The Power for Flight

NASA's Contributions to Aircraft Propulsion

Jeremy R. Kinney

Library of Congress Cataloging-in-Publication Data

Names: Kinney, Jeremy R., author.
Title: The power for flight : NASA's contributions to aircraft propulsion / Jeremy R. Kinney.
Description: Washington, DC : National Aeronautics and Space Administration, [2017] | Includes bibliographical references and index.
Identifiers: LCCN 2017027182 (print) | LCCN 2017028761 (ebook) | ISBN 9781626830387 (Epub) | ISBN 9781626830370 (hardcover)) | ISBN 9781626830394 (softcover)
Subjects: LCSH: United States. National Aeronautics and Space Administration–Research–History. | Airplanes–Jet propulsion–Research–United States–History. | Airplanes–Motors–Research–United States–History.
Classification: LCC TL521.312 (ebook) | LCC TL521.312 .K47 2017 (print) | DDC 629.134/35072073–dc23
LC record available at *https://lccn.loc.gov/2017027182*

Copyright © 2017 by the National Aeronautics and Space Administration. The opinions expressed in this volume are those of the authors and do not necessarily reflect the official positions of the United States Government or of the National Aeronautics and Space Administration.

National Aeronautics and Space Administration
Washington, DC

Table of Contents

Dedication — v
Acknowledgments — vi
Foreword — vii

Chapter 1: The NACA and Aircraft Propulsion, 1915–1958 1

Chapter 2: NASA Gets to Work, 1958–1975 49

Chapter 3: The Shift Toward Commercial Aviation, 1966–1975 73

Chapter 4: The Quest for Propulsive Efficiency, 1976–1989 103

Chapter 5: Propulsion Control Enters the Computer Era, 1976–1998 139

Chapter 6: Transiting to a New Century, 1990–2008 167

Chapter 7: Toward the Future 217

Abbreviations — 233
Bibliography — 239
About the Author — 273
Index — 275

Dedication

To Cheryl and Piper

Acknowledgments

Any author is in debt to many who help with the research and writing process. I wish to thank series editor Dr. Richard P. Hallion for asking me to participate in this project and for providing help and understanding at critical moments during my research and preparation of the final manuscript. Tony Springer of the National Aeronautics and Space Administration (NASA) Aeronautics Research Mission Directorate offered stalwart help and thoughtful counsel as he oversaw this series and my participation in it.

I received invaluable assistance from other NASA staff members. At Headquarters in Washington, DC, I would like to thank archivists Jane Odom, Colin Fries, and John Hargenrader of the History Program Office and Gwen Pitman of the Photo Library. At the Glenn Research Center, Robert Arrighi and Marvin Smith of the History Office and Chief Dhanireddy R. Reddy and Deputy Chief Dennis Huff of the Aeropropulsion Division were of extraordinary help. James Banke served as a thoughtful commentator on the manuscript. Bob van der Linden, Howard Wesoky, Melissa Keiser, and Allan Janus of the Smithsonian Institution's National Air and Space Museum in Washington, DC, provided important support. For assistance with photographs held by industry, Matthew Benvie of General Electric, Marie Force of Delta Airlines, Mary E. Kane of Boeing, and Judy Quinlan of Northrop Grumman facilitated access.

The research and writing of history is a communal effort, and historians necessarily stand on the shoulders and exploit the work of others who have gone on before. I have acknowledged in the text and the bibliography the authors of several critical previous works that have addressed the National Advisory Committee for Aeronautics (NACA)/NASA legacy in aircraft propulsion. This book would not have been possible without the foundation provided by these individuals. Nevertheless, readers should realize that all errors of fact, interpretation, or omission are solely my own.

My wife, Cheryl, has been a devoted supporter and a beloved taskmaster as I worked to manage both personal and professional schedules. We welcomed our beautiful daughter, Piper, during the writing of this book, and I greatly love and appreciate them both for so unselfishly letting me dedicate myself to this project over many months, days, and evenings.

Foreword

The *New York Times* announced America's entry into the "long awaited" Jet Age when a Pan American (Pan Am) World Airways Boeing 707 airliner left New York for Paris on October 26, 1958. Powered by four turbojet engines, the 707 offered speed, more nonstop flights, and a smoother and quieter travel experience compared to newly antiquated propeller airliners. With the Champs-Élysées only 6 hours away, humankind had entered into a new and exciting age in which the shrinking of the world for good was no longer a daydream.[1] Fifty years later, the *New York Times* declared the second coming of a "cleaner, leaner" Jet Age. Decades-old concerns over fuel efficiency, noise, and emissions shaped this new age as the aviation industry had the world poised for "a revolution in jet engines."[2] Refined turbofans incorporating the latest innovations would ensure that aviation would continue to enable a worldwide transportation network. At the root of many of the advances over the preceding 50 years was the National Aeronautics and Space Administration (NASA).

On October 1, 1958, just a few weeks before the flight of that Pan Am 707, NASA came into existence. Tasked with establishing a national space program as part of a Cold War competition between the United States and the Soviet Union, NASA is often remembered in popular memory first for putting the first human beings on the Moon in July 1969, followed by running the successful 30-year Space Shuttle Program and by landing the Rover Curiosity on Mars in August 2012. What many people do not recognize is the crucial role the first "A" in NASA played in the development of aircraft since the Agency's inception. Innovations shaping the aerodynamic design, efficient operation, and overall safety of aircraft made NASA a vital element of the American aviation industry even though they remained unknown to the public.[3] This is the story of one facet of NASA's many contributions to commercial, military, and general aviation: the development of aircraft propulsion technology, which provides the power for flight.

NASA's involvement in the development and refinement of aircraft propulsion technologies from 1958 to 2008 is important for three reasons. First, at the most basic level, NASA's propulsion specialists pushed the boundaries of the design of power plants for both subsonic and supersonic flight. Innovations that emerged from NASA programs included ultra-high-bypass turbofans; advanced turboprops; and refined systems reflecting the desire for more efficient, quieter, cleaner, and safer engines. The second reason explains how NASA achieved that success. The Agency played a major role as an innovator, facilitator, collaborator, and leader as it interacted with industry and other Federal agencies, primarily the Federal Aviation Administration (FAA) and the Department of Defense (DOD). NASA's involvement in aircraft propulsion as, in the words of longtime propulsion specialist Dennis Huff, a "technology broker" highlights the continual presence of the Federal Government in the creation of technology.[4] The third reason is that, as a result of NASA's efforts, the U.S. aircraft propulsion industry has led the world consistently in the development of new technology with improved performance, durability, environmental compatibility, and safety.[5] Overall sales of military, commercial, and general aviation engines accounted for 25 percent of the entire aviation industry's revenues for 2006.[6]

NASA has four major aeronautical centers that deal with aircraft propulsion issues based on their collective expertise: Langley Research Center in Virginia, Ames Research Center and Armstrong Flight Research Center in California, and Glenn Research Center at Lewis Field in Ohio. Glenn is NASA's primary propulsion facility.[7] Glenn's research facilities include five wind tunnels, the Aero-Acoustic Propulsion Laboratory, the Engine Research Building, the Propulsion Systems Laboratory, and the Flight Research Building.[8] More importantly, it is the specialists of Glenn and the other Centers who have served at the core of the Administration's work in aircraft propulsion. The work of all propulsion researchers at NASA falls under the programs of the Aeronautics Research Mission Directorate, with an overall goal to advance breakthrough aerospace technologies.

Airplanes incorporate synergistic technologies that embody four primary systems: aerodynamics, propulsion, structures, and control. The development of these internal systems into an overall practical and symbiotic system has been at the core of the airplane's success over the course of the 20th century. Aircraft designers must maintain a balance among lift, drag, thrust, and weight. In other words, without an equal balance among the four forces of flight, where the wings and propulsion system must generate enough lift and thrust to overcome the weight and drag of an airplane's structure, the airplane is incapable of flight.[9]

The purpose of an airplane's propulsion system is to create thrust, the force that propels an airplane through the air. The combination of a propeller and an internal combustion piston engine was the first practical system and remains in widespread use to this day. A propeller is an assembly of rotating wings, or blades, which converts the energy supplied by a power source into thrust to propel an airplane forward the same way a wing generates lift to make an airplane rise upward. Replacing the piston engine with a gas turbine to drive a propeller resulted in the turbine propeller, or "turboprop," engine. The propeller and its power source are the most efficient at moving a large mass of air for thrust at speeds of up to 500 miles per hour.

The second type of propulsion system, the jet engine, which is another type of gas turbine, emerged during World War II and serves as the dominant propulsion system for high-performance military and commercial aviation since it is most efficient at speeds of over 500 miles per hour. A jet engine takes in air, compresses it, mixes it with vaporized fuel, ignites it, and pushes it out to create thrust. The main parts of a jet engine that accomplish that process are the inlet, compressor, burner, turbine, and exhaust nozzle.

There are different types of gas turbine engines to suit the specific needs of the various types of aircraft. The oldest configuration is the turbojet, which is a pure jet that produces a lot of thrust at the expense of high fuel consumption. The addition of a large, enclosed, multiblade fan to a turbojet harnessed higher efficiencies while developing the high thrust of the turbojet. The fan created a secondary airstream that bypassed the rest of the engine and contributed to the overall production of thrust. The bypass ratio—the correlation between the mass flows of air traveling in those two pathways—is a gauge of propulsive efficiency. The widespread introduction of turbofans in the 1960s represented a dramatic jump in efficiency for jet-powered aircraft. Supersonic fighter aircraft feature afterburners for short bursts of extra speed. The injection of fuel into the hot exhaust stream produces additional thrust at the cost of high fuel consumption for increased engine power at takeoff, climb, and cruise. In turboprop and turboshaft engines, the turbine section takes energy from the exhaust gas stream to turn a propeller or rotor in addition to the compressor.

Propulsion technology is more than just piston engines; propellers; gas turbines; and individual components such as compressors, turbine blades, and disks. Support technologies, called accessories, include control apparatus; oil; fuel; and hydraulic pumps, lubricants, and fuels. Moreover, as you will see in this book, there are interrelated technical goals rooted in efficiency, noise, and emissions. Issues related to airframe integration, primarily engine nacelle placement and inlet and exhaust design, also can affect propulsion systems.

This is a survey of NASA's work in aircraft propulsion from its origins as the National Advisory Committee for Aeronautics (NACA) to the early 21st century. It stands as a point of departure rooted in an extensive body of work that addresses the topic, and it is supported by primary source material. It introduces NASA's role in the technology while taking into account economic, political, and cultural dimensions. In these pages, you will meet members of a national aeronautical community that shaped aircraft propulsion. The dramatic development and use of aircraft propulsion technology were the result

The Power for Flight

of a communal response to challenges and concerns that tell us much about the priorities, goals, and determination of a society that needed engines and related systems for military, commercial, and general aviation.

The chapters in this book survey six major eras and themes from NASA's involvement in the development of aircraft propulsion. Chapter 1 presents the history of aircraft propulsion through the story of the NACA, from the early flight period to the early days of the Cold War. Originally dedicated to the piston engine–propeller combination, the NACA shifted its focus during the emerging turbojet revolution. The Committee's work in high-speed flight continued until its dissolution in 1958. The newly created NASA and its support of military high-speed and commercial subsonic flight during the 1960s and 1970s is the subject of chapters 2 and 3. NASA's propulsion program stood at the intersection of military, industrial, and academic research as it worked to refine the military airplane and first addressed public concerns that persist today over the place of the commercial jetliner in American life. The first national programs for a commercial supersonic transport (SST) serve as the bridge between the two worlds. The establishment of the Aircraft Energy Efficiency Program of the 1970s and 1980s, presented in chapter 4, reflected NASA's desire to nurture and, in some cases, reinvent turbofan and turboprop technology during a chaotic period of oil embargoes and escalating fuel prices.

While the propulsion focus at NASA Glenn is at the center of this book, another NASA Center figured prominently in the development of new propulsion-related technologies. Chapter 5 discusses the flight research programs dedicated to digital engine controls and thrust vectoring at Dryden Flight Research Center (now the Neil A. Armstrong Flight Research Center) from the late 1960s to the 1990s. Chapter 6 documents NASA's late-20th-century efforts to direct its own research programs in efficiency, noise, and emissions and to participate in joint endeavors that complemented the work of other Government programs. Chapter 7 addresses the shift in focus for NASA's aircraft propulsion efforts and what the future might bring. This book concludes with a brief discussion of NASA's achievements in aircraft propulsion in the context of the Agency's first 50 years.

Endnotes

1. Paul J.C. Friedlander, "Jet Age Prospect," *New York Times* (October 26, 1958): 25.
2. Matthew L. Wald, "A Cleaner, Leaner Jet Age Has Arrived," *New York Times* (April 9, 2008): H2.
3. Robert G. Ferguson, *NASA's First A: Aeronautics from 1958 to 2008* (Washington, DC: NASA SP-2012-4412, 2013), pp. 3–4.
4. Dennis Huff, telephone conversation with author, August 29, 2013.
5. James Banke, "Advancing Propulsive Technology," in *NASA's Contributions to Flight, Vol. 1: Aerodynamics*, ed. Richard P. Hallion (Washington, DC: NASA SP-2010-570-Vol 1, 2010), p. 735.
6. Aerospace Industries Association, *Aerospace Facts and Figures 2008* (Arlington, VA: Aerospace Industries Association of America, 2008), p. 8.
7. NASA Glenn has had several different names over the years. During the NACA period, it was known as the Aircraft Engine Research Laboratory (AERL, 1941), the Flight Propulsion Research Laboratory (1947), and the Lewis Flight Propulsion Laboratory (1948). With the creation of NASA, the laboratory became Lewis Research Center (1958). NASA modified the name of the Center on March 1, 1999, to honor former Mercury astronaut and Ohio Senator John H. Glenn. This study will use the appropriate name according to the historical period being discussed.
8. Glenn's other fields of expertise are power, communications, and microgravity science.
9. A helpful guide to understanding the operation of aircraft engines for the nonspecialist is *Pushing the Envelope: A NASA Guide to Engines* (Cleveland: NASA Glenn Research Center EG-2007-04-013-GRC, 2007).

Shown above is a Republic XP-47M Thunderbolt fighter, complete with Hamilton Standard propeller and Pratt & Whitney radial engine, installed in the Altitude Wind Tunnel at the Aircraft Engine Research Laboratory in September 1945. (NACA)

CHAPTER 1
The NACA and Aircraft Propulsion, 1915–1958

The primary American civilian Government agency concerned with aeronautical research and development from the early flight era to the advent of the Space Age following the shock of Sputnik on October 4, 1957, was the National Advisory Committee for Aeronautics (NACA). According to the Naval Appropriation Act of March 3, 1915, the NACA possessed total freedom to "supervise and direct the scientific study of the problems of flight, with a view to their practical solution," as well as a responsibility to "determine the problems which should be experimentally attacked" in the United States. Furthermore, the act allowed the NACA to "direct and conduct research and experiment in aeronautics" at laboratories placed under its control.[1] From its creation in 1915, the NACA exemplified the Government's commitment to continued aeronautical progress. Acting as a coordinator for the military, the aviation industry, and research universities, the NACA set the pace of American aeronautics.

The core structure of the NACA was the committee framework. Inherent in the structure of the Committee were the specialist subcommittees dedicated to specific disciplines within aeronautics, which included groups addressing power plants, propellers, lubricants and fuels, and other topics that dealt with fundamental challenges in the development of propulsion technology. Their formation reflected the identification of areas that required further research and development before they reached a level of maturity that facilitated practical commercial and military use.[2]

In addition to conducting fundamental research in propulsion technology, the NACA's central role in disseminating its and the aeronautical community's information was present in the propulsion sphere, too. NACA publications in the form of technical reports, notes, and memoranda featured the Committee's research, contracted research, and translations of foreign articles of interest to American aeronautical engineers.[3] The committee went one step further by continuing to sponsor the critical *Bibliography of Aeronautics* initiated by Paul Brockett of the Smithsonian Institution to cover the period 1909 to 1932.[4]

The NACA and the Beginnings of Its Propulsion Research

Known for his effective leadership of the NACA in terms of promoting its overall role in fundamental research during his tenure, George Lewis started his career in aircraft propulsion. He earned his master's degree in mechanical engineering from Cornell University in 1910 and taught engineering as a professor at Swarthmore College until 1917. Lewis joined the Clarke Thomson Research group, a private foundation established in Philadelphia in 1918 for the promotion of "the advancement of the science of aviation" with a particular focus on propulsion systems. As a member of the Power Plants for Aircraft subcommittee, he authored a technical report on aircraft engine valves before joining the NACA in 1919.[5] Clarke Thomson made his fortune manufacturing electric trolley cars and established the research group during World War I.[6]

Engines: The Heart of the Airplane

To better understand internal combustion engine problems, Langley power plant engineer Marsden Ware and his colleagues created the NACA Universal Test Engine in 1920. Ware was a 1918 graduate of the mechanical engineering program at Rensselaer Polytechnic Institute in Troy, NY. The single-cylinder test engine featured a 5-inch bore, a 7-inch stroke, and a head assembly that facilitated a wide variation of compression ratios and lift and timing of the intake and exhaust valves, as well as the capability to connect a number of accessories such as magnetos. Ware's initial investigations centered on increasing horsepower through increased compression ratio and altered valve timing rather than the more intuitive increase in throttle settings.[7]

During the 1920s and 1930s, the next steps in the development of aircraft propulsion technology were an unknown. The dominant propulsion system in the United States consisted of the reciprocating piston engine—specifically the radial, air-cooled configuration—and the propeller.[8] In 1940, Langley Memorial Aeronautical Laboratory's Power Plants Division focused on the "conventional, incremental" approach to the reciprocating engine. There was no place for new and unconventional systems of powering aircraft. The division focused on improving the cooling properties of the air-cooled radial engine, specifically on the improved design of cylinder fins, baffles, and shrouds.[9] It was the point of view of former United Aircraft Corporation President Eugene E. Wilson that, since the piston engine was a well-known quantity in the 1920s, the NACA really had no way to advance the state of the art through fundamental research.[10]

Nevertheless, in September 1934, Langley opened the Aircraft Engine Research Laboratory. Designed primarily by Carlton Kemper,

Addison Rothrock, and Oscar W. Schey, the facility included dynamometers, equipment for fuel-spray research, and a two-stroke cycle test bed. Their research focused on increasing the power and efficiency of engines. The NACA's limited but important work on designing air-cooled cylinder fins, examining fuel behavior, and addressing the relationship between octane and high-compression engines took place there.[11]

Propellers: Rotating Wings with a Twist

While the NACA may not have had much to offer in terms of innovating new power plants, its research programs in the development of propeller technology set the standard for technical excellence and Government-university collaboration. The NACA-sponsored propeller program conducted by two Stanford University professors is an excellent case study of that relationship.[12]

Until the completion of Langley Memorial Aeronautical Laboratory in June 1920, the NACA coordinated and conducted all of its experimental research through contracts with research universities. One of the earliest and most consistently funded programs was propeller research. The NACA's first annual report in 1915 acknowledged the lack of consistent propeller data regarding efficiency as one of the general problems facing American aeronautics. In its efforts to refine, develop, and perfect the propeller, the NACA enlisted the help of Government researchers and university professors from across the Nation but concentrated its main effort at Stanford University near Palo Alto, CA. As the 1920s progressed, the NACA became involved with aeronautical research, including propeller research, at major American universities such as the Massachusetts Institute of Technology (MIT), New York University, the University of Michigan, and the California Institute of Technology.[13]

Despite regular use in airships, however, propeller development through the late 18th and 19th centuries was more the result of empiricism than established theory.[14] In France in 1885, Russian-born Stefan Drzewiecki devised a theory for calculating propeller performance based on measured airfoil data that, had they been used by the aeronautical community, would have greatly affected propeller development. Known today as the blade-element theory, Drzewiecki's theory considers the propeller to be a warped airfoil, each of whose segments represents an ordinary wing as the segments travel in a helical path. Drzewiecki was the first to calculate the forces on blade segments to find the thrust and torque output for the entire propeller as well as innovating the use of airfoil data to determine propeller efficiency. Drzewiecki published various papers and texts beginning in 1885 and ending with *Théorie Générale de l'Hélice Propulsive* in 1920.[15]

Along with being the first successfully to develop a practical flying machine, Wilbur and Orville Wright were the first to address the propeller from a

The Power for Flight

Figure 1-1. While they emphasized their Flyer's control system as their unique contribution to the development of flight, the Wrights' propellers were equally revolutionary. (National Air and Space Museum, Smithsonian Institution, NASM 9A05000)

theoretical and overall original standpoint. The Wrights came to the conclusion during the winter of 1902 and 1903 that a propeller was not a screw, but a rotary wing, or airfoil, which generated aerodynamic thrust to achieve propulsion. With that concept established, they built upon the revolutionary wind tunnel experiments they used in designing the wings for the 1903 Flyer. The Wrights successfully designed propellers that were efficient enough to transfer power from their 12-horsepower internal combustion engine to achieve powered flight. The Wrights created the world's first true airplane propeller and a theory to calculate its performance that would be the basis for all propeller research and development that followed.[16]

Individual investigators, most notably Gustave Eiffel in France and D.L. Gallup of Worcester Polytechnic Institute in the United States, continued propeller research and development in the 1910s. The Europeans who worked with Drzewiecki's theory found it to be largely unreliable but still effective in designing propellers of 70 to 80 percent efficiency. Thus, a significant knowledge of *how* to design a quality propeller existed by 1916. What did not exist was an effective collection of propeller data to aid designers in creating theoretically feasible and aerodynamically efficient propellers.[17]

The NACA's first annual report in 1915 acknowledged the lack of consistent propeller data as one of the general problems facing American aeronautics. The need for "more efficient propellers," able to retain their efficiency over a variety of flight conditions, was a primary concern. Identifying the need, the NACA suggested a solution. Unaware of the Wright brothers' findings, the Committee acknowledged the existence of "competent authorities" on marine-propellers who would be able to transfer their expertise to the refinement of propeller design.[18] The "competent authorities" that the Committee alluded to were professors William F. Durand and Everett P. Lesley, the two individuals responsible for the NACA's propeller studies at Stanford University.

An 1876 graduate of the United States Naval Academy and professor emeritus from Cornell University's prestigious Sibley College of Engineering, William F. Durand (1859–1958) served as the head of Stanford's mechanical engineering department beginning in 1904. A noted authority on marine-propellers, he became interested in aeronautics, specifically propellers, in 1914. His influential article of the same year, "The Screw Propeller: With Special Reference to Aeroplane Propulsion," which appeared in the *Journal of the Franklin Institute*, secured his charter membership in the NACA and bridged the gap between marine and aeronautical engineering. The article also inaugurated a prestigious career in aeronautics. Durand held the NACA chairmanship from 1916 to 1918, membership and the secretary's position on the influential President's Aircraft Board in 1925, and a charter trusteeship with the Daniel Guggenheim Fund for the Promotion of Aeronautics beginning in 1926.[19]

Everett P. Lesley was an equally important member of the Stanford propeller research team. He received a master's degree in naval architecture from Cornell University and served for two years at the Navy's Experimental Towing Tank. Lesley came to Stanford's mechanical engineering department in 1907 with a considerable knowledge (like Durand) of marine-propellers. One historian characterized Lesley as a versatile engineer with an "outstanding ability to make things work in the laboratory," a quality crucial to the success of the Stanford propeller tests.[20]

Durand proposed at the NACA's first meeting in 1915 that the committee sponsor extensive propeller investigations at Stanford University. In doing so, he initiated a 13-year relationship between the two institutions.[21] Stanford University received its first contract for propeller research in October 1916; unfortunately, the specific amount of this initial contract is not available. Durand personally participated in the awarding of the NACA's research contracts to Stanford University. His selection reflected the NACA's belief that the most qualified individuals, no matter their affiliation to the committee, should have the opportunity to conduct research.[22] The objective of the initial research and the 11 experiments that followed, however, involved the refinement of engineering practices that benefited overall airplane design, not just propeller design. There were differences between strict propeller design and the need to select and incorporate efficient propellers into airplane design. Propeller design involves the meticulous creation of an efficient airfoil, while airplane design requires that work to be already predetermined.[23] Perceiving the propeller as a major component within the technical system of the airplane reflected the Committee's desire to fulfill its goal of working toward the practical solution of the overall problems of flight.

Later, in June 1918, Durand illustrated the technical systems approach to airplane design before the Royal Aeronautical Society of Great Britain. In his delivery of the Annual Wilbur Wright Memorial Lecture, entitled "Some Outstanding Problems in Aeronautics," Durand succinctly voiced the NACA's position on propeller refinement within the broader sphere of airplane design. He defined a typical powered, heavier-than-air flying machine as "an airplane-motor-propeller combination" in which each of the three components was totally dependent on the others. The propeller was responsible for one crucial function in this relationship: converting the motor's energy into propulsive power to enable the aircraft's wings and fuselage to generate lift.[24] Durand and the NACA clearly believed that propeller research at Stanford University would significantly contribute to the development of American aeronautical technology.

Before embarking upon in-depth research, Durand and Lesley first had to oversee the construction of the Stanford University Aerodynamical Laboratory during the fall and winter of 1916–1917. Funding for laboratory construction came from the initial propeller research contract of October 1916. Intent on starting experiments by the spring of 1917, the Stanford professors wanted the facility completed immediately to expedite research. They designed an Eiffel-type wind tunnel with a 5.5-foot throat and a 55-mile-per-hour maximum test stream speed. The specific instruments the Stanford professors incorporated into the laboratory were dynamometers for calculating thrust and torque, a revolution counter, and an airspeed meter.[25]

The interaction between the NACA and Stanford University from 1915 to 1917 indicates the existence of a Government-research subrelationship. As the NACA Chairman, Durand influenced the shaping of a national aeronautical research and development policy that stressed the overall development of the airplane through a technical system approach. As university researchers, Durand and Lesley gained prestige from performing and publishing research for the Government. The university itself benefited from the new facilities. Individual leadership, university research, and direct Federal funding strengthened the subrelationship between the NACA and Stanford University.

Durand and Lesley conducted a broad-based study of propeller performance entitled "Experimental Research on Air-Propellers" for the NACA from 1917 to 1922. The most important contribution of this groundbreaking series of experiments was the establishment of a standard table of propeller coefficients available to designers through mathematical calculation and wind tunnel studies. During the 5 years of testing, the knowledge of and expertise in calculating propeller performance grew incrementally and created new avenues of experimentation.[26]

Durand and Lesley's initial goal for the 1917 experiments involved the development of a series of design constants and coefficients derived from wind tunnel tests on 48 standard propeller model shapes. They intended to use the results as a final check against propeller data obtained from other aeronautical laboratories, the Drzewiecki theory, and full-flight experiments. By cross-checking these methods and the methods of other researchers, Durand and Lesley hoped to establish a standard methodology for continuing aeronautical research. The results of the tests, expressed in graphical form, encouraged Durand and Lesley to assert that their model propeller data contributed to the refinement of propeller design by a significant amount.[27]

For 1917, the NACA appropriated $4,000 for Durand and Lesley's propeller research at the newly completed Stanford Aerodynamical Laboratory. This figure was 56 percent of the total budget of $7,100 for the Committee's special reports for the entire year, indicating the significance the Committee placed on propeller studies. That budget was second only to the $68,957 awarded for the construction of the Committee's research laboratory at Langley. The total NACA appropriation for 1917 was $87,515.70.[28] Of the NACA's total appropriations awarded up to the middle of 1918, 38 percent of it was for the Stanford propeller studies.[29]

The NACA authorized continued propeller research at Stanford University for the summer and autumn of 1918. Confident in the success of their previous study, Durand and Lesley continued with their standard model propeller studies as well as experimenting with a variable-pitch propeller.

Speaking before the Aeronautical Society of Great Britain earlier in June 1918, Durand identified this propeller configuration as important to the overall refinement of airplane efficiency. He believed that development of a workable variable-pitch propeller was "of the highest order of importance" and "outstanding as one of the appliances for which the art of aerial navigation is definitely in waiting."[30]

Paralleling Durand's opinions was the NACA's public request for assistance in the development of variable-pitch propellers. In the June 1, 1918, issue of *Aviation and Aeronautical Engineering,* the NACA's Special Sub-Committee on Engineering Problems reported that no significant progress had been made in the area. The committee's call to arms regarding the development of a variable-pitch propeller exemplifies the overflow of Government-sponsored university research into the private sector and indicates the overall importance of the new configuration.[31]

Earlier in the spring of 1919, the NACA reorganized its infrastructure in a way that further cemented the bonds of the Government-research subrelationship. The Committee abolished all of its subcommittees and created three technical and three administrative committees. The three technical committees, Aerodynamics, Power Plants, and Aircraft Construction, monitored all research for the NACA. The Committee on Aerodynamics specifically retained direct control of all aeronautical research at Stanford University and Langley Memorial Aeronautical Laboratory later in 1920. Of the administrative groups, the Governmental Relations committee worked to coordinate between Federal agencies, and the Committee on Publications and Intelligence aimed to make the NACA an overall source of technical information. The Bureau of Standards, the Army's Engineering Division, and the Navy's Bureau of Construction and Repair provided their reports to the NACA as a courtesy for increased dissemination of knowledge.[32]

Durand and Lesley collaborated in a reexamination of their previous experimental propeller research in 1922. Having gained a better understanding of the intricacies of propeller theory, they synthesized their previous reports in a new report to provide systematic propeller design data in a usable form.[33] Noted aerodynamicist Max Munk asserted that the Stanford University propeller study was "the most perfect and complete one ever published." He judged the experiments were "selected and executed in the most careful way" and that Durand and Lesley's methodology was "excellent."[34]

Durand and Lesley's "Experimental Research in Air-Propellers" expanded from one general model propeller study into four more complementary analyses of propeller efficiency. What they do illustrate, however, is the increasing complexities of engineering research and the extent to which the Federal Government would support further inquiry. The combined "Experimental Research on Air-Propellers" became the point of departure not only for the NACA's propeller research, but for the ever-growing Government-research relationship between Stanford University and the Committee.

As Durand and Lesley's experiments in collecting the systematic data necessary for propeller design progressed, they diverted their attention toward a new avenue of propeller specialization: vertical flight. Earlier in 1918, Durand had accepted the chairmanship of the NACA's subcommittee for Helicopters, or Direct-Lift Aircraft.[35] Not until 1921 did the Stanford Aerodynamical Laboratory investigate model propellers for use in helicopters.

The "Experimental Research on Air-Propellers" study was successful because the model propeller families were of already-established designs. Durand and Lesley were simply testing the viability of a certain methodology that confirmed predetermined calculations. In their efforts to experiment with new and unproven propeller designs, they acknowledged the need for further evaluation of the correlation among model propeller tests, airfoil theory, and full-flight testing.[36] What resulted was an increasing sophistication of the NACA-sponsored research at Stanford University.

During the spring and summer of 1924, comparative experiments between standard model propeller testing and full-flight testing resulted in a collaborative effort between the NACA's Langley Memorial Aeronautical Laboratory and the Stanford Aerodynamical Laboratory. Lesley traveled to Virginia to conduct the full-flight tests while Durand directed the model propeller tests at Stanford.[37]

Rather than discouraging airplane designers in their search for aerodynamically efficient propellers, the comparative testing made them aware of the inconsistencies of applying model propeller data to full-scale designs.[38] An NACA report issued the same year attested that researchers "can never rely absolutely" upon model data until they verify the data through full-flight tests.[39]

The comparative study between model and full-scale propellers rekindled another aeronautical problem that Durand first identified in his June 1918 address to the Aeronautical Society of Great Britain. He believed that the "widest and most important outstanding problem in connection with airplane propulsion" was the aerodynamic relationship between the propeller and the airplane itself.[40] As stated before, the NACA and Durand did not consider the propeller experiments to be of singular value, but one of importance to overall airplane design. Furthermore, aircraft structures, the fuselage, and wings directly affected the performance of propellers in flight conditions. In the last series of Stanford propeller studies conducted by Durand and Lesley from 1923 to 1929, the professors independently researched the integration of propellers into airplane design.[41, 42]

Aeronautical engineering knowledge concerning the inclusion of the propeller into overall airplane design had grown dramatically by 1930. Durand and Lesley's methods added sophistication to an increasingly complex field of propeller design. In one respect, their research influenced the NACA's 1925 decision to improve its propeller research facilities at the Langley Memorial Aeronautical Laboratory. Their exposure of the inadequacies of theoretical and model propeller testing convinced the committee that the hybridization of full-flight testing with wind tunnel testing was necessary.[43] The completion of Langley's Propeller Research Tunnel (PRT) in 1927 marked the end of the

Stanford Aerodynamical Laboratory's importance as the center of the NACA's propeller research.[44]

As has been discussed, the interaction between the Stanford Aerodynamical Laboratory and the NACA during the period 1915 to 1930 clearly illustrated the Government-research subrelationship within an embryonic military-industrial-research complex. Acting on Government mandate, Durand and Lesley pursued research in the hope of advancing the technical development of American aeronautics. The partnership between Stanford and the NACA precipitated a growing interdependency between the Federal Government and academe. As a result of this Government-research subrelationship, the level of knowledge of incorporating propeller design into overall airplane design matured during the 1930s.

As the engineers of the Langley Laboratory designed and constructed the tools for continued work in aeronautics, they were able to make their own investigations into basic propeller research. Early tests of blade profiles in the Variable-Density Tunnel (VDT), however, proved unsatisfactory. What the NACA needed was the ability to test the aerodynamic properties of full-scale propellers. This need led to the opening of the PRT.[45]

After the opening of the PRT, Langley researchers utilized other tunnels such as the 24-Inch Jet Tunnel. During World War II, Langley was the center of the NACA's study in improving propellers with high thrust at high speeds. The majority of the work on propellers took place in the 8-Foot High Speed and 16-Foot High Speed tunnels, the latter under the direction of John Stack. Melvin N. Gough conducted a simultaneous flight research program. Theodore Theodorsen and his colleagues in the Physical Research Division investigated vibration and flutter.[46]

The NACA's principal contribution to propeller development in the 1930s was improved propeller efficiency at high speeds. The RAF-6 and Clark Y airfoils proved sufficient through the 1920s and 1930s. As aircraft speeds increased, shock waves and compressibility decreased efficiency. Langley sponsored three programs addressing propeller efficiency conducted by Fred Weick in the PRT, Eastman N. Jacobs in the VDT, and John Stack in 24-Inch High Speed Tunnel. During the 1930s, propellers realized efficiencies of 80 to 85 percent for aircraft that cruised at 300 to 350 miles per hour (mph). The NACA recognized that efficiencies dropped to 70 percent as speeds increased and made propeller development a major focus.[47] The NACA announced a new family of airfoils, the 16-series, which resulted from using the thin airfoil theory. Thin airfoils facilitated faster and more efficient propeller blades. Distribution to the Army, Navy, and manufacturers ensured that the 16-series became the new choice for high-speed propellers.[48]

NACA researchers also worked to refine the aerodynamic properties of a propeller blade along its entire length with airfoil sections called cuffs. Blade design reflected a compromise where most of the blade was an airfoil, but the portion where it attached to the hub, called the root, was round for structural strength. In 1939, Langley researchers in the Full-Scale Tunnel investigated a single-engine fighter that was theoretically capable of 400 mph with the right propeller but could not yet reach that speed with a conventional propeller root-shape. To each blade root, they attached airfoil-shaped cuffs that covered over 45 percent of the blade. The increased blade area enabled the fighter to reach 400 mph at 20,000 feet. Langley carried on with extensive research on cuffs that allowed for the modification of existing blade designs.[49] The North American P-51 Mustang fighter was modified with blade cuffs for increased performance, and cuffed propellers likewise improved the cooling of radial engine designs such as the Republic P-47 Thunderbolt and—very significantly, because of its notorious cooling and engine fire problems—the four-engine Boeing B-29 Superfortress long-range bomber.

Figure 1-2. The use of propeller cuffs on fighters like the P-51 Mustang maximized the aircraft's overall performance. (National Air and Space Museum, Smithsonian Institution, NASM 7A35592)

11

In 1941, John Stack and his colleagues at the 8-Foot High Speed Tunnel began work toward the development of a propeller that would be efficient at 500 mph at 25,000 feet. They believed that blades of varying widths and combinations and based on the 16-series airfoils were the key. They determined that an 11½-foot dual-rotation propeller, comprising two tandem three-blade propellers, would exhibit 90 percent efficiency when coupled with an engine of 2,800 horsepower. The blades were called paddle blades, for the increased production of thrust came from their having a larger chord length from the leading to the trailing edge of the blade. Continued wind tunnel and flight tests revealed promising areas of research into expanding propeller performance, centering on blade airfoil sections that would offer higher critical speeds, that is, the point where the drag of the blade began to rapidly increase, reducing propeller efficiency. George Gray regarded the 16-series as the "first family when it comes to speed" due to its superiority to other types of blade sections. Using wider, or paddle, blades and expanding the number of blades increased the overall area for producing thrust. Dual contrarotation, the use of two sets of propellers connected to one engine and rotating in opposite directions (one clockwise, the other counterclockwise), alleviated torque roll, the sideways direction imparted by single propellers (which could roll an airplane on its back if a pilot too-rapidly manipulated the throttle at low speeds and high power settings), and maximized the energy of both propellers. (Though used on some aircraft, such as later Supermarine Seafire fighters in Britain, the postwar Fairey Gannet antisubmarine aircraft, and a variety of Soviet-era transports and bombers, the dual-contrarotating propeller has always been more of an exception to conventional design than a mainstream design element). Langley researchers revealed that new and efficient propellers could have up to eight blades.[50] By the end of the war, the NACA had provided important avenues for the continued refinement of the airplane propeller.

While the NACA made no contributions to the mechanical design of propellers during the interwar period, its research staff did succeed in reducing revolutions per minute (rpm) while increasing horsepower, speed, and efficiency. At the beginning of the 1930s, two- and three-blade propellers using RAF-6 and Clark Y airfoils generated speeds between 150 and 250 mph for engines with 500 and 1,100 horsepower at 1,800 to 1,500 rpm with an overall efficiency of 83 percent. Work during the period 1935–1941 resulted in four-blade propellers utilizing an NACA 2409-34 airfoil that was able to absorb 1,600 horsepower at a speed of 350 mph at 1,430 rpm with efficiency of 87 percent. At the end of World War II, a six-blade dual-rotation propeller with NACA 16-508 blades was able to absorb 3,200 horsepower at 900 rpm at a speed of 500 mph with an overall efficiency of 90 percent.[51]

Other Propulsion-Related Technologies

Ever mindful of new areas to investigate, the NACA kept evaluating the state of aeronautical technology, especially as it pertained to propulsion. The Subcommittee on Aircraft Fuels and Lubricants focused on another challenge facing aviation: high-octane fuels and engine knock. In an internal combustion engine, a spark plug ignites a fuel-air mixture that is compressed at the top of the cylinder by the piston. The resultant explosion, characterized as a flame front with an accompanying increase in temperature and pressure, pushes the piston down. Efforts to improve engine performance, primarily increased compression, by operating at higher rpms, or supercharging, introduced the possibility of "knock," the uncontrolled combustion of fuel-air mixture in an internal combustion engine that led to mechanical damage and unsafe temperatures in the cylinders and pistons.[52] Cearcy D. Miller's invention of a high-speed photography process capable of up to 40,000 frames per second captured the combustion process and permitted the determination of the exact moment engine knock began.[53]

The introduction of high-octane aviation fuel offered new challenges to the operation of aircraft engines. Fuels with high anti-knock properties facilitated higher compression ratios and leaner fuel-air mixtures that provided increased power for brief periods, primarily during takeoff and conditions warranting War Emergency power, or 100 percent of the engine's output. Those conditions also rapidly exceeded the engine's cooling capability. Engine designers turned to an old trick, the injection of water along with the fuel-air mixture, to achieve direct cooling of the cylinders and pistons. The water simply absorbed heat, evaporated, and exited through the exhaust as steam as it offered a 15- to 25-percent surge in power as it increased knock resistance.[54]

NACA researchers worked to refine the process of water injection. They determined that the optimum amount of water for injection was 1 pound for every 2 pounds of fuel. Full-scale tests with that mixture showed that a 2,100-horsepower engine could be boosted to 2,800 horsepower for brief periods of up to 6 minutes. The research also revealed that water injection prevented knock by speeding the movement of the flame front through the cylinder.[55]

Two propulsion projects the NACA pursued in the 1920s and 1930s were the Roots blower, or supercharger, and the diesel engine. Commonly used in industrial and automotive applications, a Roots blower consisted of two cycloidal rotors that pumped air into an engine's intake upon each revolution. The compressed air increased performance as an airplane flew at higher altitudes, restoring a level of power normally seen only at lower, denser altitudes. Marsden Ware dedicated a considerable amount of research to the 88-pound device and found it to be rugged and smooth in operation; he advocated its expanded use by aircraft engine makers.[56]

The engine community did not see potential in the Roots blower. Pratt & Whitney engineer Luke Hobbs remarked that the Roots was bulky, and heavy. Additionally, it operated at a high temperature, which distorted the structure and leaked air, which in turn affected the needed compression to force air into the engine.[57] Despite those problems, U.S. Navy Lieutenant C.C. Champion, Jr., flew an experimental Navy Wright XF3W-1 Apache fighter equipped with the NACA-designed Roots supercharger to a world-record altitude of 38,419 feet on July 25, 1927. The Roots served as the first stage to the Pratt & Whitney Wasp engine's own geared centrifugal supercharger, which became an industry standard. The use of the Roots blower for Champion's flight and subsequent record-breaking altitude flights by fellow naval aviator Lieutenant Apollo Soucek in the XF3W-1 in 1929–1930 became the only significant examples of the use of noncentrifugal, nonturbine superchargers in aircraft.[58]

The NACA conducted exhaustive research directed toward the development of aircraft diesel engines during the late 1920s and through the 1930s. The work centered on injection-system development and combustion-chamber design. All work was fundamental and used single cylinders for research and not actual engines. There was hope for flying diesels such as the 1931 Collier Trophy–winning design by Packard and advanced designs from Europe, but with the advent of high-octane fuels, conventional spark-ignition engines offered better performance. There were virtually no diesel engines in widespread use by the outbreak of World War II. Despite the extensive and pioneering research from 1927 to 1937, the NACA's diesel engine program, in the words of one of America's leading propulsion specialists, C. Fayette Taylor, "found no practical application."[59] The primary focus on diesel engines throughout the most of the 1930s proved to be a costly diversion.[60]

The NACA had a role in continued refinements to the piston engine. The injection of water or a water-alcohol mixture into the cylinder to cool the combustion chamber enabled greater compression ratios and, for the most part, eliminated the problem of engine knocking. The reduction in temperature increased performance, especially when used with modifications such as turbo- and supercharging. Pratt & Whitney initiated the development of water injection shortly before World War II in collaboration with the Materiel Division at Wright Field. The NACA followed up with research.[61] Immediately after the war, water-alcohol injection proved especially beneficial for military transports and commercial airliners at takeoff and for fighter aircraft that needed short bursts of extra speed and power.[62]

Propulsion Integration: Beginnings of Engine-Airframe Matching

Up to World War II, the NACA was known primarily for its pioneering fundamental work in aerodynamics, specifically drag reduction. The NACA model of engineering methodology centered on experimental parameter variation, a systematic process of elimination based on repetition and the variation of parameters until an ideal solution for an engineering challenge was found. The application of that knowledge to making a foundation technology of the aeronautical revolution—in this case, the radial, air-cooled piston engine—aerodynamically feasible became a hallmark of the NACA's work. The American aeronautical community, recognizing the engine's light weight for the horsepower produced and its simplicity placed a major emphasis on the radial engine. New and fast "express" aircraft like the Lockheed Vega benefited greatly from the inclusion of a 450-horsepower Pratt & Whitney Wasp in their design. Unfortunately, the protruding cylinders of the radial engine—a "wheel-like" engine that is both broad and flat in the airstream—created a great deal of drag for an otherwise streamlined airplane.[63]

Fred E. Weick and his colleagues at Langley addressed the fundamental problem of incorporating a radial engine into aircraft design in the PRT. The wind tunnel's 20-foot opening allowed them to test full-size aircraft structures. Their pioneering work on a new engine-encircling structure, called the NACA cowling, simultaneously reduced drag and improved engine cooling.

The NACA cowling arrived at the right moment to increase the performance of new aircraft, and it became a standard design feature on radial-engine aircraft, whether high-speed commercial express aircraft, military fighters and bombers, or general aviation designs. Famous aviator Frank Hawks flew his scarlet-red Texaco Lockheed Air Express, with an NACA cowling installed, from Los Angeles to New York nonstop in a record time of 18 hours and 13 minutes in February 1929. Tests of a Curtiss AT-5A Hawk fighter with an NACA cowling increased its top speed from 118 mph to 137 mph, equivalent to adding 83 horsepower to the engine. The National Aeronautic Association recognized that the NACA's contribution to overall aircraft design was so great that the association awarded the Committee its first Collier Trophy in 1929 for its innovative work.

The combination of aerodynamic streamlined design, radial engines with NACA cowlings, variable-pitch propellers, retractable landing gear, and other innovations resulted in the "modern airplane."[64] The Douglas DC series was the most successful of these new aircraft. The first, the DC-1, debuted in July 1933 with the major innovations of the aeronautical revolution, including an advanced NACA-designed airfoil and radial engines covered with NACA

The Power for Flight

Figure 1-3. This image shows NACA cowling #10 in the Langley PRT. (NASA)

cowlings. It led to the DC-2 of May 1934, which carried 14 passengers while cruising at 212 mph. The follow-on DC-3 of December 1935 carried 21 people and became the most popular and reliable propeller-driven airliner in aviation history.

The AERL and World War II

Surrounded by representatives of both the military and industry, the NACA's Director of Aeronautical Research, George Lewis, broke ground for the Committee's new Aircraft Engine Research Laboratory (AERL) at Cleveland on January 23, 1941. As he drove a special pick with a nickel-plated head into the ground, the NACA transferred its highly successful model of fundamental research into the field of aircraft propulsion, all to benefit the American aviation industry.[65] War raged in Europe and Asia, with the potential of American involvement becoming increasingly certain. The creation of the AERL reflected the widely held belief that the United States needed to retain superiority in aeronautical technology vis-à-vis Europe in the late 1930s, especially in propulsion. After American entry into World War II in December, the NACA and

The NACA and Aircraft Propulsion, 1915–1958

Figure 1-4. The Douglas DC-1 and other "modern" aircraft benefited greatly from NACA innovations. (Rudy Arnold Photo Collection, National Air and Space Museum, Smithsonian Institution, NASM XRA-8489)

the AERL embarked upon a widespread program—not to innovate through basic research, but to evaluate, develop, and refine existing piston engine and propeller technology for the war effort.

The AERL had its origins in the late interwar period and reflected the work of a new member of the Committee, George Mead. Mead was a legendary engine designer known for his Pratt & Whitney Wasp and Hornet radial engines. A 1916 graduate of the Massachusetts Institute of Technology (MIT), he began to work with aircraft engines as an experimental engineer at the Wright-Martin Company in New Brunswick, NJ, during World War I. He also served as engineer-in-charge of power plant research at McCook Field before becoming chief engineer at Wright Aeronautical in Paterson, NJ. In 1925, he cofounded Pratt & Whitney Aircraft with Frederick B. Rentschler and assumed the title of vice president and chief engineer. With the creation of the United Aircraft and Transport Corporation in 1929 from the nucleus of Pratt & Whitney, he rose to a technical leadership position within America's leading manufacturer of airframes, engines, and propellers. After retiring from United Aircraft in 1939, he accepted President Roosevelt's appointment to the NACA. His fellow members quickly elected him vice chairman, and he assumed leadership of the Power Plants Committee.[66]

Acting upon Charles Lindbergh's recommendations, Mead formed the Special Committee on New Engine Research Facilities to establish the design

17

of a new NACA aeronautical laboratory in Cleveland, OH. The centerpiece of the new facility would be an Altitude Wind Tunnel (AWT), which simulated an altitude of 30,000 feet at 490 mph and allowed focused testing of engines, superchargers, and propellers individually or as complete propulsion systems. Other facilities included wind tunnels and laboratories capable of investigating model and full-scale engines; fuels and lubricants; and components including superchargers, carburetors, instruments, and fuel and ignition systems. Ultimately, the new propulsion laboratory would be at the intersection of the work of the NACA, Government, industry, and the military. The NACA was to fill the gap left open by industry with an emphasis on development and production, not fundamental research, rooted in national defense and ensuring the future of commercial aviation.

Before the Subcommittee of the House Committee on Appropriations, NACA Chairman Vannevar Bush requested $8.4 million for the construction of a new laboratory focused on aircraft propulsion. His appeal was not an easy proposition that won ready acceptance. There was considerable resistance, centered on the nature of research versus development, as well as the cost, from Congress and engine manufacturers. The House committee members misunderstood the difference between fundamental research and engineering development and questioned the need for additional Government funding. The manufacturers felt the money was better spent in the form of direct grants that allowed them to focus on developing their specific products. The NACA endeavored to take competition and exclusion out and introduce fundamental engineering into the equation.[67] The Dunkirk evacuation and the fall of France in June 1940 added additional impetus to the creation of a dedicated aircraft engine research laboratory. The First National Defense Appropriations Act of June 1940 authorized the new laboratory.

Cleveland was a leading center of aviation. It was home to Thompson Products, the maker of automotive and aircraft engine parts, primarily the ever-crucial intake and exhaust valves found in all aircraft engines. Every September, hundreds of thousands of people swarmed the grandstands and displays of the National Air Races. The city on the Lake Erie shore was also a water, rail, road, and air transportation hub connecting the East Coast, the Midwest, and the West. Thompson Products president Frederick C. Crawford led the effort to bring the NACA to Cleveland. The city made 200 acres adjacent to the airport available to the Federal Government for just $500. On November 25, 1940, Cleveland civic leaders proudly announced the city's selection as the site for the new flight propulsion laboratory.[68]

The NACA's activities in Cleveland during World War II were far from fundamental. World War II resulted in a change in focus. The NACA in Cleveland focused on improving the performance and reliability of existing engines produced by Wright Aeronautical, Pratt & Whitney, and Allison, which was an impetus created by Chief of the Army Air Forces, General Henry H. "Hap" Arnold. He was a supporter of the laboratory from the outset and took the opportunity to redress what he believed to be the failure of aircraft manufacturers to produce high-performance military fighter engines.[69]

Four Langley Power Plants Division sections, led by Oscar Schey, Benjamin Pinkel, Addison Rothrock, and Charles Stanley Moore, arrived in Cleveland. Schey was chief of the Supercharger Division. A graduate in mechanical engineering from the University of Minnesota, he joined the NACA in 1923, where he became well known for his work on the Roots supercharger. His promotion of the use of valve overlap and fuel injection in piston engines to reduce supercharger requirements were innovations ignored in the United States but enthusiastically received in Nazi Germany.[70] The Thermodynamics Division, led by another propulsion expert, Benjamin Pinkel, focused on turbosupercharger research, but their engine exhaust redesign increased the power available to high-performance

Figure 1-5. The Aircraft Engine Research Laboratory in Cleveland, OH, supported the United States aviation production program during World War II. Much of the development work involved straightforward evaluation of propeller-and-engine combinations on torque stands to determine their overall power. (NASA)

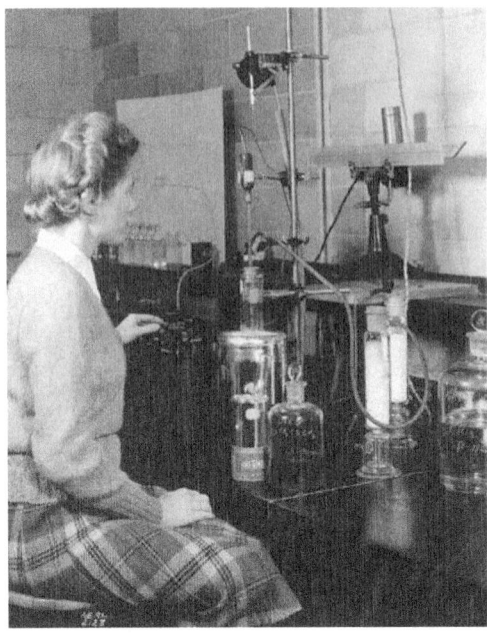

Figure 1-6. The NACA employed a large number of women at its laboratories. This researcher is testing the chemical properties of fuels and lubricants at the Aircraft Engine Research Laboratory in 1943. (NASA)

aircraft, including the Merlin-powered North American P-51 Mustang long-range fighter.[71]

The Aircraft Engine Research Laboratory's first new research program, beginning in October 1942, involved the Allison V-1710 V-12 engine. The V-1710 was the only high-performance liquid-cooled inline engine available in the United States in the late 1930s. Curtiss selected it to power its P-40 fighter, which first flew in October 1938. After December 7, 1941, American pilots found the P-40 unable to outmaneuver more advanced German and Japanese fighters in combat, especially at high altitudes, though it was a rugged and otherwise very useful aircraft, particularly for low-altitude operations. The increased refinement of the V-1710 was a full effort by all four divisions at the AERL. Schey's division focused on the supercharger; Addison Rothrock and his colleagues in the Fuels and Lubricants Division investigated knock limitations in the cylinder heads;[72] Pinkel's group worked to improve cooling; and Charles Stanley Moore's Engine Components Division addressed the refined fuel-air distribution in the Bendix-Stromberg pressure carburetor.

The teams at the AERL succeeded in getting more power out of the V-1710, and later turbosupercharged variants powered the futuristic Lockheed P-38 Lightning, which was capable of speeds approaching 400 mph at altitudes of up to 30,000 feet. Nevertheless, the NACA researchers saw the V-1710 as a flawed design that wasted their efforts during the chaotic early days of World War II. Exchanging the V-1710 on the North American NA-73 for a British 1,650-horsepower Rolls-Royce Merlin engine with a two-stage supercharger made the resultant P-51 Mustang the fastest and highest-flying Allied piston fighter to enter operational service during the war, with speeds approaching 450 mph and at altitudes of up to 40,000 feet.[73]

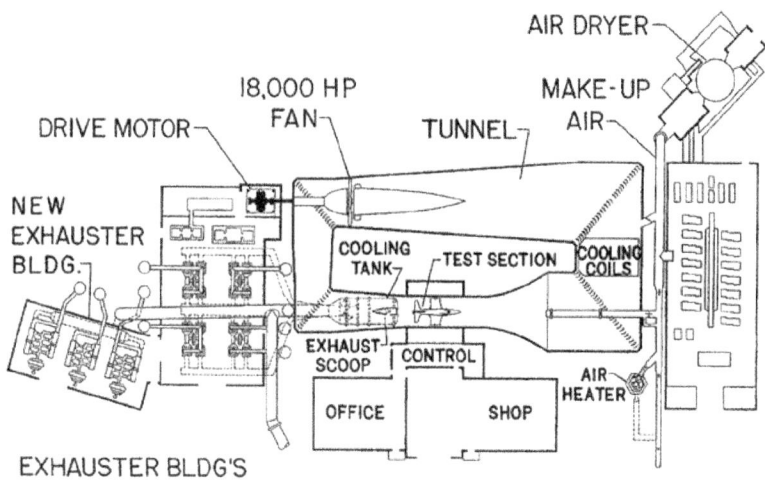

Figure 1-7. This schematic drawing of the Altitude Wind Tunnel at the AERL indicates the industrial scale of the NACA's work in aircraft propulsion research. (NASA)

The AWT was the centerpiece facility of the AERL.[74] At the cost of $6 million, it was unrivaled in its capability to test full-scale engines and propellers in simulated altitude conditions. The researchers in Cleveland addressed the next major engine development problem facing the American aviation production program, the Wright R-3350 Duplex-Cyclone radial engine, when the AWT opened in May 1944. Four turbosupercharged 18-cylinder R-3350s, each rated at 2,200 horsepower, powered the Boeing B-29 Superfortress, then the most advanced airplane in the world, with a highly refined streamlined design, all-metal monoplane construction, retractable landing gear, a pressurized cabin, and an advanced electronically based centralized defensive weapon system incorporating remotely operated turrets fired by gunners in sighting cupolas. Capable of carrying 16,000 pounds of bombs and cruising at 235 mph at altitudes of up to 30,000 feet, the B-29 was also the only strategic bomber that could reach Japan from American air bases in the Pacific. But it had a drawback: the R-3350, one of the most complex piston engines ever produced, experienced overheating and catastrophic engine fires due to rushed development. Tests in the AWT led to improved exhaust turbines and the elimination of high-altitude fuel-vaporization problems, which increased the B-29's payload capacity by 5½ tons and permitted improved high-altitude operation.[75] Beginning in November 1944, units of the U.S. Army Air Forces' 20th Air Force initiated the strategic bombing of Japanese cities, culminating in August 1945 with the atomic bombing of Hiroshima and Nagasaki.

The journal *Aviation* argued that the history of the NACA and the success of American aviation were intertwined during the late 1930s and 1940s. In terms of the global air war during World War II, the editors proclaimed that the Committee was the "force behind our air supremacy" in early 1944.[76]

Gas Turbines Usher in a Second Aeronautical Revolution

Early on August 27, 1939, test pilot Erich Warsitz took off from the Rostock airfield near the Baltic Sea in the world's first gas turbine–powered, jet-propelled airplane, the Heinkel He 178. The power plant was a Heinkel HeS 3 centrifugal-flow turbojet, which was capable of 838 pounds of thrust and could propel the small silver airplane at speeds of up to 360 mph. Watching the flight was Hans von Ohain, the engine's young inventor, and Ernst Heinkel, the speed-obsessed sponsor of the project.[77] The 5-minute flight, which took place just 5 days before the Nazi invasion of Poland that signaled the outbreak of World War II, ushered in the second aeronautical revolution and the next great age in aviation history, the Jet Age. Britain followed with its own jet aircraft, the experimental Gloster E.28/39, which first flew on May 15, 1941, in Cranwell, England. By the late summer of 1944, both Nazi Germany and Great Britain introduced operational jet fighter aircraft (the Messerschmitt Me 262 and the Gloster Meteor I), and the world's air forces scrambled to catch up. The United States flew its first jet aircraft, the Bell XP-59A Airacomet, on October 1, 1942, at Muroc Dry Lake (now Edwards Air Force Base) in California—but its engines, though built by General Electric (GE), were derivatives of a design by Britain's Frank Whittle, whose first flightworthy engine had powered the Gloster E.28/39.[78]

The invention of the jet engine and the requisite engineering to make it and the aircraft that followed viable equaled the achievement of the Wright brothers and constituted a second revolution in aeronautics. As a new and revolutionary type of propulsion system, the jet engine allowed airplanes to fly higher and faster than ever before. The reaction of the aeronautical community to that new technology resulted in a generation of new airplanes with remarkably new capabilities.

Unlike many aeronautical innovations, the turbojet engine was not of American origin. Simultaneous events in Great Britain and Germany in the 1920s and 1930s brought about the creation of this new propulsion technology by two pioneers: Sir Frank Whittle and Dr. Hans von Ohain. As a Royal Air Force officer with an engineering background, Frank Whittle sought an alternative to the piston-engine-propeller combination and theorized, using Newton's third law of physics, that a gas turbine could be used to produce jet propulsion. His patent for a gas turbine–powered jet propulsion concept in

The NACA and Aircraft Propulsion, 1915–1958

Figure 1-8. This image shows America's first jet: the Bell XP-59A Airacomet. (Bell Helicopter Textron via National Air and Space Museum, Smithsonian Institution, NASM 90-6683)

1930 went unnoticed by the British Government as well as the international aeronautical community. After receiving his advanced degree in mechanical sciences at Cambridge, Whittle found private support to develop his invention and founded Power Jets, Ltd., in March 1936. The first complete engine, the W.U., or Whittle Unit, ran on a test stand in April 1937, becoming the first jet engine in the world to successfully operate in a practical fashion. The British air ministry contracted Power Jets to build a flying engine, while Gloster Aircraft received another contract to build a jet-propelled airplane. The Gloster E.28/39's W.1X engine produced 860 pounds of thrust at speeds of up to 338 mph. The Royal Air Force introduced the Gloster Meteor into operational service in July 1944, making it the first and only turbojet-powered airplane to serve with the Allies during World War II.[79]

A young Ph.D. with a degree in physics from Göttingen University in Germany, Hans von Ohain, patented an aeronautical gas turbine engine in November 1935. He and a friend constructed a promising, but small, working demonstration model with private funds. With a letter of introduction from his mentor at Göttingen, von Ohain met with aircraft manufacturer Ernst Heinkel, a self-confessed high-speed enthusiast, who quickly gave the young physicist a job developing what became Germany's first jet engine. The collaboration proceeded rapidly. Von Ohain moved from hydrogen to gasoline as the fuel to make the engine practical for flight. Heinkel engineers designed a purpose-built airframe for the new engine. The Heinkel He 178 flew on August 27, 1939, making it the world's first gas turbine–powered, jet-propelled airplane in history to fly. Von Ohain's HeS 3B engine generated 838 pounds of thrust and propelled the He 178 at speeds of up to 360 mph.[80]

The success of the He 178 and von Ohain's engine encouraged the German air ministry to pursue the development of jet-propelled aircraft during World War II. Junkers Motoren Werke began work on a new design, the Jumo 004, the world's first practical axial-flow jet engine, under the direction of Dr. Anselm Franz, the head of the company's supercharger group. The axial-flow compressor design consisted of an alternating series of rotating and stationary blades, where the overall flow path essentially moved along the axis of the engine. The 004 series powered the Messerschmitt Me 262A-1a Schwalbe ("Swallow")—the first practical jet airplane—that first flew on July 18, 1942. Able to fly well over 500 mph, the Me 262 was a spectacular symbol of what jet aircraft could and would do over the next 50 years of flight.[81]

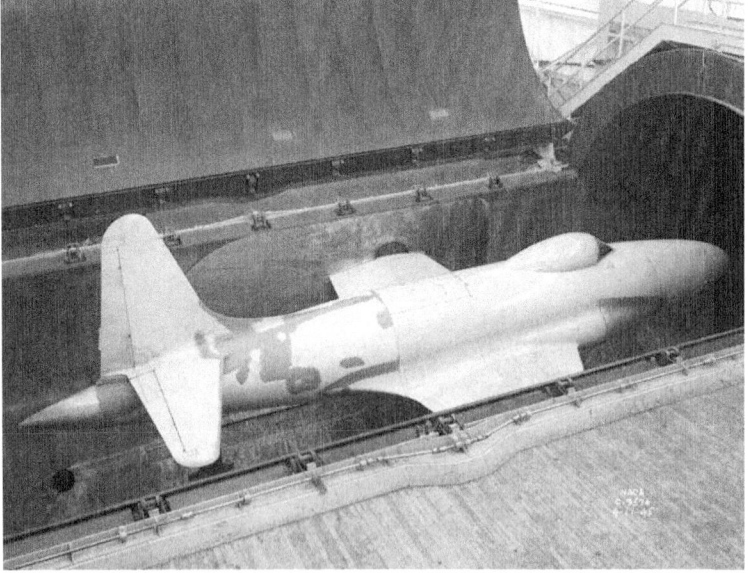

Figure 1-9. The NACA created the Altitude Wind Tunnel to test full-scale piston engines and propellers in simulated altitude conditions. The first tests of a turbojet engine took place in February 1944. During the spring of 1945, researchers installed a full-size Lockheed YP-80A Shooting Star fuselage with a GE I-40 engine for high-altitude evaluation. The tests led to data that predicted the engine thrust at all altitudes, which contributed to the operational success of America's first practical jet fighter. (NASA)

With its impressive record in making the airplane better through the interwar period, the NACA failed to recognize the significance of jet propulsion, the single most important development in aviation during the second half of the 20th century, and trailed far behind Great Britain and Germany. On the question of jet propulsion, Mead, the head of the influential Power Plants Committee, remarked in December 1942, "I doubt whether such a revolutionary change in propulsion could be developed in time to be of use in this war...."[82]

Catching Up with Europe

By the time the United States entered World War II, it was already behind in the "race" with Great Britain and Germany to develop a workable jet aircraft. The American aeronautical engine industry underwent a transition to meet the challenge of engineering the new technology. In the process, one established company, Wright Aeronautical, left the business, while another, Pratt & Whitney, persevered, and a new one, GE, rose to the challenge. By 1960, America was a leading member of a new aeronautical gas turbine industry, thanks, in large measure, to the NACA's work.

The NACA was completely unaware of the impending "turbojet" revolution. Its researchers believed that the continued evolution of the piston engine was the future direction of propulsion technology. The design and operation of the AERL reflected that practical purpose.[83] The American aeronautical community concentrated on present and immediate needs that reflected the dominance of the piston engine and propeller as the primary propulsion system during the interwar period. When faced with fighting a global aerial war, the American military embraced that standard and wedded it to large-scale production programs to avoid the strategic mistake of putting too much emphasis on new technologies that would only be practical in the long term and after costly research and development, not immediately and on aerial battlefields around the world.[84]

The two major American manufacturers of aeronautical engines were Wright Aeronautical and Pratt & Whitney. Both were pioneers in the field and were responsible for the dominance of the piston engine during the interwar and World War II periods. Wright's innovative designs included the 610-horsepower V-1400 racing engine that powered famed aviator Jimmy Doolittle's Schneider Trophy–winning Curtiss R3C-2 Racer in 1925; the 225-horsepower J-5 Whirlwind that powered Lindbergh's *Spirit of St. Louis* in 1927; and the four 2,200-horsepower R-3350 turbosupercharged radials of the Boeing B-29 Superfortress, which spearheaded the American strategic bombing campaign against Japan in 1944–1945, serving as the world's first atomic-armed bomber as well. Former Wright Aeronautical employees created Pratt & Whitney in 1925 to design and market the Wasp radial engine, which, like the Wright J-5, was a key marker in the emergent technology leading to the modern commercial and military aviation in the 1920s and 1930s. The need for aircraft that could fly faster than current technology, such as the combination of the powerful and established radial engine with the constant-speed propeller, would allow was becoming more urgent almost daily.

The impetus for aeronautical gas turbines came from within the aviation community. Airframe manufacturer Lockheed Aircraft Corporation began work on its L-1000 turbojet in 1940 in the first serious American attempt to work with the new technology. Nathan C. Price came to Lockheed in the late 1930s to develop steam turbines for aircraft but then turned to a gas turbine instead. His hiring coincided with the company's realization that a new type of power plant was needed to attain radically higher speeds. Lockheed vice president of engineering Hall L. Hibbard committed the company to the project since he believed that no engine manufacturer would build the engine.[85]

Lockheed developed the L-1000 for the L-133, an entirely new aircraft that relied entirely upon jet propulsion for power and control through small jets in the wingtips. Lockheed designers predicted that the new airplane would reach 625 mph at 50,000 feet. Lockheed was ready to develop both the engine and airframe by 1941 and formally submitted its plans to the Army Air Forces in 1942. Discussions continued until May 1943, when the Army Air Forces told Lockheed that other companies had been working on other jet engine designs since 1941. The Army Air Forces awarded Lockheed a long-term contract in mid-1943 and designated the L-1000 the XJ-37. Lockheed transferred Price, his staff, and the design to the Menasco Manufacturing Company in 1945.[86]

Northrop Aircraft began work on a turboprop aircraft during the late 1930s. A gifted Czech engineer, Vladimir H. Pavlecka, who brought his enthusiasm for using a gas turbine to drive a propeller from Europe, directed the effort. But his imaginative design, the aptly named Turbodyne, bogged down in a series of mismanaged Government contracts throughout the 1940s, though it was successfully flown in the nose of a Boeing B-17 flying test bed.[87]

Aircraft engine manufacturer Pratt & Whitney experimented with a turboprop in 1940. Engineer Leonard S. Hobbs and researcher Andrew Kalitinsky of MIT designed an engine, called the PT-1, that featured a free-piston reciprocating diesel compressor and a turbine wheel geared only to the propeller. Pratt & Whitney undertook the program as an experimental development effort that would generate design knowledge; consequently, the PT-1 was not meant for production. By 1945, the private long-term venture cost $3.3 million ($32 million in modern currency).[88]

Official American development of aeronautical gas turbine engines was more reactionary than innovative during the late 1930s and early 1940s. Due to recent German developments, the American military was very interested in rockets by 1938. General Arnold asked the NACA to investigate the issue in 1941. Members of the newly created Special Committee on Jet Propulsion, headed by the venerable William F. Durand, included the NACA, the military and naval air organizations, the Bureau of Standards, and the leading engineering universities. The Committee's industrial representatives were not aircraft engine manufacturers, but makers of industrial and marine-turbines: Westinghouse, Allis-Chalmers, and General Electric (Schenectady).[89]

Arnold requested that the leading aircraft engine manufacturers, Pratt & Whitney, Wright Aeronautical, and Allison, be excluded for two reasons. First, their exclusion would prevent them from opposing any new developments that would offset their primacy in the aeronautical marketplace. Second, exclusion prevented them from diverting financial and engineering resources away from the conventional engines that the Army was using to fight the war once the United States entered World War II. The companies received no information about gas turbine development before 1945.[90]

Durand urged the development of jet over rocket propulsion within the Committee, which recommended that the U.S. Government issue contracts with Westinghouse, Allis-Chalmers, and GE's turbine group at Schenectady. This recommendation was quite remarkable since the mainstream aeronautical community still believed that jet propulsion was not practical in 1941.

The three companies submitted very different designs. Westinghouse and Allis-Chalmers worked directly with the Navy. The former submitted a turbojet—called the 19A—that became the only original 1942 American design to fly before the end of the war. The small engine produced 1,200 pounds of thrust and flew as a booster for a Goodyear FG-1 Corsair (a co-produced derivative of the better-known Chance Vought F4U-1 Corsair) in January 1944. The improved 19B generated 1,365 pounds thrust and was the primary power plant for the McDonnell FH-1 Phantom in January 1945, the first Navy pure jet fighter to land aboard an aircraft carrier. Allis-Chalmers produced a turbine-driven ducted fan, but it suffered from a slow development program. The Army's cooperation with GE's turbine group at Schenectady resulted in the TG-100 turboprop. All three engines represented what would become the standard configuration for aeronautical gas turbines: an axial rather than centrifugal compressor driven by a turbine wheel.[91]

The most well-known American aeronautical gas turbine program was the importation and development of Whittle's design. Army representatives working with the Royal Air Force in England learned of the British turbojet program early in 1941. Great Britain's dire situation regarding the threat of an imminent

Nazi invasion convinced the British Government to assist the United States in jump-starting its gas turbine program. General Arnold inspected the Whittle engine and saw the flight of the Gloster E.28/39. Wright Field engineering officer Colonel David Keirn arrived in England in August 1941. He returned with the W.1X and drawings of the W.2B production engine on October 1, 1941, for delivery to GE. The British Government sent an early Whittle engine and drawings of the latest design to the Supercharger Division of General Electric at West Lynn, MA, in October 1941. GE's expertise in interwar turbine and turbosupercharger development made it the obvious choice to develop the American military's first jet engine. A number of British engineers, including Frank Whittle, followed to give their input. GE's improved centrifugal-flow Whittle turbojet engine, the I-A, generated 1,250 pounds of thrust.

On October 1, 1942, test pilot Robert M. Stanley took off from Muroc Dry Lake in America's first jet airplane, the Bell XP-59A Airacomet. Propelled by two GE I-A engines, the Airacomet reached a speed of 390 mph in the skies over the high desert of California. Just a few months earlier, the world's first

Figure 1-10. Captured German jet technology, especially the Junkers Jumo 004B axial-flow turbojet, was of considerable interest to NACA researchers at the Aircraft Engine Research Laboratory in the immediate post–World War II period. (NASA)

practical jet airplane, the Messerschmitt Me 262, had first flown in Germany at speeds of up to 540 mph. As America's first foray into jet aircraft, the Airacomet was a humble start since it exhibited performance equal to only the best piston-engine propeller-driven fighters. It appeared that the United States was behind in the development of a new and revolutionary technology.[92]

During the summer of 1943, Keirn received permission to inform the NACA of GE's work on the turbojet engine. That information directly led to the construction and staffing of the Jet Propulsion Static Laboratory with Kervork K. Nahigyan as the new section's head in September 1943. The unremarkable one-story building surrounded by a barbed-wire fence and located at the edge of the Cleveland airport runway featured "spin pits" that absorbed flying debris created by failed compressors.[93]

There were initial steps to work of equal value in the development of gas turbine engines and engine superchargers. In 1938, the NACA precipitated work on axial-flow compressors, the configuration found in all of today's jet engines. This work, initiated by Eastman N. Jacobs and Eugene W. Wasielewski, was followed up by considerable work at Langley. Jacobs was a highly influential researcher due to his work on symmetrical section laminar-flow airfoils; the application of airfoil theory, a hallmark of the NACA's interwar work, to the design of multistage compressors led to a more comprehensive understanding of compressors.[94]

Jacobs's engine received full endorsement by Durand's Special Committee above the other ideas submitted by the spring of 1941. Its ducted fan configuration consisted of a piston engine and a two-stage axial compressor. Air entered the duct and was compressed, mixed with atomized fuel, and ignited in the combustion chamber. The heated gas exited through a high-speed nozzle to propel the engine forward. The engine was a modified version of the power plant used by Secondo Campini to power the Caproni N.1 and quickly became known as "Jake's Jeep."[95] There were many development problems related to the axial-flow compressor, but it was very clear that the NACA believed that the configuration was the correct path.[96]

The rapid advance of pure gas turbine engines and the flight of the XP-59A caused the Army to cancel the ducted fan engine originated by Eastman Jacobs at Langley in February 1943, though it lingered officially until the Durand Committee canceled it on April 15, 1943. The failed Jeep played a role in future turbojet development in the United States as a point of departure for axial-flow compressor research, but it contributed nothing further.[97] The axial compressor of the first turbojet developed by the GE Schenectady group, the 3,750-pound-thrust TG-180 (J35), benefited from the NACA's research, but the compressor first flew in a Republic XP-84 Thunderjet in February 1946.[98]

The NACA's failure to anticipate the pure turbojet developments in Germany and Britain prior to the Second World War seriously damaged its reputation with the military, particularly with Hap Arnold's Army Air Forces. Afterward, due to its failure to anticipate jet propulsion before 1941 and the distraction of the failed Jacobs ducted fan during the critical early years of the war, Arnold effectively took the NACA out of the leading edge of aircraft propulsion by giving GE the task of developing the Whittle engine. Instead, the NACA was given the task of testing already-developed engines from GE and Westinghouse in its Static Test Laboratory.[99]

All in all, the NACA's Cleveland engine laboratory had a remarkable blind eye when it came to jet propulsion. No consideration was given to the field at all in the development of the facility. The great potential significance of the jet engine was enunciated by George Lewis in August 1943.[100] Afterward, by March 1944, a complete transformation toward work on jet propulsion had occurred at the agency.[101] Meanwhile, in October 1943, Kervork Nahigyan and his staff in the Static Laboratory built and tested the first afterburner, a direct byproduct of the work on the burner for Jake's Jeep.[102]

Abe Silverstein and his group adapted the AWT to test jet engines and acquired hands-on experience in jet technology. During the AWT facility's first runs in February 1944, NACA researchers used the AWT to support new turbojet technology, specifically the 1,600-pound-thrust GE I-16 turbojet installed in a complete Bell P-59 Airacomet fuselage. The Cleveland laboratory's use of the AWT produced a succession of turbojet advancements that resulted in a surge of thrust capabilities in the late 1940s and early 1950s. In June 1944, Lockheed's YP-80A Shooting Star became the first jet aircraft completely manufactured in the United States and the first U.S. aircraft to fly faster than 500 mph. The Altitude Wind Tunnel was used in the spring of 1945 to study the performance of the Shooting Star's two 3,750-pound-thrust GE I-40 engines at high altitudes. An attempt to forecast thrust levels at altitude, based on sea-level measurements, was successful, and a performance curve was created to predict the I-40's thrust at all altitudes. The production P-80 fighter with the I-40 engine proved a great success in the early days of the U.S. Air Force, spawning two important derivatives, the T-33 trainer and the F-94 interceptor.[103]

Lewis was the only facility capable of testing gas turbines in a comprehensive manner during the immediate postwar period. The NACA researchers there would build upon this experience to become the U.S. Government's experts in jet propulsion technology immediately after the war.

Figure 1-11. This image shows America's first jet engine: the General Electric I-A Engine. (National Air and Space Museum, Smithsonian Institution, NASM 9A11739)

NACA Research Transitions from Piston to Jet

During the early days of the Cold War, the aeronautical propulsion industry was at a crossroads. Would it continue with propeller-driven piston engine aircraft or go with the new jet technology? In 1945, the piston engine was the main power plant for aviation. Within 3 years, there were multiple alternatives.[104] The NACA's propulsion research in Cleveland transitioned from wartime development troubleshooting to long-term fundamental research. In December 1945, a laboratory plan divided agency research into nine categories with varying emphasis: turbojets (20 percent); turboprops (20 percent); continuous ramjets (12.5 percent); intermittent ramjets (5.5 percent); rocket engines (4 percent); reciprocating engines (13 percent); compound engines (reciprocating engine and turbosupercharger) (15 percent); icing research (5 percent); and "engines for supersonic flight" (5 percent). The priorities reflected the overall shift toward fundamental research, with clear lines of responsibility between the NACA and industry, which reflected overall the National Aeronautical Research Policy approved in March 1946.[105]

The Power for Flight

The NACA renamed the AERL as Lewis Flight Propulsion Laboratory in September 1948 to honor George W. Lewis (1882–1948), the NACA's first Executive Officer and Director of Aeronautical Research. In a lecture before the Industrial College of the Armed Forces that fall, the NACA's Director of Aeronautical Research, Dr. Hugh L. Dryden, identified the existence of a "technical revolution" in the immediate postwar period. In 1945, the propeller-driven four-engine Boeing B-29 Superfortress bomber and the Lockheed straight-wing P-80 Shooting Star turbojet fighter were on the cutting edge of military aircraft technology. The swept-wing and turbojet-powered Boeing B-47 Stratojet and North American F-86 Sabre fighter loomed on the immediate future's horizon in 1949. Dryden outlined its characteristics, which consisted of two areas, aerodynamics and propulsion, and the central purpose, speed. The revolution required new airframe configurations and a replacement for the reciprocating piston engine.[106]

Dryden continually justified the need for innovation in turbojet engine research and, by extension, for Lewis Laboratory. The work mirrored the state of the art in technology to keep up with Soviet developments, especially as they pertained to the air war over Korea. Dryden conceded that American manufacturers were capable of producing first-class turbojet, ramjet, turboprop, and rocket engines, but they had yet to make one that was fuel-efficient and made entirely with materials sourced in the Western Hemisphere. What would make that possible would be a vision rooted in the accomplishments and experience of the past combined with ingenuity and research in science and technology.[107] It is important to realize that Dryden was talking about military, not civil, aircraft, since he believed that "at the present time…civil aviation looks to be a very unimportant phase."[108]

The transonic region between Mach 0.75 and Mach 1.25 constituted a particular area of interest in the early years of the Jet Age. Wind tunnel test technology had not yet adequately caught up with the speed potential of the jet engine, necessitating the use of specialized research aircraft that effectively used the sky as a laboratory. Along the way, a number of difficulties were discovered that required solutions. Tests of the new Convair YF-102 delta-wing jet fighter indicated unexpectedly high drag rise, a problem overcome by an extraordinarily gifted NACA researcher, Richard Whitcomb, who postulated the transonic area rule. Modified with a so-called "wasp waist" fuselage that had a higher fineness ratio than the original design's, the production F-102 became a great success and a Cold War mainstay of the Air Force's Air Defense Command. Whitcomb's revolutionary ideas about transonic drag rise and fuselage shaping influenced the design of all subsequent transonic high-performance airplanes.

But if concepts such as the area rule and other advanced aerodynamic shaping ideas were to be implemented, they required power plants capable of propelling aircraft into the flight regimes where they could work. In Lewis's case, its research on turboprops, jet engines, ducted fans, compressors, cooled turbines, stable afterburning, and advanced propulsion concepts such as ramjets and rockets greatly advanced aircraft capabilities into the postwar era.[109]

The NACA conducted as well as sponsored important research in aircraft engine technology during the post–World War II period.[110] The arrival of German scientists in the United States created an influx of advanced knowledge, which included Ernst Eckert's groundbreaking work on heat transfer. Laboratory staff also quickly built up an unequaled expertise in aircraft engine testing that placed them between the hardware-oriented industry and theoretical academic researchers. That caused considerable tension regarding the issue of the dissemination of information generated by Government researchers and the proprietary rights of manufacturers. Nevertheless, Lewis researchers excelled at gaining a broader understanding of turbojet design and sharing that information. The pivotal 1956 "Compressor Bible," rooted in the work on Jake's Jeep, was made available on a confidential basis to industry.[111] By the 10th anniversary of the opening of Lewis, Hugh Dryden confidently announced, "The United States no longer trails in jet propulsion."[112]

These developments did not guarantee success for America's aircraft engine industry. Wright Aeronautical and its parent corporation, Curtiss-Wright, emerged from World War II in a sound economic state, but their leaders favored a dividend for shareholders rather than investing heavily in gas turbine research. Curtiss-Wright executives decided that a spare-parts, maintenance, and repair business for Wright radial engines would be the primary focus of its aeronautical engine program. By 1960, the company had devolved into a subcontractor supplying aircraft subassemblies and component parts for its former competitors—a cautionary tale about the failure to innovate and keep up with the times.[113]

Pratt & Whitney had more success. Encouraged by the U.S. Navy, the company bought the license for the Rolls-Royce Nene turbojet in 1948 and manufactured its J42 and J48 variants, which powered the Grumman F9F Panther carrier-based jet fighter (which fought Soviet MiG-15s powered by the Klimov VK-1, the Soviet variant of the Nene).[114] Pratt & Whitney oversaw the introduction of the most important turbojet engine of the early postwar era, the J57. The new engine, the first capable of generating 10,000 pounds of thrust, powered America's frontline aircraft of the Cold War, including the B-52 Stratofortress intercontinental bomber, the North American F-100 Super Sabre, and the Convair F-102 Delta Dagger. With over 21,000 J57s in service by 1960, Pratt & Whitney was one of the world's leading gas turbine engine manufacturers.[115]

The Power for Flight

Despite the emphasis on military aircraft, the development of turbojet engines for commercial use was reaching a new level of importance. The Pratt & Whitney JT3s that powered the pioneering Boeing 707 and Douglas DC-8 airliners as they opened the age of mass jet travel were adaptations of the highly successful J57 military engines. Lewis Associate Director Abe Silverstein and Physics Division Chief Newell D. Sanders proclaimed in April 1956 that the "age of the jet transport has arrived."[116] Although these engines were designed for speed above all other criteria, they caused specific problems in the areas of noise, reliability, and safety, among other issues, when compared to piston engines.

Toward the Turboprop

The apparent primacy of the jet in the immediate years after World War II led one popular aviation writer to ask, "Has the propeller a future?"[117] The propeller community conducted numerous wind tunnel and flight research experiments in conjunction with similar investigations by industry to see

Figure 1-12. Besides the turbojet, NACA researchers in Cleveland, OH, investigated other new propulsion technologies. The first turboprop flown in the United States, the GE TG-100A, underwent tests in a streamlined nacelle to determine the performance characteristics of the compressor and turbine in the Altitude Wind Tunnel in late 1946. (NASA)

if they could provide the answer. One focus was on developing supersonic propellers capable of taking long-range transport aircraft into the transonic regime between Mach 0.8 and 1.2.[118] At the Langley Aeronautical Laboratory in Virginia, NACA engineers created a "propeller research airplane" by installing an Aeroproducts propeller in the nose of a McDonnell XF-88B test bed (the predecessor to the far more capable F-101A Voodoo). The flight program, which ran from 1953 through 1956, revealed a propeller design that was 79 percent efficient at a speed of Mach 0.95.[119]

Such promising programs appeared to be futile in the early days of the Jet Age. At a 1949 NACA conference on transonic aircraft design, a Hamilton Standard engineer, having just reported on the results of wind tunnel tests on a supersonic propeller for a U.S. Air Force contract, remarked that when there was a choice between a propeller and a jet engine, "even if the propeller is good, it is not wanted."[120] The aeronautical community faced many developmental problems with supersonic propellers in the 1950s, and the propellers seemed unnecessary if jet technology provided equal or better performance. A major challenge was reducing the noise that resulted from the shock waves at the blade tips. The four-blade supersonic Aeroproducts propeller on the Air Force's experimental Republic XF-84H propulsion test bed, which offered mediocre performance overall, generated such high-intensity noise and resonance effects that it rendered bystanders sick.[121] The NACA, which targeted foreseeable fundamental aeronautics problems facing America, lost interest and disbanded the longstanding Subcommittee on Propellers for Aircraft in 1957 as the four-decade history of the organization came to an end. Propellers as a major research area disappeared in the early days of NASA (established in 1958).[122]

Establishment of the Propulsion Systems Laboratory

The Lewis Laboratory in Cleveland struggled to keep up with the rapid pace of aircraft engine development in the wake of World War II. With a war mentality in full effect, the NACA marshaled its resources to help develop and refine new aircraft, missile, and rocket engines for the American military. Design work began in 1948 on a facility with expanded capability as tensions between East and West increased during the early days of the Cold War. Two 14-foot-diameter and 24-foot-long test chambers, called Nos. 1 and 2, combined the static sea level test stands with the complex Altitude Wind Tunnel, which recreated actual flight conditions on a larger scale. The new Propulsion Systems Laboratory (PSL) opened 4 years later in 1952 as full-out war raged on the Korean Peninsula. The PSL was the only facility in the United States capable of operating increasingly powerful and complex large, full-size aircraft and rocket propulsion systems in simulated altitude conditions. The ability to

The Power for Flight

control the test environment was important in the advancement of the aircraft engine systems and placed the researchers of the PSL at the cutting edge of propulsion development in the 1950s.[123]

PSL No. 1 hosted exclusively turbojet tests as the first truly powerful and American-designed axial flow designs emerged. The Air Force requested the help of the NACA's Cleveland laboratory in improving the afterburner performance of the prototype GE XJ79-GE-1 afterburning turbojet in 1957. The J79, destined to become the iconic engine of the early Mach 2 era (it propelled such mainstays as the Lockheed F-104A Starfighter, the McDonnell F4H-1 Phantom II, the Convair B-58A Hustler, and the North American A3J-1 Vigilante), featured innovative variable stators designed by GE engineer Gerhard Neumann. The variable stators helped maintain efficient compression of the airstream during all flight regimes as the air was progressing to the engine face. PSL tests simulating speed and altitude conditions of Mach 2 at 59,400 feet revealed that modification of the fuel system and flame holder increased combustion efficiency by 19 percent, reduced pressure drop, and lowered fuel consumption by 10 percent.[124] The combination of GE's J79 with the Air Force's Lockheed F-104A Starfighter supersonic interceptor in 1958 won the two companies a shared Collier Trophy for 1958.

Figure 1-13. This image shows the General Electric J79 in PSL No. 1. (NASA)

Early Ramjet Research

With the Cold War pushing the need for expanded nuclear and conventional capabilities, the United States embraced captured Nazi technology and repurposed it in its struggle against the Soviet Union. One area in which German scientists (and Soviet ones as well) had made significant advances was in the design of ramjet engines for aircraft and missile systems. A ramjet, consisting of an inlet, combustion chamber, and exhaust nozzle, is the simplest and lightest form of high-speed propulsion within the atmosphere. The NACA had a long heritage of conducting ramjet research dating to the early postwar era, launching a variety of small ramjet test vehicles from a modified Northrop P-61C Black Widow mother ship, and using a special test pylon installed on the P-61 so that it could test special "two dimensional" ramjet configurations. (One of the P-61 project pilots was future X-15 research pilot and Gemini-Apollo astronaut Neil A. Armstrong.) The designers of the surface-launched Boeing IM-99 Bomarc pilotless interceptor missile and the North American SM-64 Navaho supersonic intercontinental cruise missile incorporated hybrid propulsion systems for higher performance. Rockets boosted these missiles into the air; once the missiles were up to speed, ramjets propelled them to their

Figure 1-14. In this image, a Marquardt RJ43 ramjet engine for the Boeing IM-99 BOMARC pilotless interceptor missile is being prepared for testing in PSL No. 1 in 1954. (NASA)

targets. PSL researchers investigated ramjets for these platforms from 1954 to 1956. The Bomarc relied upon two 28-inch-diameter Marquardt RJ43 ramjets to propel it above Mach 2. PSL researchers evaluated the responsiveness of the fuel control system and the pneumatically actuated shock-positioning control unit that was crucial to supersonic operation.[125] For the Navaho's twin 48-inch-diameter Pratt & Whitney XRJ47s, they investigated ignition, burner and flame holder designs, fuel flow control, and overall engine performance as they ran the engine at Mach 2.75 and simulated altitudes between 58,000 and 73,000 feet.[126] The advent of faster-response intercontinental ballistic missiles, budgetary problems, and difficulties in making ramjets practical made the Navaho short-lived, but the Bomarc became a standard surface-to-air missile for the United States and Canada, tasked with defending the North American continent against incoming Soviet nuclear bombers.[127]

Aircraft Nuclear Propulsion

Researchers at Lewis also investigated a new frontier for aviation: nuclear propulsion. Abe Silverstein believed that it was a new opportunity to extend the range and speed of aircraft and to collaborate with the Atomic Energy Commission (AEC) and the Nuclear Energy for the Propulsion of Aircraft (NEPA) program initiated in 1946. That interest coincided with the opening of the Plum Brook Station and its nuclear research reactor in Sandusky, OH, in 1956. Silverstein believed that Lewis could contribute to studies related to the effects of radiation on materials and reorganized the Lewis research groups to reflect that focus. Unfortunately, the impracticality of a nuclear-powered airplane, owing to the excessive weight of the airborne reactor and environmental concerns, led to the cancellation of the program in 1961. The Federal Government invested $1 billion in the failed project, which included both GE and Pratt & Whitney. As interest in a nuclear-powered airplane waned, NEPA and Lewis interest shifted to the development of a nuclear rocket for space travel instead, a subject beyond the scope of this study.[128]

Endnotes

1. Annual Report of the National Advisory Committee for Aeronautics, 1915 (hereafter "NACA AR," followed by year) (Washington, DC: Government Printing Office, 1916), p. 9.
2. Alex Roland, *Model Research: The National Advisory Committee for Aeronautics*, vol. 1 (Washington, DC: NASA SP-4103, 1985), pp. 29–30.
3. For examples from German publications, see A. Heller, "The 300 h.p. Benz Aircraft Engine," translated from *Zeitschrift des Vereines Deutsche Ingenieure* (1920), NACA Technical Note No. 34 (1921) (hereafter "NACA TN," followed by the number and year); and Otto Schwager, "Notes on the Design of Supercharged and Over-Dimensioned Aircraft Motors," translated from *Technische Berichte*, vol. 3, NACA TN 7, 1920.
4. The *Bibliography of Aeronautics* appeared in separate volumes during the period 1921–1936.
5. NACA AR 1918, pp. 25–26. See E.M. Nutting and G.W. Lewis, "Air Flow Through Poppet Valves" (Washington, DC: NACA Technical Report No. 24, 1918) (hereafter "NACA TR," followed by the number and year), p. 25.
6. Luke Hobbs, interview by Walter T. Bonney, October 27, 1971, file 001018, p. 11 of transcript, NASA Historical Reference Collection, NASA History Program Office, NASA Headquarters, Washington, DC (hereafter "NASA HRC").
7. Marsden Ware, "Description of the NACA Universal Test Engine and Some Test Results," NACA TR 250, 1920; James R. Hansen, *Engineer in Charge: A History of the Langley Aeronautical Laboratory, 1917–1958* (Washington, DC: NASA Scientific and Technical Information Office, 1987), p. 423.
8. Edward Constant identifies this paradigm as "normal technology" that reflects the status quo in the history of technical systems. Edward W. Constant, *The Origins of the Turbojet Revolution* (Baltimore: Johns Hopkins University Press, 1980), pp. 10–11.
9. Virginia P. Dawson, *Engines and Innovation: Lewis Laboratory and American Propulsion Technology* (Washington, DC: NASA SP-4306, 1991), p. 43; Oscar Schey, Benjamin Pinkel, and Herman H. Ellerbrock, Jr., "Correction of Temperatures of Air-Cooled Engine Cylinders for Variation in Engine and Cooling Conditions," NACA TR 645, NACA AR 1939.
10. Hobbs interview.

The Power for Flight

11. See NASA TN 634 and TRs 634, 644; and Hansen, *Engineer in Charge*, p. 450.
12. For a history of the program from the viewpoint of the development of engineering methodology, see Walter G. Vincenti, "Air-Propeller Tests of W.F. Durand and E.P. Lesley: A Case Study in Technological Methodology," *Technology and Culture* 20 (1979): 712–751.
13. "Problems," NACA AR 1915, pp. 13, 15; Roland, *Model Research*, pp. 33, 46; Roger E. Bilstein, *Orders of Magnitude: A History of the NACA and NASA, 1915–1990* (Washington, DC: NASA SP-4406, 1989), p. 4.
14. J.L. Nayler and E. Ower, *Aviation: Its Technical Development* (London: Peter Owen/Vision Press, 1965), p. 156; R.T.C. Rolt, *The Aeronauts: A History of Ballooning, 1783–1903* (New York: Walker and Co., 1966), pp. 82, 203. Pioneer French aeronaut Jean-Pierre Blanchard used a hand-cranked propeller on the first flight across the English Channel in 1785. Jean Baptiste Marie Meusnier, the "father of the dirigible," initiated the trend toward using propellers in his airship designs during the same period.
15. Walter G. Vincenti, "Air-Propeller Tests of W.F. Durand and E.P. Lesley: A Case Study in Technological Methodology," *Technology and Culture* 20 (1979): 718–719; Fred E. Weick, *Aircraft Propeller Design* (New York: McGraw-Hill Book Co., 1930), pp. 37–38.
16. Tom D. Crouch, *A Dream of Wings: Americans and the Airplane, 1875–1905* (New York: Norton, 1981; reprint, Washington, DC: Smithsonian Institution Press, 1989), pp. 33, 294; Peter L. Jakab, *Visions of a Flying Machine: The Wright Brothers and the Process of Invention* (Shrewsbury, England: Airlife, 1990), pp. 184, 194–195.
17. "Work of the Committee," NACA AR 1915, p. 12; Vincenti, "Air-Propeller Tests," pp. 718–719.
18. "Problems," NACA AR 1915, pp. 13, 15; Vincenti, "Air-Propeller Tests," p. 720. Vincenti asserts that there are inherent similarities in marine-propeller and air-propeller research, specifically regarding methodology. For a contemporary view of the relation between aeronautical and marine engineering, see Jerome C. Hunsaker, "Aeronautics in Naval Architecture," *Transactions of the Society of Naval Architects and Marine Engineers* 32 (1924): 1–25.
19. Frederick E. Terman, "William Frederick Durand (1859–1958)" in *Aeronautics and Astronautics: Proceedings of the Durand Centennial Conference Held at Stanford University, 5–8 August 1959*, ed. Nicholas J. Huff and Walter G. Vincenti (New York: Pergamon Press, 1960), pp. 4–7. Durand's life spanned the growth of the NACA and American

aeronautics. In 1933, Durand left the NACA—only to return for the duration of World War II to head the Committee's jet propulsion investigations.
20. Vincenti, "Air-Propeller Tests," p. 722. This article was later incorporated into Vincenti's *What Engineers Know and How They Know It: Analytical Studies from Aeronautical History* (Baltimore: Johns Hopkins University Press, 1990).
21. Terman, "William Frederick Durand," p. 7.
22. Roland, *Model Research*, pp. 33–34.
23. "General Problems," NACA AR 1916, p. 14; Vincenti, "Air-Propeller Tests," pp. 718, 721–722.
24. William F. Durand, Wilbur Wright Memorial Lecture, "Some Outstanding Problems in Aeronautics," NACA AR 1918, pp. 33, 39.
25. William F. Durand, "Experimental Research on Air-Propellers," NACA TR 14, 1917, pp. 87–88, 91. Even though Lesley is denied equal authorship, Durand acknowledged Lesley's importance to the tests and completion of the report within the body of the text. Vincenti, "Air-Propeller Tests," p. 722. Vincenti describes the Eiffel-type tunnel as having a "free-jet test seam inside a closed room, tapering intake and exit channels, and return flow within the surrounding building."
26. Vincenti, "Air-Propeller Tests," pp. 727, 729.
27. Durand, "Experimental Research on Air-Propellers," pp. 83, 85; "Summaries of Technical Reports," NACA AR 1917, p. 27.
28. "Financial Report," NACA AR 1917, p. 30.
29. Roland, *Model Research*, p. 46.
30. William F. Durand and E.P. Lesley, "Experimental Research on Air-Propellers, II," NACA TR 30, 1918, p. 261; Durand, "Some Outstanding Problems," pp. 31, 41–42.
31. "Problems in Propeller Design," *Aviation and Aeronautical Engineering* 4 (June 1918): 108.
32. The final administrative group was the Personnel, Buildings, and Equipment committee. Roland, *Model Research*, p. 74.
33. William F. Durand and E.P. Lesley, "Experimental Research on Air-Propellers, V," NACA TR 141, 1922, p. 169; "Report of the Committee on Aerodynamics," NACA AR 1922, p. 39.
34. Max M. Munk, "Analysis of W.F. Durand's and E.P. Lesley's Propeller Tests," NACA TR 175, 1923, p. 291.
35. NACA AR 1918, p. 13.
36. Vincenti, "Air-Propeller Tests," pp. 732–733.

37. William F. Durand and E.P. Lesley, "Comparison of Tests on Airplane Propellers in Flight with Wind Tunnel Model Tests on Similar Forms," NACA TR 220, 1925, p. 273.
38. Vincenti, "Air-Propeller Tests," p. 734.
39. D.W. Taylor, "Some Aspects of the Comparison of Model and Full-Scale Tests," NACA AR 1925, pp. 265–266.
40. Durand, "Some Outstanding Problems," pp. 41–42.
41. E.P. Lesley and B.M. Woods, "The Effect of Slipstream Obstructions on Air-Propellers," NACA TR 177, 1923, pp. 313, 332; Vincenti, "Air-Propeller Tests," p. 737.
42. William F. Durand, "Interaction Between Air-Propellers and Airplane Structures," NACA TR 235, 1926, pp. 107, 109. This would be the next-to-last test conducted by Durand on behalf of the NACA.
43. Vincenti, "Air-Propeller Tests," p. 739.
44. For more information on the importance of the Propeller Research Tunnel in the subsequent aeronautical revolution of the late 1920s and 1930s, see Hansen, *Engineer in Charge*.
45. Fred E. Weick and Donald H. Wood, "The Twenty-Foot Propeller Research Tunnel of the National Advisory Committee for Aeronautics," NACA TR 300, 1929; George Gray, *Frontiers of Flight: The Story of NACA Research* (New York: Knopf, 1948), p. 208.
46. Gray, *Frontiers of Flight*, pp. 208–209.
47. Ibid., p. 208.
48. John Stack, "Tests of Airfoils Designed To Delay the Compressibility Burble," NACA TN 976, 1944; Gray, *Frontiers of Flight*, pp. 210–211.
49. Gray, *Frontiers of Flight*, pp. 212–213.
50. Ibid., pp. 213–215.
51. Ibid., p. 216.
52. Stephen L. McFarland, "Higher, Faster, and Farther: Fueling the Aeronautical Revolution, 1919–1945," in *Innovation and the Development of Flight*, ed. Roger D. Launius (College Station, TX: Texas A&M University Press, 1999), p. 101; Robert Schlaifer and Samuel D. Heron, *Development of Aircraft Engines and Aviation Fuels* (Boston: Harvard University Graduate School of Business Administration, 1950), p. 568.
53. Miller authored a series of reports on this topic that appeared during the period 1941 to 1946. See Cearcy D. Miller, "A Study by High Speed Photography of Combustion and Knock in a Spark-Ignition Engine," NACA TR 727, 1942; Dawson, *Engines and Innovation*, p. 24.
54. McFarland, "Higher, Faster, and Farther," p. 117.

55. Ernest F. Fiock and H. Kendall King, "The Effect of Water Vapor on Flame Velocity in Equivalent CO-O$_2$ Mixtures," NACA TR 531, 1935; Addison M. Rothrock, Alois Krsek, and Anthony W. Jones, "The Induction of Water to the Inlet Air as a Means of Internal Cooling in Aircraft Engine Cylinders," NACA TR 756, 1943; McFarland, "Higher, Faster, and Farther," pp. 117–118.
56. Marsden Ware, "Description and Laboratory Tests of a Roots Type Aircraft Engine Supercharger," NACA TR 230, 1920, pp. 451–561.
57. Luke Hobbs, interview by Walter T. Bonney, October 27, 1971, file 001018, NASA HRC, p. 11.
58. C. Fayette Taylor, *Aircraft Propulsion: A Review of the Evolution of Aircraft Piston Engines* (Washington, DC: Smithsonian Institution Press, 1971), pp. 72–73.
59. Taylor, *Aircraft Propulsion*, p. 60. For more on the NACA's diesel studies, see Arthur W. Gardiner, "A Preliminary Study of Fuel Injection and Compression Ignition as Applied to an Aircraft Cylinder," NACA TR 243, 1927; A.M. Rothrock, "Hydraulics of Fuel Injection Pumps for Compression-Ignition Engines," NACA TR 396, 1932; and A.M. Rothrock and C.D. Waldron, "Effects of Air-Fuel Ratio on Fuel Spray and Flame Formation in a Compression-Ignition Engine," NACA TR 545, 1937.
60. Dawson, *Engines and Innovation*, p. 24.
61. Addison M. Rothrock, Alois Krsek, and Anthony W. Jones, "The Induction of Water to the Inlet Air as a Means of Internal Cooling in Aircraft Engine Cylinders," NACA TR 756, 1943.
62. Taylor, *Aircraft Propulsion*, p. 67; Jack Connors, *The Engines of Pratt & Whitney: A Technical History* (Reston, VA: American Institute for Aeronautics and Astronautics [AIAA], 2010), pp. 143–144.
63. Hansen, *Engineer in Charge*, pp. 123–140.
64. Donald W. Douglas, "The Development and Reliability of the Modern Multi-Engine Air Liner," *Journal of the Royal Aeronautical Society* 40 (November 1935): 1042.
65. Dawson, *Engines and Innovation*, p. 1.
66. Cary Hoge Mead, *Wings over the World: The Life of George Jackson Mead* (Wauwatosa, WI: Swannet Press, 1971), pp. 214–219.
67. Dawson, *Engines and Innovation*, p. 9.
68. Ibid., pp. 11–13.
69. Ibid., p. 19.
70. Dawson, *Engines and Innovation*, pp. 24–25; Hansen, *Engineer in Charge*, p. 422.
71. Dawson, *Engines and Innovation*, p. 25.

72. A.M. Rothrock and Arnold E. Biermann, "The Knocking Characteristics of Fuels in Relation to Maximum Permissible Performance of Aircraft Engines," NACA TR 655, 1940, pp. 267–288.
73. Dawson, *Engines and Innovation*, pp. 26–27.
74. The AWT was dismantled in 2008.
75. Mead, *Wings over the World*, pp. 156–157; Dawson, *Engines and Innovation*, pp. 27–31.
76. "NACA: The Force Behind Our Air Supremacy," *Aviation* 43 (January 1944): 175.
77. Margaret Conner, *Hans von Ohain: Elegance in Flight* (Reston, VA: AIAA, 2001), pp. 91–93.
78. The Gloster E.28/39 and the first Bell XP-59A are on exhibit in national museums, the E.28/39 in Britain's Science Museum, South Kensington, London, England, and the XP-59A in the National Air and Space Museum of the Smithsonian Institution, Washington, DC.
79. Frank Whittle, "The Birth of the Jet Engine in Britain," in *The Jet Age: Forty Years of Jet Aviation*, ed. Walter J. Boyne and Donald S. Lopez (Washington, DC: Smithsonian Institution, 1979), pp. 3–16, 20.
80. Hans von Ohain, "The Evolution and Future of Aeropropulsion Systems," in *Jet Age*, ed. Boyne and Lopez, pp. 29–34, 36.
81. Anselm Franz, "The Development of the 'Jumo 004' Turbojet Engine," in *Jet Age*, ed. Boyne and Lopez, pp. 69–74.
82. Mead quoted in Mead, *Wings over the World*, p. 270.
83. Dawson, *Engines and Innovation*, p. 19.
84. James O. Young, "Riding England's Coattails: The U.S. Army Air Forces and the Turbojet Revolution," in *Innovation and the Development of Flight*, ed. Roger D. Launius (College Station, TX: Texas A&M University Press, 1999), p. 271.
85. J.R. Kinney, "Starting from Scratch?: The American Aero Engine Industry, the Air Force, and the Jet, 1940–1960," AIAA Report 2003-2671, July 2003, p. 3.
86. Ibid., p. 3.
87. Ibid., p. 3.
88. Ibid., p. 3.
89. Ibid., p. 3.
90. Ibid., pp. 3–4.
91. Ibid., p. 4.
92. Dawson, *Engines and Innovation*, p. 41; Kinney, "Starting from Scratch?," p. 4.
93. Dawson, *Engines and Innovation*, p. 42.

94. John T. Sinnette, Oscar W. Schey, and J. Austin King, "Performance of NACA Eight-Stage Axial-Flow Compressor Designed on the Basis of Airfoil Theory," NACA TR 758, 1943.
95. Macon C. Ellis, Jr., and Clinton E. Brown, "NACA Investigation of a Jet-Propulsion System Applicable to Flight," NACA TR 802, 1943.
96. Dawson, *Engines and Innovation*, pp. 48–49.
97. Ibid., pp. 42, 55, 57.
98. Dawson, *Engines and Innovation*, p. 54; Kinney, "Starting from Scratch?," p. 5.
99. Dawson, *Engines and Innovation*, p. 57; Roland, *Model Research*, p. 185.
100. Minutes of the Meeting of the Special Committee on Jet Propulsion, August 18, 1943, National Archives and Records Administration (NARA) record group (RG) 255, p. 11; Dawson, *Engines and Innovation*, p. 57.
101. Report to the Executive Committee, March 16, 1944, NACA Executive Committee Minutes, NARA RG 255, box 9.
102. Dawson, *Engines and Innovation*, p. 58.
103. Langley Research Center, "World War II and the NACA," fact sheet FS-LaRC-95-07-01, July 1995, available online at *http://www.nasa.gov/centers/langley/news/factsheets/WWII.html* (accessed May 1, 2012).
104. Hugh L. Dryden, "Research and Development in Aeronautics," October 4, 1948, Publication No. L49-25 (Washington, DC: Industrial College of the Armed Forces, 1948), p. 27.
105. Dawson, *Engines and Innovation*, p. 70.
106. Dryden, "Research and Development in Aeronautics," p. 27.
107. Hugh L. Dryden, "Jet Engines for War," January 23, 1951, NASA HRC file 40958, pp. 5, 6.
108. Dryden, "Research and Development in Aeronautics," p. 16.
109. Dhanireddy R. Reddy, "Seventy Years of Aeropropulsion Research at NASA Glenn Research Center," *Journal of Aerospace Engineering* 26 (April 2013): 202.
110. Taylor, *Aircraft Propulsion*, p. 67; Connors, *The Engines of Pratt & Whitney*, p. 95.
111. Irving A. Johnsen and Robert O. Bullock, eds., *Aerodynamic Design of Axial-Flow Compressors* (Washington, DC: NASA SP-36, 1965), pp. iii, 1–8; Dawson, *Engines and Innovation*, pp. 127–144.
112. Dryden, "Jet Engines for War," p. 3.
113. Kinney, "Starting from Scratch?," p. 5.
114. Connors, *The Engines of Pratt & Whitney*, pp. 201–212.

115. Kinney, "Starting from Scratch?," p. 5.
116. Abe Silverstein and Newell D. Sanders, "Concepts on Turbojet Engines for Transport Application" (paper presented to the Society of Automotive Engineers [SAE] Aeronautic Meeting, New York, April 10, 1956), p. 1, NASA HRC, file 41567.
117. William Winter, "Has the Propeller a Future?," *Popular Mechanics* 89 (February 1948): 171.
118. Eugene C. Draley, Blake W. Corson, Jr., and John L. Crigler, "Trends in the Design and Performance of High-Speed Propellers," in *NACA Conference on Aerodynamic Problems of Transonic Airplane Design: A Compilation of Papers Presented, September 27–29, 1949*, NASA TM-X-56649, pp. 483–498; John V. Becker, *The High Speed Frontier: Case Histories of Four NACA Programs, 1920–1950* (Washington, DC: NASA SP-445, 1980), pp. 135–138.
119. An Allison XT38 turboprop engine powered the Aeroproducts propeller. Jerome B. Hammack, Max C. Kurbjun, and Thomas C. O'Bryan, "Flight Investigation of a Supersonic Propeller on a Propeller Research Vehicle at Mach Numbers to 1.01," NACA RM L57E20, 1957, pp. 5–6; Jerome B. Hammack and Thomas C. O'Bryan, "Effect of Advance Ratio on Flight Performance of a Modified Supersonic Propeller," NACA TN 4389, September 1958, p. 5.
120. Thomas B. Rhines, "Summary of United Aircraft Wind Tunnel Tests of Supersonic Propellers," in *NACA Conference on Aerodynamic Problems of Transonic Airplane Design: A Compilation of Papers Presented, September 27–29, 1949*, NASA TM-X-56649, p. 448.
121. Stephan Wilkinson, "ZWRRWWWBRZR: That's the Sound of the Prop-Driven XF-84H, and It Brought Grown Men to Their Knees," *Air & Space* (July 2003), available online at *http://www.airspacemag.com/how-things-work/zwrrwwwbrzr-4846149/* (accessed on January 12, 2016).
122. Becker, *The High Speed Frontier*, p. 136.
123. "New Altitude Test Facilities Aid Improvements of Turbojets," NACA AR 1952, pp. 4–6; Robert S. Arrighi, *Pursuit of Power: NASA's Propulsion Systems Laboratory No. 1 and 2* (Washington, DC: NASA SP-2012-4548, 2012), p. 15. Arrighi's work is the point of departure for discussing the PSL.
124. Harry E. Bloomer and Carl E. Campbell, "Experimental Investigation of Several Afterburner Configurations on a J79 Turbojet Engine," NACA RM E57I18 (September 23, 1957).

125. R. Crowl, W.R. Dunbar, and C. Wentworth, "Experimental Investigation of Marquardt Shock-Positioning Control Unit on a 28-Inch Ramjet Engine," NACA RM E56E09 (May 18, 1956).
126. George Vasu, Clint E. Hart, and William R. Dunbar, "Preliminary Report on Experimental Investigation of Engine Dynamics and Controls for a 48-Inch Ramjet Engine," NACA RM E55J12 (March 16, 1956).
127. Robert S. Arrighi, "History: Propulsion Systems Laboratory No. 1 & 2, Ramjets and Missiles (1952–1957)," December 3, 2012, *http://pslhistory.grc.nasa.gov/Ramjets%20and%20Missiles.aspx* (accessed September 25, 2013).
128. Dawson, *Engines and Innovation*, pp. 184–185. Also, see Mark D. Bowles, *Science in Flux: NASA's Nuclear Program at Plum Brook Station, 1955–2005* (Washington, DC: NASA SP-2006-4317, 2006).

NASA maintained an active research program with its Lockheed YF-12 Blackbirds from 1967 to 1979, which included propulsion-focused investigations. The Blackbird at top carries the experimental "coldwall" heat transfer pod on a pylon beneath the fuselage in 1975. (NASA)

CHAPTER 2
NASA Gets to Work, 1958–1975

The Soviet launch of the Sputnik satellite into Earth's orbit on October 4, 1957, initiated a major shift in American aeronautical research and development. The National Air and Space Act of July 1958 dissolved the NACA and created the National Aeronautics and Space Administration (NASA) the following October. A primary goal of NASA was to create and then manage America's civilian space program, which would enable the United States to compete with the Soviets in putting the first humans in space and, ultimately, on the Moon. The Cold War–infused space race was on. The first "A" in NASA, aeronautics, dealt with improving flight in the atmosphere. It worked in competition against the "S" in NASA, the space program. In the opinion of Congressman George P. Miller, the latter clearly overshadowed the former during the Agency's first decade.[1] Nevertheless, NASA's work in aircraft propulsion during the 1960s and 1970s reflected the Agency's contributions to military high-speed flight and subsonic commercial aviation, which included the first in-depth studies into improved fuel economy and the growing public concern over engine noise and emissions.

The NACA's seminal legacy in aeronautics changed dramatically with the creation of NASA. Quickly, the personnel, tools, and techniques used to investigate aircraft propulsion challenges became enlisted in the space race. The research conducted in the PSL at Lewis, created to evaluate gas turbine and rocket engines for flight in the atmosphere, is a case in point. After Sputnik in 1957 and the creation of NASA in 1958, the Cleveland facility shifted its focus. PSL researchers made important contributions to the Pratt & Whitney RL-10 liquid-fueled rocket that powered the Centaur and Saturn upper-stage rockets. They also worked on the first stage of the Apollo program's Saturn V rocket. They investigated the distinctive contoured nozzle for the groundbreaking Rocketdyne F-1 engines and the failed 260-inch solid rocket motor alternative.[2]

The 1960 staff and resource priorities for the Lewis Research Center reflected the new focus on space. Advanced propulsion research investigating chemical rockets, nuclear propulsion, and electric propulsion and power generation represented 35, 20, and 14 percent, respectively, of the work conducted in

Cleveland. Fundamental research into fluid mechanics, heat transfer, instrument and computing research, and radiation physics constituted 24 percent of the laboratory's efforts. Finally, air-breathing engine research in support of advanced military projects such as the North American XB-70 Valkyrie long-range supersonic nuclear bomber and its GE J93 engines represented only 7 percent.[3]

Crossing the Hypersonic Frontier: The X-15

NASA continued the high-speed programs of the NACA in the form of the X-15 flight research program (1959–1968), which investigated hypersonic flight at five or more times the speed of sound at altitudes reaching into space. Launched from the wing of a Boeing B-52 mother ship, the X-15 was a true "aerospace" plane with performance that went well beyond the capabilities of existing aircraft powered by air-breathing engines within and beyond the atmosphere. North American Aviation of Los Angeles, CA, had a special challenge in designing the X-15. For propulsion, a Reaction Motors XLR99 rocket engine produced 57,000 pounds of thrust. At hypersonic speeds, the air traveling over an airplane generated enough friction and heat that the outside surface of the airplane reached a temperature of 1,200 degrees Fahrenheit. North American used titanium as the primary structural material and covered it with a new, high-temperature nickel alloy called Inconel-X. The X-15 relied upon conventional controls in the atmosphere but used reaction-control jets to maneuver in space. The long, black research airplane, with its distinguishing cruciform tail, became the highest-flying airplane in history. In August 1963, the X-15 flew to 67 miles (354,200 feet) above Earth at a speed of Mach 6.7, or 4,534 miles per hour. Overall, the 199 flights of X-15 program generated important data on high-speed flight and provided valuable lessons for NASA's space program.[4] The use of rocket power to propel an air-launched aircraft into the hypersonic range was successful, but it also illustrated the need for other forms of propulsion for practical high-speed flight.

NASA's Participation in the National SST Program, 1961–1971

NASA's work in high-speed commercial aviation, centering on ever-faster airliners, culminated with the ill-fated supersonic transport (SST). With the subsonic jet airliner an everyday technology in the 1960s, the next step was building an SST. Achieving supersonic commercial flight in the 1960s was viewed by many as the next logical triumph of American civil aviation, proof of the United States' enduring technological superiority in aerospace. America

Figure 2-1. This image shows the North American X-15 research airplane. (National Air and Space Museum, Smithsonian Institution, SI 77-14083)

had revolutionized international air transport in the piston era, and then in the early turbojet era. Now, many saw extending that dominance into supersonic civil air transport service as the next logical step.

The Secretary of the Department of Defense (DOD), Thomas S. Gates, Jr., and the administrators of NASA and the FAA, T. Keith Glennan and Elwood R. Quesada, respectively, issued a joint recommendation in October 1960 to initiate a national program for the development of a commercial SST. The reasons were many. The creation of an SST was in the national interest because it would guarantee American leadership in commercial aviation, which was vital to the Nation's economy, security, and prestige. The technical foundation was there with the long tradition of supersonic research in the United States that had culminated with the recent B-58 and B-70 bomber programs. The research and development cost of any SST would be beyond the capabilities of a single company and airline to support. The sheer magnitude of the project required Government leadership, funding, and technical expertise combined with the participation of industry. They predicted that supersonic transports, "either of foreign or of United States origin," would dominate worldwide commercial air

The Power for Flight

Figure 2-2. The failed Boeing 2707 and its four GE4 turbojets represented NASA's first work toward a supersonic commercial airliner. (NASA)

routes as early as 1970.[5] Working toward the creation of a practical American SST as part of a Nationwide effort was the kind of challenge NASA endeavored to take on, especially since it was already doing that with the space program.

DOD, NASA, and the FAA—with the support of President John F. Kennedy—initiated a design competition between the leading aircraft manufacturers. The Government chose Boeing's Model 2707 design in December 1966, and General Electric received the contract for its four engines, designated GE4. The engines for the new SST could not be military power plants. They had to be as fuel-efficient, quiet, and reliable as standard airliner engines at both supersonic and subsonic speeds. The latter included takeoff, landing, and loitering in a holding pattern.[6]

Almost immediately, a national debate began that centered on the cost of development, predicted to total $5 billion, and the environmental impact. Many groups objected to the prospect of experiencing frequent sonic booms, which had been an area of particular interest for NASA (and continues to this day).[7] On Wednesday, December 2, 1970, the Senate voted unanimously to regulate public exposure to sonic booms by prohibiting flights of civilian SSTs over the United States. Senate Bill S. 4547 went further to ensure that when SSTs became a reality, they would operate in compliance with noise limitations established by the FAA.[8] Others expressed grave concerns over expected exhaust pollution expelled from the four GE4 turbojets and their theoretical contribution to possible deterioration of the ozone layer. Environmental concerns aside, the cause for the program's demise was the fact that it was not commercially viable. The research and development costs weighed against actual aircraft and engine orders from commercial airlines rendered the 2707 economically unsustainable. The Senate canceled funding for the program by a vote of 51 to 46 in March 1971.[9]

The Valkyrie and the SST

NASA enlisted the Air Force's experimental North American XB-70 Valkyrie into its collection of research airplanes as part of the national SST program.[10]

Figure 2-3. The XB-70 was the world's largest experimental aircraft when the Air Force and NASA partnered to use the canceled bomber as a flying laboratory to generate data for future supersonic aircraft. The controllable internal geometry of the inlets maintained efficient airflow to the six YJ93 turbojets. (NASA)

The Valkyrie was the epitome of the phrase "higher, faster, and farther." North American engineers had designed it as a strategic nuclear bomber, capable of cruising at altitudes of over 70,000 feet and at speeds greater than Mach 3. The largest and heaviest supersonic airplane ever flown, it underwent 9 long years of development related to the refinement of its aerodynamic configuration, its heat-resistant structure, and its afterburning turbojet propulsion system at a then-year cost of $1.5 billion. Journalist Keith Wheeler proclaimed that "no sky has carried anything like the XB-70."[11]

Six innovative General Electric YJ93 turbojet engines, situated in a central bay underneath the fuselage, produced thrust equal to two-thirds of the power needed to propel the nuclear carrier U.S.S. *Enterprise*. Awarded a $115 million development contract, GE engineers at the Evendale plant worked to double the power without increasing the weight under the direction of Bruno Bruckmann. Building on previous company experience and lessons learned from the J79 program, they introduced rare alloys, variable-pitch stators to increase the intake of air for more power on demand, and special techniques to offset overheating and vibration. The operating environment of the YJ93 was extreme. The engine itself ran at its optimum level with air at a pressure of 30 pounds per square inch, heated to 650 degrees Fahrenheit, and moving slower than the speed of sound. Yet the mission of the XB-70 placed the engines in an environment of low pressure, atmospheric temperatures of –65 degrees Fahrenheit, and flight speeds of 2,000 mph. The solution was high-speed inlet ducts 60 feet in length that heated the air, compressed it, and slowed it down to 350 mph before it entered the front face of the engine compressor. One YJ93 "monster engine" generated 30,000 pounds of thrust, with a thrust-to-weight ratio of 6:1, which meant it was capable of pushing forward 6 pounds for every pound it weighed.[12]

Before squadrons of Valkyries were able to join the ranks of the Strategic Air Command, Congress canceled the program in 1961, citing development costs and questions regarding the feasibility of its being a strategic nuclear bomber, given the rapid development of Soviet surface-to-air missiles as well as air-to-air interceptors capable of traveling above Mach 2. Congress did allow the construction of two airframes for research and development purposes, something they later denied when canceling the funding for the SST a decade later. NASA's Flight Research Center (subsequently Dryden and now the Armstrong Flight Research Center) envisioned the XB-70 as a flying test bed for addressing the problems faced by the national SST program. NASA and the Air Force entered into a joint $50 million assessment program in March 1966. The tragic collision with a Lockheed F-104 Starfighter that resulted in the loss of NASA's chief research pilot, Joe Walker; Valkyrie copilot Carl Cross; and the second XB-70A aircraft on June 8, 1966, altered that partnership. NASA took over the entire program and focused on acquiring SST-related flight data involving such issues as sonic boom ground overpressures, autopilot performance as the aircraft transited varying atmospheric pressures and dynamic conditions, and cruise efficiencies and performance. Much research was done on analyzing inlet performance. During the transition from subsonic to supersonic flight, the presence of airflow distortion and turbulence induced compressor stall. The other specter was the reduction of noise. Experimentation on the XB-70A led to better understanding of propulsion and airframe integration, especially regarding mixed compression inlets, used on the design of other supersonic aircraft. Flights of the XB-70 also contributed to the growing body of information available to the National Sonic Boom Program, which confirmed the impracticability of allowing SSTs to operate over the continental United States.[13]

Supersonic Cruise Aircraft Research

Advocates of the SST moved past its cancellation to create a new program to continue research and development, albeit on a much smaller scale. NASA established the Supersonic Cruise Aircraft Research (SCAR) Program in late 1971. Langley managed the overall project. In the new NASA model, contracts went out to manufacturers and the NASA Centers supplemented their work with focused studies. The SCAR propulsion effort at Lewis consisted of two programs.

The first program, the main engine-related project, was the variable-cycle engine (VCE). It addressed the noise and emissions problems of the GE4 by exhibiting the best characteristics of both a turbojet and a turbofan. Turbojets were most efficient at supersonic speeds but were loud and inefficient at subsonic speeds. Conversely, turbofans offered subsonic efficiency and lower noise but were less efficient at higher Mach numbers. A VCE offered both configurations. Both Pratt & Whitney and GE received development contracts, and the interplay between Lewis, the military, and manufacturers resulted in two different engine designs.[14] Reflecting the importance of computer simulation and modeling embodied in the emerging field of computational fluid dynamics (CFD) to predict aerodynamic behavior, Lewis researcher Larry Fishbach facilitated the use of a new design code with the Naval Air Development Center. This code, called the Navy-NASA Engine Program, simulated the proposed technical scenarios for the engine designs.[15]

The second program, the Experimental Clean Combustor Program (ECCP), emerged in response to the anticipated introduction of Environmental Protection Agency (EPA) airport emissions standards (and is discussed in more depth in chapter 3). Tests of the experimental engines took place from 1978 until the termination of SCAR in 1981. Economic inflation, Federal budget

The Power for Flight

reductions, and the need for NASA to keep the Space Shuttle Program funded outweighed any justifications for another Government-funded supersonic commercial airliner program.[16]

Nevertheless, the propulsion research of SCAR proved to be longstanding. Environmental research revealed that an advanced SST's effect on the ozone layer, which was a major concern during the 1960s, would be less harmful than previously believed. (The true culprit proved to be chlorofluorocarbons used as the delivery medium in aerosol spray cans.) The VCE program accelerated the capability of Pratt & Whitney and GE to use and experiment with advanced design codes, materials, and structures while incorporating the new parameters of noise and emissions reduction into their engines. As historian Eric Conway noted, the VCE was successful "solely on its merits as a technology program" despite the fact that it did not lead to a flight-ready engine.[17]

NASA and the Lockheed Blackbird Family

The Lockheed SR-71 Blackbird reconnaissance airplane was the fastest piloted aircraft with air-breathing engines in history when it entered service in the U.S. Air Force in 1966. It could fly higher and faster than any Soviet fighter or missile by cruising at Mach 3 near the upper edge of Earth's atmosphere at altitudes above 85,000 feet. The sinister and futuristic-looking Blackbird featured a sleek delta wing that spanned 55 feet, a fuselage 100 feet long, and a height of 18 feet. The Blackbird received its name from the special paint covering its outside surfaces. The paint and the titanium alloy structure underneath allowed the skin of the airplane to withstand high-speed aerodynamic heating caused by the friction of the air passing over the surface, absorb radar signals, and serve as camouflage in the dark sky at high altitudes. The initial variant was the A-12, the Central Intelligence Agency's (CIA's) replacement for the Lockheed U-2 spyplane. The single-seat A-12 gave rise to several two-seat derivatives, one of which was the M-12, a mother ship for the GTD-21 reconnaissance drone; the YF-12A interceptor; and the SR-71 strategic reconnaissance aircraft. Created by the highly successful Advanced Development Projects division of Lockheed Aircraft—better known since World War II as the "Skunk Works" in an homage to the moonshine still featured in Al Capp's *Li'l Abner* comic strip—the SR-71 served on the frontlines of overhead atmospheric reconnaissance through the Cold War and afterwards, until it was finally retired in 1998.

Two conventional Pratt & Whitney J58 turbo-ramjet engines, each generating 30,000 pounds of thrust, powered the Blackbirds. They had to operate across a wide range of speeds and conditions in flight, from a takeoff speed of over 200 mph to a maximum cruise of 2,200 mph, or Mach 3.3. Innovations

Figure 2-4. Originally designed as a fighter during the Cold War, the YF-12 proved more important as a NASA research airplane. (NASA)

based on advanced thermodynamic design enabled the J58 to generate all-out thrust at high Mach speeds and operate efficiently at high temperatures. The Skunk Works team designed a complex air inlet and bypass system for the engines to deter supersonic shock waves from moving inside the engine intake and causing flameouts.[18]

NASA facilities were significant to the design and development of the Blackbirds. Lockheed went to Ames Research Center at Moffett Field, CA, to evaluate the critical relationship between the airframe and the engine inlet system. After the Blackbird was publicly announced in 1964, NASA saw an opportunity to use it as a platform for flight research into the high supersonic (greater than Mach 3) regime. The Agency and the Air Force entered into a joint flight research program at Dryden Flight Research Center using the canceled fighter variant of the Blackbird, the YF-12A, on December 10, 1969. Subsequently, the program added a modified SR-71A flight-test aircraft, redesignated the "YF-12C," to the study effort. For the propulsion component of the program, the goal was to establish a baseline of engine performance data for present and future use to serve as a validation for computer predictions and wind tunnel tests. Dryden managed and coordinated the overall program and was responsible for developing a cooperative control system. Lewis conducted analyses on the inlet designs, performed full-scale tests in the 10- by 10-foot wind tunnel, performed engine calibration tests, and developed a new control system. Ames took charge of the design, analysis, and testing of all wind tunnel models.[19]

Besides investigating better inlets and controls, boundary layer noise, heat transfer under high Mach conditions, altitude performance at supersonic speeds, and overall airframe-propulsion system interaction, the YF-12A/C flight research program investigated the problem of inlet "unstart." The phenomenon resulted from improperly matched airflow where internal pressure forced the internal standing shock wave to "pop" out of the inlet. The resultant loss of thrust induced exaggerated yaw, pitching, and rolling and threatened to destroy the airplane. The flightcrews imposed a program of deliberate unstarts to familiarize themselves with recovery procedures and to influence an improved inlet spike design on the production SR-71. Overall, the YF-12 program generated a wide range of data on variable-cycle engine and mixed-compression inlet operation for the benefit of future supersonic aircraft design.[20]

As the YF-12 program gained momentum at Dryden, NASA worked to make supersonic aircraft quieter at Lewis. The evaluation of inlets and exhaust nozzles began with ground tests by the Wind Tunnel and Flight Division. Researchers placed microphones at intervals around the test rig to measure the direction and level of the noise. The downside was that the ground reflected the sound waves and compromised the test results. Flight tests of a modified Convair F-106 Delta Dart took place at Selfridge Air Force Base in Michigan. The installation of a GE J85 turbojet under each of the F-106's delta wings resulted in a generalized engine-airframe combination similar to that of a future supersonic aircraft. During the tests, the research pilot flew as low as 300 feet with the main J75 engine at idle so that a microphone tape recorder station on the ground could obtain good noise signals. NASA engineers installed experimental nozzles on the J85s for each test run.[21]

Fixing the F-111: Overcoming Classic Engine-Airframe Mismatch

The General Dynamics F-111 supersonic all-weather multipurpose tactical fighter bomber was the product of an ill-considered 1961 Department of Defense plan initiated by Secretary Robert McNamara. The Tactical Fighter Experimental, known widely as TFX, would fulfill both an Air Force supersonic strike aircraft requirement and a Navy fleet-defense interceptor requirement. That these two requirements were basically incompatible did not prevent the program from being advanced by Secretary McNamara, who saw a chance to

Figure 2-5. NASA contributed greatly to improving the design of the General Dynamics F-111A. (USAF)

achieve both joint-service "commonality" and acquisition savings. Not surprisingly, the program proved to be seriously flawed; the Navy variant, the F-111B, was so heavy that it never entered service, and the Air Force variant, the F-111A, was underpowered and proved unable to meet the original performance requirement. Eventually, the F-111 design was made to work, and the final F-111F "Aardvark" was a remarkably successful strike aircraft, as exemplified by its performance in 1991 in the first Gulf War, together with an electronic warfare variant, the EF-111A "Sparkvark."

One of the challenges of the F-111 program was that it incorporated a new innovation for American military aircraft: a variable-geometry, or swing, wing. When fully extended, it facilitated short takeoffs and landings. When fully swept, it enabled the F-111 to attain speeds of up to Mach 2.5, exceeding 920 mph at less than 200 feet altitude. The variable-sweep wing had first appeared on the Bell X-5 research airplane, which the Air Force and the NACA had extensively tested in the 1950s. Though attempted unsuccessfully on Grumman's XF10F-1 Jaguar naval fighter in the mid-1950s, the F-111 represented its first practical application. Other American and foreign swing wing aircraft included the B-1 strategic bomber, the F-14 fighter, and the MiG-23. From the beginning, a multitude of problems centered on making one airplane meet the requirements of two services and their disparate missions beset the TFX program, along with continuous technical challenges. In November 1962,

The Power for Flight

Secretary McNamara selected the General Dynamics design instead of a rival design offered by Boeing due to the former's greater commonality between the Air Force and Navy designs, but he did so over the objections of the program's evaluation committee (and later by Congress). General Dynamics moved forward with the development of the new F-111A for the Air Force and the F-111B for the Navy.

The F-111 was the first combat airplane in the world powered by afterburning turbofans. Still new to aviation, turbofans exhibited different airflow characteristics from those of turbojets and were sensitive to pressure distortion between the bypass duct and the engine core. For optimum operation, the inlet airflow of a turbofan had to be uniform or the engine would experience "compressor stall," where abnormal airflow led to a dramatic reduction in operating pressure that affected overall power, often referred to as "flameout." The F-111 designers chose quarter-round inlets and placed them under the wing roots next to the fuselage for optimum performance for the low-level mission. They reduced drag but introduced airflow disturbances into the inlet. The extreme range of operational mission profiles for the F-111—ranging from the supersonic dash in dense air at sea level to the high-speed cruise at extreme altitudes, with a wide range of speeds and flight attitudes in between—exacerbated the problem. The two Pratt & Whitney TF30 engines were highly susceptible to stalling, something General Dynamics discovered even before the first F-111 test flight in December 1964.[22]

NASA had not contributed to the initial inlet design of the F-111. NASA's focus, through John Stack and other aerodynamicists at Langley, was initially on the variable-sweep wing and overall aerodynamic refinement of the aircraft for the supersonic low-level nuclear mission. Langley and Ames actively supported the development program in many areas, which amounted to the most extensive wind tunnel support ever provided for one aircraft by NASA or the NACA.[23] Langley engineers were finding indications of problems with the inlet, but these were a byproduct of their other work, and their concerns were lost in the bureaucratic hustle that characterized the development of the F-111 in the 1963–1965 period. The staff at Lewis was fully engrossed in programs supporting the space program and ballistic missile development.[24]

That changed, however, when the inlet-engine compatibility problems arose. Immediately, NASA was asked to investigate the problem. In March and April 1965, Ames conducted an investigation into the use of vortex generators, small fins used to control airflow, to minimize distortion in the F-111 supersonic inlet system. NASA researchers shared the results with General Dynamics' engineers, who left the meetings with detailed design drawings. The F-111/TF-30 Propulsion Program Review Committee, with members representing General Dynamics, Pratt & Whitney, the Air Force, and the Navy,

NASA Gets to Work, 1958–1975

as well as researchers from Ames and Lewis, met in September and October 1965 to discuss engine control problems, inlet distortion evaluation, engine development, and an estimated stall-free envelope.[25]

The solution put forth by General Dynamics was the so-called Triple Plow I inlet, which the Air Force approved for production in early 1967. The design modified the splitter plate between the front of the intake and the fuselage to divert turbulent boundary-layer air that hugged the fuselage, incorporated hydraulically extended engine cowls, and integrated 20 vortex generators into each inlet. Pratt & Whitney introduced a less-temperamental TF30 engine. The modifications improved performance, but they still did not fully satisfy the Air Force engineers in the F-111 Systems Program Office at the Aeronautical Systems Division of Air Force Systems Command, located at Wright-Patterson Air Force Base, OH. They worked through the late 1960s and offered a better solution in the form of the Triple Plow II inlet, characterized by an enlarged inlet duct and major structural changes. It provided a full-capability flight Mach/maneuvering envelope free of compressor stalls.[26] The final cost to fix the inlet-engine compatibility problem was over $100 million.[27]

Only the Air Force's F-111 went into service, beginning with the F-111A in 1967, because the Navy's variant, the F-111B, grew too heavy and was too underpowered to fly from aircraft carriers and serve in the fleet air defense role. After solving the structural problems, the Air Force went on to operate successive strike, strategic bomber, and electronic warfare versions of the F-111 in Southeast Asia, North Africa, and the Middle East through 1998. The strategic bomber variant, the FB-111A, flew into the 1990s. The Royal Australian Air Force, the only foreign customer for the aircraft, operated the F-111C from 1973 to 2010.

Overall, the F-111 experience was an extremely cautionary tale for the American aerospace industry on the issue of commonality and on the problems of airframe-engine integration. For the long term, the research conducted to solve the inlet-engine compatibility problems related to the F-111 proved beneficial to the development of later American military aircraft, including the Grumman F-14 Tomcat, McDonnell Douglas F-15 Eagle, General Dynamics F-16 Fighting Falcon, and Rockwell B-1 Lancer.[28]

Developing and Refining Advanced Military Aircraft Engines

The Military and NASA Lewis's Propulsion Systems Laboratory (PSL)

Since its creation, NASA supported engine development for American aircraft. In 1966, with much of the work on the Apollo program already accomplished,

The Power for Flight

Figure 2-6. A research technician checks the installation of a J85 engine in PSL No. 2 in 1974. (NASA)

NASA's PSL refocused attention upon the jet engine, which continued to grow in size, sophistication, complexity, and performance. Ironically, it involved a bit of catch-up: with the researchers and facilities in Cleveland committed to space during the 1956–1966 focus, the PSL had been relatively subordinated in air-breathing engine development by the Air Force's Arnold Engineering Development Center in Tennessee. This facility, inspired by the German Kochel high-speed tunnel complex discovered at the end of the Second World War, had opened in 1951 and quickly gained a reputation for analytical excellence using a variety of tunnels, shock tubes, and other research tools. As well, the PSL had been somewhat supplanted (though to a lesser degree) by the growing capabilities of industrial research facilities maintained by the aeropropulsion industry itself. Lewis Research Center created a new Airbreathing Engine Division in 1966. The organization had at its disposal the Propulsion Systems Laboratory, a new Quiet Engine Test Stand, the 10- by 10-Foot Supersonic Wind Tunnel, and the Agency's Convair F-106 Delta Dart research aircraft to study turbofans.

To better determine the baseline performance characteristics of any engine undergoing evaluation, the Lewis engineers chose the compact,

4,350-pound-thrust GE J85 turbojet for calibration studies. The J85 was one of GE's most successful engines. Small and powerful at 3,000 pounds of thrust, it had the highest thrust-to-weight ratio of any American-built jet engine. GE had based its compressor on an experimental five-stage NASA design, which included high-stage loading technology.[29] The American military used the engine in a wide range of aircraft that included the Northrop T-38 Talon and F-5 fighter family, the North American T-2 Buckeye trainer, and the Cessna A-37 Dragonfly (a trainer modified for light attack). Comprehensive testing established the desired transonic nozzle inlet temperature and pressure and gas flow rate, which allowed researchers to determine the J85's gross thrust within a margin of error of less than 1 percent. With the methodology in place, they could rationalize the testing of new nozzle and compressor designs and noise-reduction technology not only for future tests in the PSL, but also in test aircraft and other facilities, using the known characteristics of the J85 as a baseline.[30]

NASA placed new emphasis on increasing the performance of turbofans in the early 1960s, seeking as well to reduce their noise to make them commercially practical. The turbojets that powered the first commercial airliners consumed large amounts of fuel in relation to the distances they carried aircraft, which affected airline revenues and operating costs. The addition of a large, enclosed, multiblade fan to a turbojet harnessed the efficiency of the propeller while developing the high thrust of the turbojet. The unprecedented 60- to 80-percent leap in efficiency increased thrust, improved fuel economy, and reduced noise. The introduction of the new "turbofan" facilitated the success of low-cost, long-distance air transportation and military flight in the early 1960s.

The first practical American turbofan engine, General Electric's 16,000-pound-thrust CJ805-23 aft-fan, was ready for flight in 1959. It resulted from considerable pioneering research and financial expense on the part of both GE and the NACA, primarily in the newly emerging field of CFD. Unfortunately, the airplane it was destined to power, the Convair 990 airliner, was not yet ready. The delay worked against the subsequent market success of the CJ805, allowing rival Pratt & Whitney time to catch up with a rival front-fan design, the JT3D. Though initially resistant to the idea of the turbofan, Pratt & Whitney beat General Electric to the market by replacing the first three low-pressure stages of the JT3 turbojet with two fan stages, which improved thrust and fuel economy while reducing noise. The resulting JT3D generated 18,000 pounds of thrust, and the airlines quickly adopted it for the 707 and DC-8 aircraft in 1960. Operators could convert their JT3s into JT3Ds with a simple kit, an extraordinarily attractive enticement. Pratt & Whitney produced over 8,000 JT3D engines and dominated the growing market for commercial turbofan engines in the early 1960s. In 1990, Jack Parker, head of

The Power for Flight

GE Aerospace and Defense, ruefully remarked, "We converted the heathen, but the competitor sold the bibles." [31]

From that competitive beginning, turbofans became the engine of choice for both military and commercial applications, for they greatly increased range thanks to their better fuel economy. But they were not an unalloyed blessing. For the U.S. Air Force and Navy, low-bypass turbofans offered better fuel economy over turbojets while still providing high thrust and supersonic capability. Unfortunately, the extreme conditions in which military aircraft operated caused problems in overall efficiency, which affected the overall performance of the next generation of military aircraft. The 25,000-pound-thrust Pratt & Whitney TF30 afterburning turbofan slated for the Grumman F-14A Tomcat and the General Dynamics F-111 twin-engine supersonic variable-sweep aircraft suffered from decreased engine pressures at high altitudes and distorted airflow that reduced the stability of the compressor. Many F-14As were lost from engine failures, in part because a loss of power in a single engine could cause the aircraft to enter an unrecoverable spin, thanks to its widely placed engine layout. Indeed, one Grumman test pilot, Charles "Chuck" Sewell, who ejected over the Atlantic from an F-14A that experienced engine failure, quipped, "If the engines say Pratt & Whitney, the seats should say Martin-Baker," a reference to an outstandingly reliable ejection seat manufacturer. For the subsequent TF30 tests in PSL No. 1, engineers devised a system of nozzles that injected air into the test chamber airstream that simulated those

Figure 2-7. This image shows the Grumman F-14A Tomcat. (Northrop Grumman)

disturbances and enabled the engineers to chart the probability of engine stalling under various conditions.[32]

After the Vietnam War, the services had a number of new aircraft under development that required NASA analytical assistance. In the mid-1970s, the staff of the PSL and the U.S. Air Force Aero Propulsion Laboratory at Wright-Patterson Air Force Base collaborated on a Full-Scale Engine Research (FSER) program to investigate flutter, inlet distortion, and electronic controls. The project focused on two service engines: the General Electric J85-21 (a new variant of the proven J85 and used in the Northrop F-5E/F Tiger II lightweight fighter) and the Pratt & Whitney F100, a very-high-performance engine intended for the new McDonnell-Douglas F-15A Eagle and General Dynamics F-16A Fighting Falcon, two high-priority defense programs. For the J85 portion, the PSL targeted internal compressor aerodynamics and mechanical flutter caused by distortion in the airflow. The collection of data documenting the unique differences between flutter and instability led to better understanding of the phenomena in jet engines.[33]

Of the two programs, the F-15 and the F-16, the F-15 was by far the most critical, for it was essential (as was the Navy's F-14) for maintaining American air superiority in the face of rapidly changing air combat threats. The F-15 was one of the first so-called "fourth generation" jet fighters, with advanced stability augmentation (though it was not, like the slightly later F-16, a "fly by wire" design); an optimized aerodynamic and control configuration for

Figure 2-8. Shown here is the McDonnell Douglas F-15A Eagle. (Boeing)

superb agility; and two powerful new afterburning turbofan engines, namely, the Pratt & Whitney F100. Even before the aircraft took to the air, General James Ferguson, commander of Air Force Systems Command, made note of the already "substantial contribution" made by researchers at Langley, Ames, Lewis, and the Flight Research Center to the evolution of the program.[34]

NASA officials discussed with the Air Force the specific propulsion choices facing the F-15 as the design moved forward prior to its first flight in 1972. The Air Force had a choice of using either a conventional short or advanced, but yet-to-be-proven, long blade chord compressor to guarantee high performance and efficiency. The Lewis Laboratory was well suited to assist due to its experience in troubleshooting the F-14A's Pratt & Whitney TF30 with special instrumentation developed and procured for that effort. Brigadier General Benjamin Bellis, director of the F-15 System Program Office (SPO), emphasized that he wanted all the NASA help and support he could get regarding the Eagle's two Pratt & Whitney F100 afterburning turbofan engines, which were experiencing unanticipated reliability and safety issues.[35] The result was a collaborative meeting that also provided a basis for closer coordination and partnering, an outcome NASA Administrator James C. Fletcher hailed as forming the "basis for future cooperation and supportive efforts on the engine side similar to what we have on the airframe side."[36]

The Pratt & Whitney F100 was a new-generation military low-bypass turbofan that featured variable compressor and fan blades rather than the traditional static types found on earlier designs. These new engines required better and more rapidly adjusting engine control systems that handled multiple parameters and variables while increasing overall accuracy and response. The Air Force requested that NASA develop and test such a system on an F100 engine. The PSL team at Lewis offered the testing facilities; Pratt & Whitney provided the engine and design data; and Systems Control, Inc., of Palo Alto, CA, created the basic computer logic for the controls. The team began with the creation of a digital controller utilizing a linear quadratic regulator and tested it in computer simulations in 1975. Full-scale F100 testing in the PSL began in mid-1977.[37]

The amount of work taking place within the PSL led to the construction of two additional test chambers—at the cost of $14 million—that opened in February 1972. Designated Nos. 3 and 4, the structures reflected the lessons learned in Nos. 1 and 2. They were larger, at 40 feet long and 24 feet in diameter, and simpler in construction, with the shared exhaust cooler that gave the facility its distinctive Y-shape. More importantly, they enabled researchers to test engines twice as powerful as any in existence at the time, with simulated altitudes of up to 90,000 feet and speeds of up to Mach 3 in No. 3 and Mach 6 in No. 4.[38] PSL Nos. 1 and 2 continued in operation until 1979, when their

Figure 2-9. This image shows a Pratt & Whitney F100 engine installed in the much larger PSL No. 4 in 1981. (NASA)

limited capacity compared to the cost of operations rendered them obsolete. They sat unused until their demolition in 2009.[39]

Despite the success of the PSL and other propulsion programs at Lewis, the 1970s were a tumultuous time for the Center's staff and especially for Bruce Lundin, who became the Center Director in 1969. His desire to expand Lewis's role in commercial aviation faced considerable challenges. NASA overall faced budget cuts that led to the termination of permanent civil service positions in the wake of the end of the Apollo program and fragmented the research community. Continued uncertainty related to interagency competition with the new Department of Energy (DOE), along with the inability of Lundin and Headquarters in Washington to work together effectively, led to the Director's resignation in 1977. More importantly, the nature of the research changed. Rather than retaining a degree of independence rooted in basic research, Lewis managed contracts to industry or served as one piece of a national puzzle managed from Headquarters in Washington, which emphasized short-term development.[40]

Endnotes

1. George P. Miller, "NASA," in Congressional Record (hereafter Cong. Rec.), 92nd Cong., 1st sess., vol. 117, December 14, 1971, p. H12545, copy available in the NASA HRC, file 012336.
2. "1962–1972: The First Decade of Centaur," *Lewis News* (October 20, 1972): 4–5, available at *http://pslhistory.grc.nasa.gov/PSL_Assets/History/C%20Rockets/First%20Decade%20of%20Centaur%20article%20(1972).pdf* (accessed August 15, 2013); Robert S. Arrighi, "History: Propulsion Systems Laboratory No. 1 & 2, Rocket Engines (1958–1966)," December 3, 2012, *http://pslhistory.grc.nasa.gov/Rocket%20Engines.aspx* (accessed August 14, 2013).
3. Dawson, *Engines and Innovation*, p. 179.
4. Dennis R. Jenkins, *Hypersonics Before the Shuttle: A Concise History of the X-15 Research Airplane* (Washington, DC: NASA SP-2000-4518, 2000), pp. 21–44, 119.
5. Federal Aviation Agency, Department of Defense, and NASA, "A National Program for a Commercial Supersonic Transport," NASA TM-X-50927, October 1960, pp. 1–3.
6. Ibid., pp. 17–18.
7. Ira Schwartz, ed., *Third Conference on Sonic Boom Research Held at NASA, Washington, D.C., October 29–30, 1970* (Washington, DC: NASA SP-255, 1971).
8. Senate Daily Digest, December 2, 1970, p. D1216, copy available in the NASA HRC, file 012336.
9. Erik M. Conway, *High-Speed Dreams: NASA and the Technopolitics of Supersonic Transportation, 1945–1999* (Baltimore: Johns Hopkins University Press, 2005), p. 152.
10. Richard P. Hallion, *On the Frontier: Flight Research at Dryden, 1946–1981* (Washington, DC: NASA SP-4303, 1984), pp. 178–181.
11. Keith Wheeler, "The Full Story of the 2.8 Seconds That Killed the XB-70," *Life* 61 (November 11, 1966): 128.
12. Keith Wheeler, "Building the XB-70," *Life* 58 (January 15, 1966): 77, 79.
13. Richard A. Martin, "Dynamic Analysis of XB-70-1 Inlet Pressure Fluctuations During Takeoff and Prior to a Compressor Stall at Mach 2.5," NASA TN D-5826, 1970, pp. 1–2; Terrill W. Putnam and Ronald H. Smith, "XB-70 Compressor-Noise Reduction and Propulsion Performance for Choked Inlet Flow," NASA TN D-5692, 1970, p. 1; James St. Peter, *The History of Aircraft Gas Turbine Engine Development in the United States: A Tradition of Excellence* (Atlanta: International Gas

Turbine Institute of the American Society of Mechanical Engineers, 1999), p. 263; Hallion, *On the Frontier*, pp. 185–186.
14. Warner L. Stewart, "Introduction: Session III-Propulsion," in *Proceedings of the SCAR Conference Held at Langley Research Center, Hampton, Virginia, November 9–12, 1976* (Hampton, VA: NASA CP-001), pp. 337–338.
15. L.H. Fishbach and Michael J. Caddy, "NNEP: The Navy-NASA Engine Program," NASA TM-X-71857, December 1975, p. 1.
16. Conway, *High-Speed Dreams*, pp. 170–174.
17. Ibid., p. 188.
18. Smithsonian Institution National Air and Space Museum, "Lockheed SR-71 Blackbird," November 11, 2001, *https://airandspace.si.edu/collection-objects/lockheed-sr-71-blackbird* (accessed January 18, 2017).
19. James A. Albers, "Status of the NASA YF-12 Propulsion Research Program," NASA TM-X-56039, March 1976, pp. 1–2, 10; Hallion, *On the Frontier*, pp. 191, 193–194; Peter W. Merlin, "Design and Development of the Blackbird: Challenges and Lessons Learned" (AIAA Paper 2009-1522, presented at the 47th AIAA Aerospace Sciences Meeting, Orlando, FL, January 5–8, 2009), pp. 28, 30–31, 37.
20. Hallion, *On the Frontier*, p. 193; Peter W. Merlin, *Mach 3+: NASA/USAF YF-12 Flight Research, 1969–1979* (Washington, DC: NASA History Division, 2002), pp. 16–19, 43.
21. NASA, "Nozzle Design Reduces Jet Noise," September 8, 1970, NASA HRC, file 012336; H. Dale Grubb to Roman C. Pucinski, October 7, 1970, NASA HRC, file 012336.
22. Robert F. Coulam, *Illusions of Choice: The F-111 and the Problem of Weapons Acquisition Reform* (Princeton: Princeton University Press, 1977), pp. 176–180.
23. Staff of the NASA Research Centers, "Summary of NASA Support of the F-111 Development Program, Part I-December 1962-December 1965," NASA Langley Working Paper 246, October 10, 1966, p. 2.
24. U.S. Senate Committee on Government Operations, *TFX Contract Investigation (Second Series)*, part 2, 91st Cong., 2nd sess., March 25–26, April 7, 9, 14, 1970, pp. 345–348; Coulam, *Illusions of Choice*, p. 191.
25. NASA Research Centers, "Summary of NASA Support of the F-111 Development Program," pp. 23, 27–28.
26. G. Keith Richey, "F-111 Systems Engineering Case Study," Center for Systems Engineering at the Air Force Institute of Technology, March 10, 2005, pp. 64–71, *http://www.afit.edu/docs/0930AFIT14ENV125%20 2-2.pdf* (accessed January 13, 2016).

27. Coulam, *Illusions of Choice*, pp. 167, 184–186.
28. Donald L. Hughes, Jon K. Holzman, and Harold Johnson, "Flight-Determined Characteristics of an Air Intake System on an F-111A Airplane," NASA TN D-6679, March 1972, p. 1; Richey, "F-111 Systems Engineering Case Study."
29. St. Peter, *History of Aircraft Gas Turbine Engine Development*, pp. 296, 298.
30. Charles M. Mehalic and Roy A. Lottig, "Inlet Temperature Distortion on the Stall Limits of J85-GE-13," NASA TM-X-2990, 1974, pp. 1, 14–15.
31. Jack Parker, quoted in George E. Smith and David E. Mindell, "The Emergence of the Turbofan Engine," in *Atmospheric Flight in the Twentieth Century*, ed. Peter Galison and Alex Roland (Boston: Kluwer Academic Publishers, 2000), p. 143.
32. Roger A. Werner, Mahmood Abdelwahab, and Willis M. Braithwaite, "Performance and Stall Limits of an Afterburner-Equipped Turbofan Engine With and Without Inlet Flow Distortion," NASA TM-X-1947, 1970; Willis M. Braithwaite, John H. Dicus, and John E. Moss, "Evaluation with a Turbofan Engine of Air Jets as a Steady-State Inlet Flow Distortion Device," NASA TM-X-1955, 1970. Sewell quote from a conversation he had with Richard P. Hallion at Edwards Air Force Base in 1982. See also Robert S. Arrighi, "History: Propulsion Systems Laboratory No. 1 & 2, Return to Turbojets (1967–1974)," December 3, 2012, *http://pslhistory.grc.nasa.gov/Return%20to%20Turbojets.aspx* (accessed August 15, 2013).
33. "The Fan-Compressor Flutter Team," *Lewis News* (February 3, 1978), available online at *http://pslhistory.grc.nasa.gov/PSL_Assets/History/F%20Turbofan%20Engines/Full%20Scale%20Engine%20Program%20article%20(1978).pdf* (accessed August 16, 2013); Robert S. Arrighi, "History: Propulsion Systems Laboratory No. 1 & 2, Turbofan Engines (1974–1979)," December 3, 2012, *http://pslhistory.grc.nasa.gov/Turbofan%20Engines.aspx* (accessed August 15, 2013).
34. James Ferguson to Thomas O. Paine, October 17, 1969, NASA HRC, file 011664.
35. Albert J. Evans, "Visit to F-15 Systems Program Office, Wright-Patterson Air Force Base, September 13, 1971," September 21, 1971, NASA HRC, file 011664.
36. James C. Fletcher to Robert C. Seamans, Jr., October 8, 1971, NASA HRC, file 011664.
37. John R. Szuch, James F. Soeder, Kurt Seldner, and David S. Cwynar, "F100 Multivariable Control Synthesis Program-Evaluation of a

Multivariable Control Using a Real-Time Engine Simulation," NASA TP 1056, October 1977, pp. 1–4. See also W.J. Deskin and H.G. Hurrell, "A Summary of NASA/Air Force Full-Scale Engine Research Programs Using the F100 Engine" (paper presented at the 15th AIAA, SAE [Society for Automotive Engineers], and ASME [American Society for Mechanical Engineers] Joint Propulsion Conference, Las Vegas, NV, June 18–20, 1979); and B. Lehtinen, R.L. Dehoff, and R.D. Hackney, "Multivariable Control Altitude Demonstration on the F100 Turbofan Engine" (paper presented at the 15th AIAA, SAE, and ASME Joint Propulsion Conference, Las Vegas, NV, June 18–20, 1979).

38. Lewis Research Center, "The Propulsion Systems Laboratory," Brochure B-0363, March 1991, available online at *http://pslhistory.grc.nasa.gov/PSL_Assets/History/E%20PSL%20No%203%20and%204/Propulsion%20Systems%20Lab%20No.%203-4%20brochure%20(1991).pdf* (accessed August 15, 2013).

39. Robert Arrighi, "Demolition of PSL No. 1 and 2 (1980–2009)," December 3, 2012, *http://pslhistory.grc.nasa.gov/Demolition.aspx* (accessed January 3, 2013).

40. Dawson, *Engines and Innovation*, pp. 201–215.

NASA technicians prepare for a jet engine noise test on the airfield at Lewis Research Center in 1967. (NASA)

CHAPTER 3
The Shift Toward Commercial Aviation, 1966–1975

By 1970, the United States aviation industry manufactured 74 percent of all commercial aircraft in the free world, with $3 billion in revenues generated by overseas business—thanks, in large part, to NACA and NASA aeronautical research conducted from 1950 to 1970. But as NASA worked to improve military turbofans, there was a growing fear that United States was falling behind as a world leader in commercial aviation technology. Competition from state-supported manufacturers in Europe in the form of Airbus Industrie and recently Rolls-Royce, Ltd., was growing, and consequently there was a continued push in Congress to keep American aeronautics moving forward. The United States appeared to be behind in the development of new technologies such as Vertical and Short Takeoff and Landing (V/STOL), following its controversial decision to choose, largely for nontechnical reasons, to abandon its effort to develop a supersonic transport that could travel faster than Mach 2.5. The House Committee on Science and Astronautics was fully aware of the "increasing deficiencies" in the national aeronautical effort. Congressman George P. Miller (D-CA) of California was an especially vocal supporter of NASA. The Agency had the talent and expertise to confront national problems, but it needed the full support of the Government to do so. Otherwise, the future security and prosperity of the United States was in "great peril."[1]

American manufacturers GE and Pratt & Whitney dominated the turbofan market and invested considerable funding in meeting the needs of the airline industry. For them, the bottom line was increased fuel efficiency. NASA took the lead in two areas that did not affect the bottom line—noise and emissions—and aimed for considerable reduction, with engines up to 10 decibels (dB) quieter and 60 percent cleaner as the 1960s and 1970s progressed.[2] From the late 1920s to the present, commercial airlines have consistently pursued the increase of payload capacity and engine efficiency. Since the 1960s, they have had to recognize a third driving force in aviation: compliance with the FAA's noise regulations.

By the early 1970s, noise in everyday life was an increasingly widespread social concern. The World Health Organization stated in the *Washington News* that noise was "the curse of modern times and a major environmental problem."[3] The scientific measure of sound intensity, most commonly heard as noise, is the decibel. The threshold of pain starts at 120 dB, while 140 produces permanent damage. While the decibel is a specific term for the level of sound output, acoustics researchers use "PNdB," or "perceived noise in decibels," to measure the sound affecting people in the vicinity of an aircraft. In early aircraft noise research, PNdB was the subjective "measure of annoyance" that reflected the overall intensity of a sound, its frequency content, and how people responded to it. But the special circumstances of the aircraft noise problem characterized by whining jet aircraft passing overhead necessitated a better and more relevant measure and resulted in the defining of a more appropriate and relevant noise measurement unit, the effective perceived noise in decibels, or EPNdB. The EPNdB accounted for two additional factors beyond PNdB: first, it gave more importance to tones in the noise spectrum, and second, it accounted for the duration, or rise and fall, of a sound. In essence, EPNdB provided the single number that expressed the measure of the total annoyance a person experienced as an airplane flew over.

A four-engine commercial airliner of the 1960s or early 1970s generated 95 to 120 EPNdB at takeoff, which was a potential cause of hearing damage to the general populace and detrimental overall to the quality of life around airports. In a world full of "chattering jackhammers, whining motorcycles, and roaring jetliners," Representative William F. Ryan (D-NY) asserted, "the right to a quiet, peaceful environment is as basic as the right to clean air and water and pure food." The EPA led the effort to limit noise emissions as it had done for air and water, but Congress planned comprehensive legislation addressing noise in the 1970s. Congress specifically authorized the FAA to muffle "the loudest source of urban noise": low-flying jet aircraft.[4]

Early in the Jet Age, communities and environmental activists recognized that noise and emissions were significant byproducts of jet travel. While other parts of an aircraft generated noise in flight, the jet engine was the greatest source. In terms of their effect on people on the ground, jet aircraft are loudest during takeoffs and landings. At takeoffs, the exhaust gases from the engine mix with the cooler air outside the engine to create a roar. As an airliner approaches a runway to land, the pilots reduce power to slow down, which creates a hissing noise. NASA's overall gas turbine research and development program focused on increasing the fuel efficiency of high-bypass-ratio engines. Although fuel efficiency was the goal, any innovations in that area had to avoid increased engine noise, which was a factor to which the public reacted negatively beginning in the 1960s. NASA began investigating the relationship between turbofan-cycle characteristics and engine noise levels, size, and performance in 1966. Agency researchers learned that the fan generated the highest noise forward from the inlet and aft from the fan discharge ducts. NASA held the first of several conferences on aircraft noise at Langley with Government, industry, and academic participation in October 1968.[5]

The FAA and Federal Aviation Regulation 36: Regulating Aircraft Noise

A year later, in 1969, the FAA responded to congressional legislation calling for reduced aircraft noise with increased regulation; the agency consistently updated those rules over the years.[6] The FAA adopted Part 36 of the Federal Aviation Regulations (FAR), which stopped the increase of noise levels by subsonic turbojet airliners and dictated noise measurement, valuation, and level requirements for new aircraft. The FAA amended Part 36 in 1977 to establish three categories of jet aircraft that reflected the noise each class generated at takeoff, climb, and descent. Stage 1 represented the oldest and noisiest airliners, primarily the Boeing 707 and McDonnell Douglas DC-8 powered by four Pratt & Whitney JT3D turbofans each. Stage 2 included Boeing's 727 and 737 and the McDonnell Douglas DC-9/MD-80, which were noisy aircraft with, respectively, three and two Pratt & Whitney JT8D turbofans. New generations of turbofans powered Stage 3 aircraft, which included an updated 737 and the new Boeing 757, Airbus A319, Fokker 100, and various regional jets; these were the quietest of all jets. Overall, the FAA's goal was to see Stage 1 and 2 aircraft either retired from service in the continental United States or retrofitted to meet the quieter and more stringent Stage 3 standards.[7]

The FAA adopted Part 36 of the FAR in December 1969 and applied it to the type certification of new aircraft such as the McDonnell Douglas DC-10, Lockheed L-1011, and Boeing 747-200. The rule subsequently expanded to cover all aircraft produced after December 31, 1974. The FAA emphasized that the purpose of the new rule was "not to force the modification or retrofit of older airplanes," but rather to encourage each operator to adopt whatever means of achieving compliance was best suited to their individual economic situation. An operator could replace older airliners with new updated aircraft, retrofit the current fleet, or execute a mixture of those options.[8]

A turbofan engine generates noise simply by its operation. The fan, which pulls air into the front of the engine, produces noise in much the same way a propeller does. Additionally, each individual fan blade generates its own noise. Once past the fan, the airflow follows two paths: through the fan bypass duct surrounding the engine and through the inner core duct. Inside the fan duct, the swirling airflow caused by the fan requires stabilization with stator vanes to remove the swirl. The interaction between the two, often described as waves rolling onto a beach, produces tones heard as the distinctive piercing sound emitted by many engine designs. Additionally, turbulent airflow interacting with the stators creates rumbling broadband noise as well. In the engine's core duct, there are three more sources of noise. As the compressor rotors squeeze the airflow, rows of stators separate each rotor stage to straighten the flow, which produces more clamor. Then there is the explosive mixture of the compressed air and atomized fuel in the combustors. The resultant high-temperature and high-pressure combusted air violently interacts with the statorlike turbine to drive the fan and the compressor rotors. Finally, the two flows traveling through the fan and core ducts exhaust into the air to the rear of the engine. The mixing of the two types of exhaust with each other and the outside air generates broadband noise aptly named "jet noise."[9]

Jet Noise as Public Policy Crisis: The Quest for Solutions

Nacelle Acoustic Treatment: First Effort at a Technological Fix

NASA was already well on its way to making the Nation's airliners quieter in the 1960s.[10] Langley managed the Nacelle Acoustic Treatment program beginning in 1967.[11] NASA concluded its research program in 1970. Both Boeing and Douglas conducted research under contract to NASA. The technology demonstrated that noise at landing could be reduced 12–15 dB for the 707 and 7–10 dB for the DC-8. In an October 1970 letter to Representative Roman C. Pucinski (D-IL), NASA Assistant Administrator for Legislative Affairs H. Dale Grubb asserted that the research revealed "an immediate and practical way to reduce significantly the noise of present aircraft." The solution involved the retrofitting of nacelles with acoustic treatment. Grubb explained that the program was a quick fix to provide some relief while NASA and the aviation industry worked on more-effective noise-reduction methods, primarily the initiation of steep landing approaches, the installation of new and better fan assemblies on engines, and the design of a new "Quiet Engine."[12] Through the 1960s and 1970s, NASA offered both short- and long-term solutions to reduce noise and made those simultaneous efforts a major part of its research program for several years.

Yet, as *Forbes* noted, the FAA faced a paradox. On one hand, it needed to carry out its congressional mandate; on the other, it could not afford to jeopardize the financial stability of the airline industry. Noise-reduction programs were estimated to cost between $300,000 to $1 million per airliner, or $250 million to $1 billion overall. Moreover, FAA Administrator John H. Shaffer was not totally convinced of the economic and technical feasibility of retrofitting the Boeing 707 fleet, the workhorse airliner, with noise-suppression equipment despite NASA's successful 707 and DC-8 experiments. The 707 tests involved a complete redesign of the engine pods and the installation of over 3,000 pounds of insulation into each of the four engines. The successful reduction in noise decreased the 707's range by 200 nautical miles. If the conversions did work, it would take up to 4 years to complete the program. There was also the question of who would actually do the work and pay for it. The FAA's policy of "treading softly" thus pushed back the issuance of any effective commercial aircraft noise-reduction regulations to late 1971.[13]

NASA Administrator James C. Fletcher spoke before an Agency-sponsored conference on civil aviation near Langley in November 1971. He outlined NASA's central role in meeting the challenge of aircraft noise reduction, which was based on a growing record of success. Fletcher also echoed changing priorities in the aviation industry overall. He admitted that it was no longer enough to think in terms of more power, lift, and speed. For Fletcher, if NASA wanted taxpayers to continue to support civil aviation, the Agency had to adopt the motto "Fly Quiet!"[14]

The Growing Clamor over Aircraft Noise

By then, airport noise had become a national concern. The problem of aircraft noise was reaching a fever pitch in 1970, especially in the densely populated Northeast. Noise suits against airports increased as local communities grew larger and urban development transitioned to suburban sprawl. Representative Allard K. Lowenstein (D-NY) exclaimed that the FAA's apparent failure to enforce its own regulations resulted in a "nightmare of noise" for many of his constituents who lived near Kennedy International Airport, then the East Coast's major international civil airport. The Congressman demanded that the FAA require the immediate installation of "noise-muffling" materials on the apparent swarms of first-generation jet airliners—primarily the Boeing 707 and Douglas DC-8—that clouded the skies over Kennedy. He also took exception to an FAA ruling that addressed the reduction of noise for soon-to-be-introduced airliners like the Lockheed L-1011 and the Douglas DC-10, while another exempted the new Boeing 747 jumbo jet altogether. The older, noisier airliners would contribute to increased congestion for decades. Lowenstein and approximately 50 of his House colleagues met with FAA Administrator Shaffer to discuss the noise problem.[15] The city of Boston filed a $10.2 million noise-pollution suit against the Massachusetts Port Authority, the operator of Logan International Airport, and 19 individual airlines in September 1970. The city, represented by Mayor Kevin H. White, wanted the money to soundproof 15 nearby schools and provide reimbursement for the lost air rights over the city, the depreciated real estate value of the schools, and the fact that the area was

The Power for Flight

no longer fit for educational purposes. Some communities went so far as to ban late-night flights.[16]

The experience of congressional response to community concerns led to the creation of a champion for aircraft noise reduction. After joining the House in 1963, Representative John W. Wydler (R-NY) from Long Island met with unhappy constituents from the western part of his district near the edge of Kennedy Airport. Their passionate pleas for help resulted in his waging an 8-year, career-defining "jet noise fight" in Congress. He became a member of the House Committee on Science and Astronautics, which also dealt with aeronautics. Wydler and his colleagues saw that the first step was to support NASA research in jet noise and the development of a "quiet" engine that would yield actual hardware for use by the airline industry. The programs began in March 1964 and September 1966 respectively. Wydler was also on the committee overseeing the creation of the new Department of Transportation (DOT), which opened in April 1967 and assumed administrative control of the FAA. The Jet Aircraft Noise Control Bill passed by Congress in June 1968, legislation Wydler originally sponsored in the House, gave the FAA the power to regulate aircraft noise; that power centered on requiring the airlines to implement new noise-reduction technology regulations.[17]

Interagency Noise Study Efforts Lead to Reliance upon the EPNdB

The departments of Housing and Urban Development and Transportation sponsored a study of aircraft noise and abatement at John F. Kennedy International Airport in 1971. Sixteen miles from midtown Manhattan in southeastern Queens, the New York airport hosted 19.6 million passengers in 1968 and employed up to 40,000 people. Other studies focused on O'Hare International Airport (Chicago), Bradley International Airport (Hartford, CT), and Cape Kennedy Regional Airport (Melbourne, FL). The studies recommended the installation of noise mufflers on commercial aircraft engines, the implementation of revised takeoff and landing procedures, and the rapid development of NASA's Quiet Engine. The combination would potentially reduce the noise-contour area by 40 percent, which meant 45 percent fewer people would be affected.[18]

NASA and the FAA advocated the two-segment landing approach. For takeoff, the pilots climbed quickly away from the airport. As they accelerated, they retracted the flaps to reduce drag and gain more speed. At landing, the airliner moved toward the airport at a steep 6-degree approach; as it neared the runway, the pilots changed the attitude to the normal 3-degree glide slope. This practice decreased the amount of time the airliner spent at low altitudes over populated areas and subjected fewer people to noise. The adoption of the method required the adaptation of air traffic control procedures

Figure 3-1. A researcher takes aircraft noise readings as part of an airport environmental control study. (NASA)

and navigation equipment for improved safety in bad weather conditions, but no actual changes in the aircraft.[19]

The FAA first required airlines to retire or upgrade their noisiest Stage 1 aircraft, such as the Boeing 707 and McDonnell Douglas DC-8, by 1985. The installation of a "hush kit," or mufflers, cost an operator between $2 million and $3 million to make them Stage 2–compliant. As that deadline approached, the major airlines paid for conversion or ordered new and improved airliners that already met the FAA's noise standards.[20] NASA established the procedures for the retrofitting of equipment that reduced fan noise levels on the Boeing 707 and Douglas DC-8, the highly successful commercial airliners of the 1960s and early 1970s.[21]

Congressional lawmakers, the FAA, the airline industry, and NASA differed in their approaches on how best to go about reducing aircraft noise. The options included installing new architectural structures at the airports, modifying existing aircraft, or working toward the design and manufacture of new aircraft and engines that cost upward of $30 million each.[22]

To indicate the effect of sound on the quality of life, the FAA's noise regulations specifically referenced EPNdB and set maximum limits at specific ground locations near an airport. The installation of sound-measuring equipment determined EPNdB at takeoff, along a line parallel to the runway at takeoff called the sideline (3.25 nautical miles for three-engine aircraft, 3.5 nautical miles for four), and at landing. Landing noise was the worst because it intensified as the aircraft flew closer to the ground and required measuring equipment 1 nautical mile (1.15 miles) from the touchdown path and 370 feet below the airplane. For new and larger airliners such as the 747, the limit was 108 EPNdB. For new aircraft equivalent to the 707, the EPNdB restrictions were 104 at takeoff and 106 for the sideline and landing. Actual 707 and DC-8 aircraft exceeded those levels.[23]

Instituting Noise Abatement: The Search for an Operational Fix

Noise was a problem at many levels. Community disdain for noisy airports prevented the construction of several more specialized and quieter airports because of the challenges facing developers to even acquire land at all. Centralized, all-purpose airports promoted economic efficiency, but they also increased noise, congestion, and pollution. The expansion of short-haul services compounded the problem because the aircraft available were noisy and worked best from airports close to city centers, which were unavailable due to community resistance. NASA and DOT released their joint report, *Civil Aviation Research and Development (CARD) Policy Study*, in May 1971. The 2-year study was an effort to establish national goals and policies for aeronautics and aviation. It focused on several critical areas, including noise abatement, airway and airport congestion, and the lack of adequate low- and high-density short-haul aircraft systems.[24] The central message was that the aviation industry, with appropriate Government assistance, had to do a better job of tailoring technology to solutions that met those problems, which were complicated and interrelated. DOT and NASA recognized that paying attention to sociological, economic, and engineering factors was the key to the solution. The *Washington Post* believed that prescription made sense for both the industry and the public.[25]

The results of the CARD study lead to a number of conclusions regarding NASA research and development priorities. Aircraft noise abatement deserved the highest priority because of widespread concern for the environment and because that program's success affected the solutions to other problems.

Congestion was next because its solution involved an organized effort directed at the combination of air traffic control; runway capacity; ground control of aircraft; terminal processing; access and egress; parking; and airport location, acquisition, and development. A new short-haul system could help relieve congestion at existing airports. The CARD study acknowledged that constant improvements in technology for long-haul vehicles and their propulsion systems were essential to continued U.S. leadership.[26]

In response to the CARD study, Secretary of Transportation John A. Volpe and NASA Administrator James C. Fletcher announced the establishment of a new Joint Office of Noise Abatement in October 1971. DOT official Charles R. Foster was director, while Walter F. Dankhoff of NASA served as deputy director. The new organization retained the original DOT office with the addition of NASA personnel at Lewis. The consolidation took place as a measure to better manage the national program to address noise in current and future transportation systems.[27] To avoid duplication, NASA, DOT, and the FAA had to approve all programs jointly. Specifically, DOT and the FAA worked to gather noise information, which included retrofitting current aircraft to gain a better understanding of the nature of noise and its effect on communities, as well as to better inform regulatory functions.[28]

DOT and NASA would approach the problems in different ways. The new joint office identified four distinct methods of modifying the propulsion systems for narrow-body airliners, which included the Boeing 707, 727, and 737, as well as the Douglas DC-8 and DC-9. DOT continued with nacelle acoustic treatment and jet suppression. NASA directed its efforts toward both modifying the fans in existing engines and combining them with new acoustic nacelles and developing a new engine altogether. Independent reviews of each would determine the method endorsed and adopted by industry.[29]

Up to 1971, NASA worked in two main areas. Researchers worked to gain a better understanding of the generation and propagation of aircraft noise. They also went the extra step to investigate various techniques for suppressing the noise of current and expected subsonic commercial transports. The Langley Nacelle Acoustic Treatment and Lewis Quiet Engine programs were examples of NASA's work in that area.[30]

In a May 1972 memorandum to Edward E. David, Jr., the science advisor to President Richard Nixon, and William Morrill of the Office of Management and Budget, NASA Administrator James C. Fletcher, DOT Under Secretary James M. Beggs, and FAA Administrator John H. Shaffer outlined the joint program to address the noise generated by narrow-body airliners powered by Pratt & Whitney turbofans.

Specifically, the problem was with the JT3D-powered Boeing 707 and Douglas DC-8 and the JT8D-powered 727, 737, and DC-9. The Joint

The Power for Flight

Noise Abatement Office, with the support of Pratt & Whitney, Boeing, and McDonnell Douglas, worked to create a complementary relationship between technical feasibility and the regulatory process. Regarding the retrofitting of airliners, the office offered "baseline" and "expedited" programs. The former relied upon funding levels included in President Nixon's 1973 fiscal year budget. The latter alternative was created for the benefit of the House Subcommittee on Aeronautics and Space Technology. It used only acoustically treated nacelles as the basis for regulation while providing leeway for the airlines to invest in front-fan modification to meet increasingly stringent regulations.[31] The problem inherent in retrofitting aircraft with acoustically treated nacelles or new engines with quieter fans, or in purchasing new Quiet Engines, was the question of who would pay for the technology.

Refan: Going Beyond the Nacelle and Treating the Engine Itself

Noise absorption material inside a nacelle was not fully effective in reducing the noise of the exhaust jet pushing an airliner through the sky. That low-pitch roar, caused by the interaction of high velocity exhaust gases with the bypass and surrounding air, required a different solution. The only way to reduce that type of noise was to lower the jet velocity by replacing the fan with a larger one that used more energy from the exhaust jet and reduced its velocity.[32] NASA's Refan Program investigated the feasibility and cost of modifying the JT8D turbofan that powered the 727, 737, and DC-9 airliners to reduce noise during the period from 1970 to 1975. Going beyond the acoustic modification of a nacelle led to improved noise reduction, decreased fuel consumption, and increased efficiency in turbofan engines overall.[33]

There was considerable impetus to make the JT8D a more efficient, quieter, and cleaner engine. Introduced in February 1963 on the Boeing 727, the low-bypass JT8D was a commercial derivative of the Pratt & Whitney J52 military turbojet. The engine series covered the thrust range from 12,250 to 17,400 pounds and powered 727, 737-100/200, and DC-9 aircraft. In the early 1970s, Pratt & Whitney was well on its way toward producing more than 14,000 JT8Ds, which early on had earned the nickname "workhorse of the airline industry."[34]

Flight tests in 1975 revealed that engines with the refan engine modification generated 5 to 10 EPNdB below then-current noise levels of the JT8D-powered aircraft at all three FAR 36 measuring points. At the approach point, the noise reduction at takeoff was substantially greater with the engine modification than with acoustic treatment alone. For example, the total effect on the community around the airport would be to reduce the 90-EPNdB or louder noise levels heard during a takeoff and landing operation of a 727 airplane by 75 percent.

Acoustic treatment alone accounted for a 30-percent reduction—significant, but not sufficient to ameliorate local concern.[35]

The cost of modifying JT8D engines and installing new nacelles, of course, was more expensive than simply adding acoustic treatment to an existing nacelle. The technology included a larger-diameter single-stage fan; increased spacing between the inlet guide vanes, fan, and stator blades; an optimized number of blades; an internal mixer nozzle to reduce exhaust velocity; and a sound-absorbing lining. NASA worked with the FAA and DOT to provide the airlines with the data needed to inform their decision on whether or not to utilize refanned JT8D engine technology along with other options.[36]

Boeing continued to collaborate with NASA on the Refan Program. The manufacturer installed three JT8D-115 refans, developed by Pratt & Whitney under contract to NASA, on a 727 for flight tests at Boeing's Boardman, OR, airfield. While NASA's goal was simply to retrofit all existing engines, Boeing and Pratt & Whitney went further, identifying a possible new 727 derivative, the 727-300B. Equipped with 19,300-pound-thrust JT8D-217 refans, the new "stretched" 727-300B offered less takeoff noise and improved fuel efficiency. United Airlines wanted the refan and played an important role in the design of the -300. Boeing's sales staff sought out more purchasers for the refined airliner that cost $2 million more than the non-refan version of the 727.[37]

However, the buyers never materialized before both the older 727 and DC-9 aircraft went out of service. Nevertheless, Pratt & Whitney had considerable success reengineering the JT8D engine for use in updated versions of the McDonnell Douglas MD-80 and the twin-engine Boeing 737 introduced in 1980 and 1981 respectively.[38] The JT8D-217 and -219 introduced in the early 2000s provided Stage 3 noise compliance; steeper, faster, and quieter climb rates; enhanced short-field performance; and an approximately 10-percent increase in fuel economy over long distances.

The Subcommittee on Aviation and Transportation Research and Development of the House Committee on Science and Technology chaired by Dale Milford (D-TX) conducted public hearings on aircraft noise abatement during the fall of 1976. The resultant report recommended that the FAA sponsor NASA's increased research in noise-reduction technology. Overall, the subcommittee urged the creation of a coordinated noise-abatement program within the framework of national transportation policy. Specifically, they advised the Secretary of Transportation to consider the Sound Absorption Material (SAM) retrofit option for its short-term feasibility while taking into account the long-term benefits of the more expensive programs such as Refan and the purchase of new aircraft.[39]

In reaction to a directive by President Gerald R. Ford, the FAA issued regulations at the end of 1976 requiring that all commercial jet aircraft be in compliance with Federal noise standards by 1985. The ruling affected 75 percent of the national commercial aviation fleet, which included the Boeing 707, 720, 727, and 737; the McDonnell Douglas DC-8 and DC-9; the Convair 990; the British Aircraft Corporation BAC One-Eleven; and even the early 747-100 jumbo jet. Of those, 8 out of 10 were technically in violation of the then-current noise regulations implemented in 1969. The airlines and other commercial operators of jet aircraft had the option of either modifying or replacing jet engines that generated noise levels exceeding those specified in Part 36 of the Federal Aviation Regulations. Those levels ranged from 93 to 108 EPNdB depending on the weight of the aircraft.[40]

The next step was to retire or upgrade all Stage 2 aircraft to Stage 3, which required entirely new engines. A considerable portion of aircraft in commercial aviation, approximately 4,500 aircraft, or 70 percent of the airline fleet, were Stage 2–certified. They included the 727, the workhorse for the airlines, with about 1,750 in service in 1987. The solution was to retrofit the tri-jet's two side engines with NASA-influenced Pratt & Whitney JT8D-217 refans. The update improved fuel efficiency by 10 to 15 percent and increased range from 1,500 to 1,800 miles.[41] Pratt & Whitney introduced the derivative JT8D-200 series in October 1977 on the McDonnell Douglas MD-80 airliner. With the new 49.5-inch fan, it generated over 19,000 pounds of thrust at a bypass ratio of 1.78:1.[42]

NASA's Anti-Noise Initiatives from Quiet Engine Program Onwards

Quiet Engine Program

Congressional posturing and critiques of the modern world aside, NASA's eventual response to the noise problem was the initiation of the Quiet Engine Program (QEP). GE was the principal contractor working with the Agency. The new engine that their engineers developed had the potential to reduce subsonic jet engine noise 15–20 PNdB below the levels generated by a Boeing 707 or McDonnell Douglas DC-8, definitely increasing the quality of life for the people living within 5 to 10 miles of a major airport.[43] Once NASA and GE proved the technology, the FAA could enforce increasingly strict noise regulations based on a new generation of standardized Quiet Engines. In other words, NASA acted as the problem solver for the environmental noise movement, manufacturers, and the airline industry. William Hines of the *Chicago Sun-Times* believed that while getting the program started represented a minute portion of NASA's 1970 budget, a mere $9 million, it offered immense dividends in "peace of mind" for the country.[44]

NASA considered the Quiet Engine to be a long-term solution because the program would take a considerable amount of time before it became a reality. It also involved different partners. The Agency's role was to determine the design parameters. The manufacturers were to make it a reality and then the regulatory agencies needed to approve the engines. Ground tests began in 1972 at General Electric, with additional tests carried out at Lewis. NASA estimated that the cost of the entire program would be $300 million and that a flight engine would be ready by 1975.[45]

James J. Kramer led the QEP at Lewis from the beginning in 1966. Kramer joined the NACA in 1951 and worked in both aircraft and rocket propulsion. Before leading the program, he managed the 260-inch solid rocket program at Lewis. In 1971, he became the chief of the new Noise and Pollution Reduction Branch of the Aeronautical Propulsion Division at NASA Headquarters.[46]

The Quiet Engine was not a retrofit program. It was a technology demonstrator for the development of future engines and aircraft. GE finished and tested the first full-scale fan assembly during the spring of 1971. NASA was confident that the low-noise objectives were going to be met.[47] The first ground tests of a Quiet Engine began in late August 1971 at GE's Peebles, OH, test facility. The goal was to develop a 22,000-pound-thrust engine 15 to 20 dB quieter than the 1950s-generation engines on 707 and DC-8 airliners. To meet that goal, project engineers designed a high-bypass-ratio engine with a low-noise fan that had a lower rotational speed, a higher bypass ratio, adjusted tolerances between rotating and stationary parts, and honeycomb acoustic material installed in the flow passages to muffle sound. They were optimistic they were going to meet that goal. The most promising design, called Engine A, featured the gas generator core GE used on its CF6 and TF39 engines, which powered the Douglas DC-10 airliner and Lockheed C-5A Galaxy transport. Ground tests involved operating the engine at takeoff and landing conditions with different inlets and exhaust nozzles installed. Engineer Harry Bloomer reflected that it was a "smooth running engine" that exhibited no vibration, structural stress on the fan, or visible smoke in the exhaust. Once GE completed its part of the test program, Engine A traveled to Lewis for installation in an acoustic nacelle for complete propulsion system testing.[48]

Lewis displayed Engine A, fresh from additional tests, at its May 1972 Aircraft Noise Reduction Conference in Cleveland. Over 300 engineers from the aviation industry attended and learned of the ongoing progress of the $21 million project. Tests revealed that installation of the Quiet Engine on a four-engine airline would generate only 90 EPNdB at simulated takeoff and landing conditions. Airliners in service at that time generated 116 EPNdB at takeoff and 118 EPNdB at landing.[49]

Solving the ills of the first generation of American airliners, exemplified by the Boeing 707 and Douglas DC-8, was going to be expensive. At a press conference at NASA Headquarters in March 1972, Kramer remarked that the economic factors in the development of the Quiet Engine were substantial. Quiet Engines would be no more expensive in operation than the new generation of turbofans found on the Boeing 747 and Douglas DC-10. Retrofitting the aircraft that needed the quieter engines the most, the 707 and DC-8, was the challenge. Kramer provided an estimate from McDonnell Douglas, one of the contractors for the program, that it would cost between $5 and $6 million to take one aircraft out of service, install four Quiet Engines and nacelles, and return the aircraft to airline service. Reporter David Bresket of the *New York Daily News* responded that the airlines were not going to "retrofit anything" at that price. Kramer answered, "I agree."[50]

NASA realized that engine modification was cheaper than creating new engines. NASA awarded 4-month, noncompetitive contracts to Pratt & Whitney, Boeing, and McDonnell Douglas to develop Quiet Engine designs. Pratt & Whitney received $1.2 million to design modifications to its JT3D and JT8D turbofans, which powered virtually all narrow-body airliners operating in the United States. The engine maker's engineers concentrated on replacing two-stage fans with one-stage fans to reduce engine whine and exhaust noise. Boeing and McDonnell Douglas both received $800,000 and focused on the methods and materials of the acoustic treatment of engine nacelles to absorb fan noise. A secondary function of the contracts was the reduction of exhaust emissions. NASA estimated that the overall cost of this particular part of the program would be $5.6 million.[51]

Experimental Clean Combustor Program

Accompanying the concern over aircraft noise was growing public unease over air pollution in the United States. For aviation, the EPA's 1979 Standard Parameters addressed the levels of carbon monoxide (CO), unburned hydrocarbons (THC), oxides of nitrogen (NO_x), and smoke in aircraft engine exhaust emissions.[52] The International Civil Aviation Organization (ICAO), the United Nations agency responsible for worldwide commercial aviation regulation, released increasingly stringent NO_x emissions standards beginning in 1981.[53] In addition to the retrofit program to install acoustic materials to reduce noise in engine nacelles in the early 1970s, NASA also initiated a program to remove the distinctive black exhaust generated by jet engines. The airlines simply made the modifications at the regularly scheduled maintenance interval of 6,000 hours.[54] NASA reacted with a number of internal and contract studies to address the problem of CO, THC, and NO_x. The most comprehensive program was the Experimental Clean Combustor Program (ECCP) for turbofan engines conducted in partnership with GE and Pratt & Whitney, which had its origins in the defunct SCAR program.[55]

The ECCP was a multiyear, major contract effort. The primary program objectives were to generate clean and efficient combustor technology for the development of advanced commercial turbofan engines with lower exhaust emissions than those of current aircraft.[56] The program, which started in December 1972, consisted of three phases. First, Lewis researchers evaluated low-pollutant combustors as they investigated multiple-burning-zone combustors, improved fuel distribution and preparation, and the staging of combustor airflow. After selecting the most promising configurations, they went about refining them for optimum performance. The program concluded with the demonstration and evaluation of the new combustor innovations in full-scale CF6 and JT9D turbofans in 1976.[57]

For their engines, GE and Pratt & Whitney developed a two-stage combustion process wherein a pilot zone addressed low-power engine CO and THC emissions while a main zone regulated high-power NO_x emissions. They approached the process in different ways.[58] Pratt & Whitney developed an axial series arrangement called Vorbix (*vor*tex *b*urning and m*ix*ing) that employed multiple burning zones and improved fuel preparation and distribution.[59] GE introduced a double annular axially parallel design that featured a characteristically short combustor for multistage burning. Both approaches achieved substantial reductions in all pollutant categories, meeting the 1979 EPA standards for CO, THC, NO_x, and smoke at percentages of reduction that reached 69, 93, and 42 percent respectively.[60] While there was enough information to "declare victory" regarding reduced NO_x emissions, it would not be until the mid-1980s that the need influenced actual development on the part of the manufacturers.[61]

V/STOL → Q/STOL → QUESTOL

Despite considerable effort on the part of NASA, the United States was well behind the rest of the world in research and experimentation on V/STOL by 1970. The United Kingdom was a leader in the technology with its Hawker Siddeley Harrier reconnaissance/strike-fighter. The Harrier used a unique "bedpost" approach to vertical flight, with an engine, the Rolls-Royce Pegasus, that had twin "cold" nozzles off its compressor and twin "hot" exhaust nozzles. The four "legs" of thrust thus generated enabled the aircraft to take off, hover, and land. The nozzles could then be "vectored" aft to enable the aircraft to accelerate into wingborne (aerodynamic lifting) flight. This engine, developed by Sir Stanley Hooker, made practical the transonic V/STOL strike aircraft,

The Power for Flight

something long sought but never previously achieved. As an important example of international collaboration in the late 1960s, NASA assisted the British Government with the testing of the P.1127 Kestrel, the developmental test and evaluation forerunner of the Harrier; NASA later operated one (designated the XV-6A) at Langley Research Center.[62]

NASA's role included free-flight and transonic tunnel model testing at Langley in 1959 and 1960, led by Deputy Director John Stack, a Collier Trophy recipient for the conceptualization of both the first supersonic research airplane and the transonic slotted-throat wind tunnel. In mid-1962, NASA research pilots Jack Reeder and Fred Drinkwater became the first non-British pilots to fly the P.1127. Their report facilitated the funding to establish the Tri-Service XV-6A Kestrel Evaluation Squadron. NASA engineer Marion "Mack" McKinney validated the P.1127's ability to transition from hovering to forward flight with a one-sixth scale free-flight model in the Full-Scale Tunnel. Another model featured small maneuvering jets powered by hydrogen peroxide. The program was an international effort, with Hawker providing the basic design and NASA providing its experience with the Bell X-14 (an earlier vectored-thrust test bed) and the construction of the model; the Air Force funded all of this work.[63] In addition to military V/STOL aircraft, NASA worked toward the development of new short-haul Short Takeoff and Landing (STOL) airliner designs to alleviate travel problems in high-density areas such as the Northeast Corridor and the West Coast. NASA let competitive contracts with several companies to identify the optimum design for a practical STOL transport.[64]

Figure 3-2. The Kestrel V/STOL research aircraft. (NASA)

But V/STOL was not an unalloyed blessing. Whatever noise problems conventional jet aircraft experienced were greatly magnified by V/STOL aircraft as they took off under engine-borne lift and then transitioned to aerodynamic lift, transiting back to engine-borne lift for their descent to land. NASA was well aware of the noise problem with V/STOL aircraft. At one of the earliest NASA-sponsored conferences dedicated to the topic in 1960, Langley researchers identified the basic challenges. One thing was clear: the main source of noise from V/STOL aircraft was the different types of propulsion systems configured to power them through the air. At the time, viable options included pure helicopter, turbojet, or turbofan lift systems or the tilt-wing turboprop. Minimizing the adverse noise generated by a V/STOL aircraft fell into two areas based on commercial or military application. Regardless of whether it was a commercial aircraft flying over a housing community or a military aircraft trying to avoid detection in a war zone, each problem had its own solution. Lightly loaded rotor blades, propellers with low tip speeds, and turbofan and similar ducted fan engines would alleviate complaints from civic groups. For military missions, overall sound reduction would enable the operation of aircraft at increasingly lower altitudes.[65]

NASA's movement toward quieter V/STOL aircraft gained momentum over the course of the 1960s. Because of significantly greater restrictions on aircraft noise around city centers, researchers explored ways to reduce the noise generated by low-noise fans and sonic inlet choke devices and investigated methods of reducing the noise in augmenter wing flap systems. The latter featured a particular focus on the reduction of the "scrubbing" noise associated with the flap system of an externally blown flap vehicle. Also, low emissions were a concern. NASA fully expected to carry the program through to the construction and flight of a quiet STOL engine technology demonstrator.[66]

In response to the CARD study, NASA announced in August 1971 the initiation of the Quiet-STOL, or Q/STOL, aircraft program, a joint effort with the FAA and the U.S. Air Force to relieve aircraft noise and congestion. Manufacturers had until October 15 to submit a proposal for the design and construction of two experimental STOL transports that incorporated propulsive lift. STOL required 2,000 feet or less of runway. Propulsive lift used the jet engines to produce lift to augment aerodynamic lift generated by the wings and flaps. The aircraft were part of a NASA flight research program that would lead to an operational, environmentally friendly, economical, and safe turbofan-powered STOL transport. NASA saw the development of STOL systems as valuable to the aviation industry because they would enhance short-haul transportation, which covered 500 miles or less and accounted for 60 percent of air traffic in 1971; alleviate noise and congestion at airports; and modernize the military's tactical airlift operations. NASA also envisioned that the Q/STOL

The Power for Flight

Figure 3-3. The QSRA made over a dozen landings on the U.S.S. *Kitty Hawk* during naval flight trials near San Diego, CA, in August 1980. (NASA)

program would generate the data needed for technical development and the establishment of rules for certification and operation. Eventually, this program led to the development of a specialized test bed, the NASA Quiet Short-Haul Research Aircraft (QSRA), which flew an extended series of flight trials, including landing and taking off from the aircraft carrier U.S.S. *Kitty Hawk*.[67]

The Q/STOL program was one of several projects managed by the NASA Office of Advanced Research and Technology (OART). OART was responsible for providing technical support for the advancement of civil and military aviation. OART was also generating the technical basis for the development of quiet, nonpolluting STOL engines.[68]

At a STOL conference hosted at Ames in October 1972, Agency researchers shared their goals. They aimed to reduce aircraft engine noise, exhaust emissions, airport congestion, and other factors that stymied the growth of the national air transportation system. NASA envisioned that a quiet and clean STOL transport designed to operate from small airports would be both compatible to surrounding communities and convenient to the ever-increasing number of air travelers. The conference included a specific session on quiet STOL propulsion.[69]

NASA spent quite large sums of money on STOL programs. The short-lived Quiet Experimental Short Takeoff and Landing (QUESTOL) program of the early 1970s focused on new 50- and 100-passenger airliners designed for a revolutionary and novel aerial transportation network of short-haul routes. New and smaller (under 2,000 feet) runways, combined with aircraft capable of operating from them safely and quietly, provided a way to alleviate the aerial traffic jams over the United States, especially the Northeast Corridor connecting Boston, New York, Philadelphia, and Washington, DC. The key to low-speed performance and maneuverability was a high wing and T-tail configuration, engines mounted closer together and to the fuselage, and externally or internally blown flaps. Regarding the latter, cool thrust from a new type of turbofan with exhaust temperatures in the range of 350 degrees Fahrenheit directed over or through the trailing flaps generated more lift. All of those components led to an airliner capable of steep takeoffs and descents that reduced noise in the areas surrounding the airport and efficient cruise at 20,000 feet while carrying enough passengers to make an airline profitable.[70]

A considerable number of STOL research contracts went to Hamilton Standard, a veteran propulsion company that had pioneered the development of the variable-pitch propeller. The funding allowed the company to transition its theoretical work on variable-pitch fans of the 1960s into practical experimentation. Just as the variable-pitch propeller was more efficient and offered better performance than its fixed-pitch counterpart, a variable-pitch fan offered efficiency, safety, and maneuverability. The development of the new design, called the Quiet-Fan (Q-Fan), was also quieter at 95 EPNdB and facilitated STOL performance. The company initiated and funded its own private high-bypass, variable-pitch fan engine research and development program from 1969 to 1971. The work involved the creation of new analytical methodology for aerodynamics, structures, and acoustics and extensive work investigating the operational capabilities of variable-pitch fan engines, including the much-desired quiet and reverse-thrust capabilities. Engineers working with chief of preliminary design Richard M. Levintan began with testing a ⅓-scale model, which led to the construction of a full-size demonstrator engine based on the proven Lycoming T55-L-11 turboshaft. With QUESTOL funding from Lewis, Hamilton Standard and Lewis engineers conducted a formal test program of their 20-inch-, 4.5-foot-, and 6-foot-diameter Q-Fans from September 1972 through February 1973.[71]

NASA's Advanced Concepts and Missions Division at Ames and Hamilton Standard worked together on a Q-Fan for general aviation aircraft in the early 1970s. The researchers believed that the continued growth of general aviation through the late 1980s depended on improved aircraft safety, utility, performance, and cost. They saw the compact, low-noise Q-Fan propulsor

The Power for Flight

concept as the answer. The combination of a reciprocating or rotary combustion engine with the Q-Fan offered exciting new possibilities. They would meet the expected noise and pollution restrictions of the 1980s while facilitating a new generation of general aviation aircraft with cleaner airframe designs.[72]

QCSEE

The purpose of the focused and more successful Quiet, Clean, Short-Haul Experimental Engine (QCSEE, pronounced "Quick-See") program in 1974 was to demonstrate the advanced propulsion technology developed during the QEP for QUESTOL's 125- to 150-passenger short-haul airplane and NASA's follow-on powered-lift QSRA program to reduce airport congestion, aircraft noise, and air pollution. The U.S. Navy also voiced its interest in applying the technology to a V/STOL transport capable of operating from aircraft carriers. The $30 million QCSEE program had three specific goals: the development of environmentally friendly and economical short-haul propulsion technologies, the generation of data for use by the Government in setting future regulations, and the transfer of the technology and data to industry.[73]

NASA researchers started with investigations of thrust performance, fan design, and thrust-to-weight ratio for both over-the-wing and under-the-wing configurations. A major emphasis of the program was low noise and emissions characteristics that were to conform to the EPA's 1979 pollutants standards. The research expanded to investigate other ideas, including reduced fan-tip speeds, optimized rotor and stator ratios, and acoustic treatment techniques to reduce both high-frequency fan and turbine noise and low-frequency combustor noise.[74]

Lewis awarded GE the QCSEE contract. In terms of performance, the new engines needed to incorporate an efficient and quieter high bypass ratio and generate enough power to enable steep takeoffs and landings. Under the leadership of senior designer A.P. Adamson, GE constructed two QCSEE fan assemblies, one with conventional blades and another with variable-pitch blades developed in concert with Hamilton Standard. An F101 engine served as the core for both. To achieve the high bypass ratio, GE utilized a reduction gearbox that allowed the fan assembly to rotate at much slower and quieter speeds. The GE engineers utilized a large number of metal reinforced composites, which were cheaper and easier to fabricate, for the fan blades and casing, the gearbox housing, the front frame, and the exhaust nozzle. GE regarded its participation in QCSEE as a valuable experience.[75]

The two 20,000-pound-thrust QCSEEs were identical internally. For short-runway operation, both engines relied upon the wing flaps to deflect the jet exhaust for increased lift. The primary external difference between them was the method of mounting the engine to the wing. The GE engine employed the conventional under-wing pod installation found on virtually all multi-engine jet aircraft. NASA's design featured an unorthodox mounting on top of the wing with a distinctive half-moon nozzle, which protected people on the ground below from noise.[76]

Lewis and GE evaluated both QCSEEs, and the results in terms of noise, emissions, and fuel consumption were dramatic. They were 8 to 12 EPNdB quieter than the quietest engine then in use on the McDonnell Douglas DC-10 and Boeing 747 wide-body airliners, the GE CF6. That meant they were 16 EPNdB below FAA standards and 9 EPNdB below the stricter standards to be implemented in the 1980s. QCSEE technology offered a way to reduce that noise "footprint"—the area on the ground, directly below the aircraft, that was subjected to the takeoff and landing noise of the airliner—to approximately 1 square mile, which was 40 times smaller than the footprint of the Boeing 707. The researchers achieved the reduction in noise by slowing the velocity of the engine exhaust by increasing the bypass ratio and by other design features including the use of acoustically absorbent materials.[77]

The QCSEEs provided cleaner emissions and lower fuel consumption. The technology reduced two of the most problematic air contaminants from jet engines, CO and THC, by approximately 80 percent and 97 percent respectively. The new design principles incorporated by GE into the engine combustion system enabled the technology to meet increasingly stringent EPA standards. The QCSEE engines offered fuel savings of 10 percent. Glenn researchers and GE engineers achieved those savings by substituting lightweight, nonmetallic composite materials of equal or greater strength for much heavier metal components found in the engine cowling, frame, and fan blades.[78]

NASA asserted that engine manufacturers could incorporate the technology into their higher-thrust engines—those with a thrust in excess of 40,000 pounds—that powered the largest commercial airliners.[79] It was applied to the experimental QSRA developed jointly by NASA and Boeing that utilized the over-the-wing upper-surface blowing technology combined with four Avco-Lycoming YF-102 high-bypass turbofans during its test flight program from 1978 to 1980, but it did not achieve commercial application.[80] Nevertheless, the QCSEE program anticipated many of the technical features pursued in the design of turbofan engines in the early 21st century. Both GE and Pratt & Whitney incorporated high-bypass-ratio engines in the range of 10:1 to 12:1, low-pressure-ratio variable-pitch fans, variable-area fan nozzles, advanced acoustic liners, digital electronic controls, clean combustors, reduction gearing, and composite components into their new products.[81]

Fuels

While NASA worked to enhance the turbofan engine, the Agency's researchers also directed their focus to aircraft fuels. NASA's Research and Technology Advisory Council Committee on Aeronautical Propulsion convened an Ad Hoc Panel on Jet Engine Hydrocarbon Fuels that met at Lewis through the 1970s. Membership included NASA researchers and representatives from the Environmental Research and Development Agency, the Air Force's Aero Propulsion Laboratory at Wright-Patterson Air Force Base, and the Naval Air Propulsion Test Center headquartered in Trenton, NJ. The panel, carrying on in the tradition of the defunct NACA fuels and combustion subcommittee, met to discuss the state of the art in fuel development at NASA and other agencies, covering topics from improving burning properties to creating alternative synthetics.[82]

The panel submitted four recommendations to guide NASA's future support of an alternative fuel research program. First, the Agency was to initiate a supporting combustion research program. Second, researchers were to stay abreast of candidate alternative fuels to assess their toxicity or safety problems. Third, NASA was to carry out the research needed to provide the authoritative performance relationships as the basis for the preparation of fuel specifications. Finally, NASA was to develop and maintain an up-to-date analysis of alternative fuel economics and availability that made full use of all data sources.[83]

Lewis researchers dedicated a considerable amount of effort to investigating alternatives to traditional oil sources. Their evaluation of JP-5/Jet A fuel refined from Paraho-processed shale in a single combustor revealed no difference in combustion when compared to petroleum-derived fuel. Lewis contracted with Atlantic Richfield to perform the laboratory synthesis of jet fuel from coal and shale. In-house studies explored new synthetic crude oils, the design of a hydrotreating laboratory to process coal and shale into oil, and the determination of the effects of relaxed fuel specifications on combustor performance. The work at Lewis complemented the research conducted by the Agency's partners on the panel. Both the Air Force and the Navy were investigating the potential of coal and shale crude oil as the basis of jet fuel. The Navy conducted engine tests, while the Air Force conducted long-distance flights with a North American T-39 Sabreliner powered with JP-4 fuel refined from Paraho shale crude.[84] In January 1979, researchers at Langley compared liquid hydrogen, liquid methane, and synthetic aviation kerosene and concluded that the last was the most economical option.[85]

The Naval Energy and Natural Resources Research and Development Office, which provided NASA and its Department of Defense partners with their supplies of synthetic fuels for testing, determined that no technical barriers existed to increasing the production of fuels derived from shale, coal, and tar sands by 1980. Nevertheless, increasing the supply of jet fuel faced economic and environmental challenges. The office estimated that fuel derived from shale required an increase in production from 900 barrels per day in 1975 to 250,000 per day in 1980. The more widely available coal needed to expand from 10 barrels per day to 6,000. Tar sands, which presumably existed in quantities similar to the remaining reserves of domestic petroleum, required starting the refining process from scratch—from no barrels to 7,000 a day within 5 years.[86]

I. Irving Pinkel, a consulting engineer and former head of the Physics Division at Lewis, prepared a report for the Advisory Group for Aerospace Research and Development (AGARD) of the North Atlantic Treaty Organization (NATO) in 1975 on the future of aviation fuels based on discussions with leading experts in the NATO countries. Technology could do only so much. Advanced turbofan engine components offered an estimated 15-percent reduction in fuel consumption. Pinkel stressed that "any great gains in fuel consumption would probably be by use of other engine concepts such as turboprops." While petroleum sources continued to dwindle, Pinkel emphasized that over the course of the next half century, the American and European aviation industries would have to accept fuels very different from those available in the mid-1970s due to escalating costs and plummeting availability.[87]

Overall, the panel argued that any high-level work on aircraft fuel conservation technology had to include fuel properties, especially regarding how they were to be affected by changing resources. In other words, a national fuel effort that started at the oil refinery would be the only way to contribute to aircraft efficiency overall.[88]

Endnotes

1. George P. Miller, "NASA," Cong. Rec., 92nd Cong., 1st sess., vol. 117, December 14, 1971, p. H12545, copy available in the NASA HRC, file 012336.
2. Reddy, "Seventy Years of Aeropropulsion Research at NASA Glenn Research Center," p. 202.
3. "Too Many Decibels Are Making Us Ding-a-Lings," *Washington News* (June 8, 1972): 2.
4. Peter C. Stuart, "Congress Raises Voice To Soften Noise," *Christian Science Monitor* (March 3, 1972): n.p., copy available in the NASA HRC, file 012336.
5. Charles W. Harper, "Introductory Remarks," in *Progress of NASA Research Relating to Noise Alleviation of Large Subsonic Jet Aircraft: A Conference Held at Langley Research Center, Hampton, Virginia, October 8–10, 1968* (Washington, DC: NASA SP-189, 1968), n.p.; "NASA Noise Conference Planned," NASA News Release 77-221, October 14, 1977, NASA HRC, file 012336.
6. St. Peter, *History of Aircraft Gas Turbine Engine Development*, p. 398.
7. FAA, "Noise Standards: Aircraft Type and Airworthiness Certification," Advisory Circular 36-4C, July 15, 2003; U.S. Congress, Office of Technology Assessment, *Federal Research and Technology for Aviation*, OTA-ETI-610 (Washington, DC: Government Printing Office, 1994), pp. 78–79.
8. "FAA Issues Noise Regulations for Older Jets," Department of Transportation News Release 76-121, December 28, 1976, copy available in the NASA HRC, file 012336.
9. Glenn Research Center, "Making Future Commercial Aircraft Quieter: Glenn Effort Will Reduce Engine Noise," NASA Release FS-1999-07-003-GRC, July 1999, available at *http://www.nasa.gov/centers/glenn/pdf/84790main_fs03grc.pdf* (accessed July 7, 2013).
10. The NACA initiated noise-reduction studies in the early 1950s and, in response to the findings of President Eisenhower's airport commission, created a Special Subcommittee on Aircraft Noise in 1952. J.H. Doolittle, *The Airport and Its Neighbors: The Report of the President's Commission* (Washington, DC: Government Printing Office, 1952), p. 45; Banke, "Advancing Propulsive Technology," p. 737.
11. *NASA Acoustically Treated Nacelle Program: A Conference Held at Langley Research Center, Hampton, Virginia, October 15, 1969* (n.p.: NASA SP-220, 1969), pp. 11–12.

12. H. Dale Grubb to Roman C. Pucinski, October 7, 1970, NASA HRC, file 012336.
13. "Shhh…," *Forbes* (March 15, 1970): 31.
14. "NASA Chief Welcomes Jet Noise Challenge," NASA News Release 71-222, November 3, 1971, NASA HRC, file 012336.
15. "Lowenstein Says FAA Won't Curb Jets' Noise," *New York Times* (February 21, 1970): 55.
16. "Noise Suit," *Washington Post* (September 10, 1970): A24.
17. John W. Wydler, "Noise," in Cong. Rec., 91st Cong., 2nd sess., vol. 116, June 1, 1970, pp. E4989–E4991, copy available in the NASA HRC, file 012336.
18. William F. Ryan, "Aircraft Noise Abatement at JFK International Airport," in Cong. Rec., 92nd Cong., 1st sess., vol. 117, June 14, 1971, pp. E5783–E5784, copy available in the NASA HRC, file 012336.
19. "Press Briefing: NASA Aircraft Noise Abatement Research," *NASA News* (March 31, 1972): 6–7, NASA HRC, file 012336; Banke, "Advancing Propulsive Efficiency," p. 740.
20. Agis Salpukas, "Advances: Rebuilding Planes To Cut Noise," *New York Times* (November 18, 1987): D8.
21. St. Peter, *History of Aircraft Gas Turbine Engine Development*, p. 398.
22. Douglas B. Feaver, "Ford Pledges To Abate Airport Noise," *Washington Post* (October 22, 1976): A19.
23. William Hines, "NASA's Quiet Engine Still a Long Way in Future," *Chicago Sun-Times* (June 7, 1970); "Press Briefing: NASA Aircraft Noise Abatement Research," *NASA News* (March 31, 1972): 4–5, NASA HRC, file 012336.
24. George P. Miller, "NASA," in Cong. Rec., 92nd Cong., 1st sess., vol. 117, December 14, 1971, p. H12545, copy available in the NASA HRC, file 012336.
25. "DOT and NASA and Noise," *Washington Post* (August 29, 1971): B6.
26. C.A. Syvertson, "Civil Aviation Research and Development (CARD) Policy Study," in *Proceedings of the Northeast Electronics Research and Engineering Meeting, Boston, Massachusetts, November 2–5, 1971* (Newton, MA: Institute of Electrical and Electronics Engineers, Inc., 1971), p. 58.
27. "DOT/NASA Noise Abatement Office Established," NASA News Release 71-213, October 21, 1971, NASA HRC, file 012336. The interagency members included the Office of Management and Budget, the National Aeronautics and Space Council, and the Environmental Protection Agency.

28. Albert J. Evans to Associate Administrator for Advanced Research and Technology, "Back-Up Information Regarding Progress Status and Plans of NASA Aircraft Noise Reduction Programs," March 3, 1971, NASA HRC, file 012336.
29. James C. Fletcher, James M. Beggs, and John H. Shaffer, memorandum for Edward E. David, Jr., and Donald B. Rice, "Reduction of Noise from Existing Aircraft," February 9, 1972, NASA HRC, file 012336.
30. Evans to Associate Administrator for Advanced Research and Technology, "Back-Up Information Regarding Progress Status and Plans of NASA Aircraft Noise Reduction Programs."
31. James C. Fletcher, James M. Beggs, and John H. Shaffer, "Reduction of Noise from Existing Aircraft," May 17, 1972, NASA HRC, file 012336.
32. Gerald D. Griffin to Henry M. Jackson, July 23, 1974, NASA HRC, file 012336.
33. Jeffrey L. Ethell, *Fuel Economy in Aviation* (Washington, DC: NASA SP-462, 1983), pp. 6–7.
34. The bypass ratio for the JT8D is 0.96:1. Pratt & Whitney, "JT8D Engine," n.d. [2013], available at *http://www.pw.utc.com/JT8D_Engine* (accessed August 30, 2013); Bill Gunston, *The Development of Jet and Turbine Aero Engines*, 2nd ed. (Somerset, England: Patrick Stephens, Ltd., 1997), p. 183.
35. K.L. Abdalla and J.A. Yuska, "NASA Refan Program Status," NASA TM-X-71705, 1975; Gerald D. Griffin to Henry M. Jackson, July 23, 1974, NASA HRC, file 012336.
36. Dennis L. Huff, "NASA Glenn's Contributions to Aircraft Noise Research," *Journal of Aerospace Engineering* 26 (April 2013): 233.
37. "Boeing Studies Refanned 727 Benefits," *Aviation Week & Space Technology* (January 13, 1975): 24–25.
38. Huff, "NASA Glenn's Contributions to Aircraft Noise Research," 233.
39. House Committee on Science and Technology, *Aircraft Noise Abatement Technology*, 94th Cong., 2nd sess., 1976, S. Rept. W, pp. iii–iv.
40. Douglas B. Feaver, "Ford Pledges To Abate Airport Noise," *Washington Post* (October 22, 1976): A19; "FAA Issues Noise Regulations for Older Jets," Department of Transportation News Release 76-121, December 28, 1976, copy available in the NASA HRC, file 012336.
41. Agis Salpukas, "Advances: Rebuilding Planes To Cut Noise," *New York Times* (November 18, 1987): D8.
42. Gunston, *The Development of Jet and Turbine Aero Engines*, p. 183.
43. J.J. Kramer and F.J. Montegani, "The NASA Quiet Engine Program," NASA TM-X-67988 1972, p. 1.

44. William Hines, "NASA's Quiet Engine Still a Long Way in Future," *Chicago Sun-Times* (June 7, 1970), quoted in *NASA Current News*, June 10, 1970, NASA HRC, file 012336.
45. H. Dale Grubb to Roman C. Pucinski, October 7, 1970, NASA HRC, file 012336.
46. "Kramer Heads Jet Noise Research Program," NASA News Release 71-170, September 7, 1971, NASA HRC, file 012336.
47. Evans to Associate Administrator for Advanced Research and Technology, "Back-Up Information Regarding Progress Status and Plans of NASA Aircraft Noise Reduction Programs."
48. "First Quiet Engine Noise Tests," NASA News Release 71-156, August 27, 1971, NASA HRC, file 012336; M.J. Benzakein, S.B. Kazin, and F. Montegani, "NASA/GE Quiet Engine 'A'," *Journal of Aircraft* 10 (February 1973): 67–73; Huff, "NASA Glenn's Contributions to Aircraft Noise Research," p. 232.
49. Charles Tracy, "New Quiet Engine To Cut Jet Noise," *Cleveland Press* (May 15, 1972): 8.
50. "Press Briefing: NASA Aircraft Noise Abatement Research," *NASA News* (March 31, 1972): 10, NASA HRC, file 012336.
51. "Three Contracts Let for Quiet Engines," NASA News Release 72-166, August 15, 1972, NASA HRC, file 012336.
52. NO_x is a generic term for the mono-nitrogen oxides NO (nitric oxide) and NO_2 (nitrogen dioxide) that result from combustion in an engine.
53. Caitlin Harrington, "Leaner and Greener: Fuel Efficiency Takes Flight," in NASA's *Contributions to Flight*, vol. 1, *Aerodynamics*, ed. Richard P. Hallion (Washington, DC: Government Printing Office, 2010), p. 791.
54. Miller, "NASA," p. H12545.
55. St. Peter, *History of Aircraft Gas Turbine Engine Development*, p. 402.
56. Daniel E. Sokolowski and John E. Rohde, "The E^3 Combustors: Status and Challenges," NASA TM-82684, 1981.
57. R. Niedzwiecki and R. Jones, "The Experimental Clean Combustor Program: Description and Status," SAE Technical Paper 740485, February 1, 1974, p. 1.
58. St. Peter, *History of Aircraft Gas Turbine Engine Development*, p. 402.
59. R. Roberts, A. Peduzzi, and R.W. Niedzwiecki, "Low Pollution Combustor Designs for CTOL Engines: Results of the Experimental Clean Combustor Program" (AIAA paper 76-762, presented at the 12th AIAA/SAE Joint Propulsion Conference, Palo Alto, CA, July 26–29, 1976), p. 1.
60. Richard W. Niedzwiecki, "The Experimental Clean Combustor Program: Description and Status to November 1975," NASA TM-X-71849,

1975; R. Niedzwiecki and C.C. Gleason, "Results of the NASA/General Electric Experimental Clean Combustor Program" (AIAA paper 76-76, presented at the 12th AIAA/SAE Joint Propulsion Conference, Palo Alto, CA, July 26–29, 1976), p. 14.
61. Statement of Dr. Meyer J. Benzakein, Chair, Aerospace Engineering, Ohio State University, before the Subcommittee on Space and Aeronautics, Committee on Science, House of Representatives, March 16, 2005, available at *http://www.spaceref.com/news/viewsr.html?pid=15785* (accessed September 12, 2014).
62. Anthony M. Springer, "NASA Aeronautics: A Half-Century of Accomplishments," in *NASA's First 50 Years: Historical Perspectives*, ed. Steven J. Dick (Washington, DC: NASA SP-2010-4704, 2010), p. 199.
63. L. Stewart Rolls, "Characteristics of a Deflected-Jet VTOL Aircraft," in *NASA Conference on V/STOL Aircraft: A Compilation of the Papers Presented at Langley Research Center, Langley Field, Virginia, November. 17–18, 1960* (Langley Field, VA: Langley Research Center, 1960), pp. 171–176, available at *http://ntrs.nasa.gov/archive/nasa/casi.ntrs.nasa.gov/19630004807.pdf* (accessed August 21, 2015); St. Peter, *History of Aircraft Gas Turbine Engine Development*, pp. 351–352, 362.
64. Miller, "NASA," p. H12545.
65. Domenic J. Maglieri, David A. Hilton, and Harvey H. Hubbard, "Noise Considerations in the Design and Operation of V/STOL Aircraft," in *Conference on V/STOL Aircraft*, pp. 269–284.
66. Evans to Associate Administrator for Advanced Research and Technology, "Back-Up Information Regarding Progress Status and Plans of NASA Aircraft Noise Reduction Programs."
67. "Quiet-STOL Program Started," NASA News Release 71-146, August 4, 1971, NASA HRC, file 012336.
68. Ibid.
69. "STOL Conference at Ames," NASA News Release 72-201, October 11, 1972, NASA HRC, file 012336.
70. Ben Kocivar, "QUESTOL: A New Kind of Jet for Short Hops," *Popular Science* 200 (May 1972): 78, 80, 84.
71. R.M. Levintan, "Q-Fan Demonstrator Engine," *Journal of Aircraft* 12 (1975): 685; "Q-Fan Undergoing Aerodynamic Testing," *Aviation Week & Space Technology* (October 16, 1972): 59; Michael Wilson, "Hamilton Standard and the Q-Fan," *Flight International* 103 (April 19, 1973): 618.
72. M.H. Waters, T.L. Galloway, C. Rohrbach, and M.G. Mayo, "Shrouded Fan Propulsors for Light Aircraft" (SAE paper 730323, presented at the

Society of Automotive Engineers Business Aircraft Meeting, Wichita, KS, April 3–6, 1973).
73. C.C. Ciepluch, "A Review of the QCSEE Program," NASA TM-X-71818, 1975; Huff, "NASA Glenn's Contributions to Aircraft Noise Research," p. 232.
74. St. Peter, *History of Aircraft Gas Turbine Engine Development*, p. 398; C.C. Ciepluch, "Preliminary QCSEE Program Test Results" (SAE paper 771008, presented at Society of Automotive Engineers Aerospace Meeting, Los Angeles, CA, November 1977).
75. Robert R. Garvin, *Starting Something Big: The Commercial Emergence of GE Aircraft Engines* (Reston, Virginia: AIAA, 1998), pp. 163–164.
76. "NASA Aircraft Test Engines Promise Noise, Pollution Reduction," NASA News Release 78-138, September 11, 1978, NASA HRC, file 012336.
77. Ibid.
78. Ibid.
79. Ibid.
80. M.D. Shovlin and J.A. Cochrane, "An Overview of the Quiet Short-Haul Research Aircraft Program," NASA TM-78545, November 1978.
81. Huff, "NASA Glenn's Contributions to Aircraft Noise Research," p. 232.
82. NASA Research and Technology Advisory Council Committee on Aeronautical Propulsion, "Meeting Minutes: Ad Hoc Panel on Jet Engine Hydrocarbon Fuels," October 15, 1975, NASA HRC, file 011150, pp. 1–3.
83. Ibid, p. 4.
84. Ibid, pp. 4–6.
85. R.D. Witcofski, "Comparison of Alternate Fuels for Aircraft," NASA TM-80155, 1979; Harrington, "Leaner and Greener: Fuel Efficiency Takes Flight," pp. 815–816.
86. NASA Research and Technology Advisory Council Committee on Aeronautical Propulsion, "Meeting Minutes: Ad Hoc Panel on Jet Engine Hydrocarbon Fuels," p. 6.
87. Ibid, pp. 6–8.
88. Ibid, p. 9.

GE's E³ demonstrator of 1983 reflected advances in fuel efficiency and noise and emissions reductions that benefited from NASA support and research in turbofan development. (NASA)

CHAPTER 4
The Quest for Propulsive Efficiency, 1976–1989

In the October 1973 Arab-Israeli war, the United States assisted Israel, which was desperately fighting a defensive war, by supplying it with aircraft, tanks, ammunition, weapons, and intelligence. The war marked an important stage in the evolution of integrated air defenses, and afterward, America embarked upon an intensive program of electronic warfare research that would lead to more sophisticated antimissile attack systems and the development of the F-117 and B-2 stealth aircraft. But of more immediate consequence was the imposition of a brief yet crippling oil embargo upon the United States, undertaken by the Organization of Petroleum Exporting Countries (OPEC) in retaliation for America's having supported the Jewish state. It reduced the amount of oil available to the United States by almost 20 percent, or three million barrels a day. The resultant decrease in petroleum products and increase in prices created an energy crisis that affected the lives of everyday Americans.[1] For the airlines, the embargo resulted in a near-disastrous cut in the supply of, and an increase in the price of, jet fuel. The Nixon administration implemented mandatory Federal control of jet fuel, heating oil, and middle distillate supplies on November 1, 1973, which set supplies at lower 1972 levels. With the number of people traveling by air growing exponentially, the outcome commercial air carriers wanted least was to reduce service. For American Airlines' legendary Chairman C.R. Smith, there was "no chance" that the airlines could survive a 10-percent reduction in current schedules due to fuel rationing.[2]

With oil prices low through most of the 1960s, there seemed to be no concerns over fuel efficiency in aviation. With cheap fuel and turbofan engines, which operated at about 65 percent efficiency, the airlines were making increased revenues. That perspective changed as the OPEC embargo sent the United States and Europe reeling. No longer would energy sources be unlimited or cheap in Western societies, whose technological infrastructure depended on oil. While automakers reacted with compact, fuel-efficient cars and drivers adapted by traveling less, the airlines were in a dire situation. They could not drive away customers by reducing service or raising ticket prices as they faced an escalation of fuel prices that was estimated to be between 20 to 50 percent.[3] Successive oil crises that have stymied the West since the 1970s have caused a serious reevaluation of what constitutes performance in aviation. Specifically, fuel consumption has become a matter of economic survival.

There was a technical impetus reflecting these changes. The era of "higher, faster, and farther" reached a technical plateau during the latter half of the 20th century. Operational high-performance aircraft speeds and altitudes, which had been consistently increasing since December 1903, peaked at Mach 3.3 and 85,000 feet around 1970, with the Lockheed SR-71 Blackbird strategic reconnaissance aircraft, powered by Pratt & Whitney J58 turbo-ramjet engines. The choices designers made changed to reflect an emphasis on other parameters. Issues of electronic flight and propulsion control, aerodynamic and propulsive efficiency, fuel efficiency, lower costs, improved reliability and safety, and awareness of environmental issues related to emissions and noise augmented and often superseded the decades-old driving philosophy of aircraft design based solely on altitude, speed, and range.[4] In terms of fuel efficiency and emissions, the 1970s presented NASA with an opportunity to, in the words of one historian, "reclaim its mantle at the forefront of aeronautics research."[5]

Aircraft Energy Efficiency: Onset of a Transformational Program

Echoing that shift in emphasis, the U.S. Senate Committee on Aeronautical and Space Sciences, chaired by Senator Frank E. Moss (D-UT) and Barry M. Goldwater (R-AZ), recognized the threat to the American airline industry and the aviation industry in general. The Senate requested that NASA address the issue of increased fuel prices and anticipated oil shortages in January 1975.[6] NASA offered a solution for the fuel crisis. Speaking before the Committee, Administrator James C. Fletcher asserted that there was potential to halve the fuel requirement for commercial jet aircraft by 1985. He advocated low-drag wing shapes, lightweight composite materials, improved engine efficiencies, and modified flying procedures to reduce landing time. Fletcher argued that the implementation of those new technologies and practices on the 2,100 commercial airliners in American service would amount to $1.3 billion year, or the equivalent of 333,000 barrels a day in fuel savings.[7]

With political pressure intensifying, NASA formed the Aircraft Fuel Conservation Technology Task Force in February 1975 to explore potential options. The team formulated a 10-year plan that consolidated existing propulsion, aerodynamics, and structures programs into the Aircraft Energy Efficiency (ACEE) program, which would cost $670 million while promising to reduce fuel consumption by 50 percent, or over 600,000 barrels of oil a day.[8] The Committee on Aeronautical and Space Sciences held hearings, which included considerable debate for and against the program, during the fall and winter of 1975. In the end, the program received approval and recommendation for implementation during fiscal year 1976. The reasons for approval were many. Increased fuel efficiency would stimulate the U.S. aviation industry and give both air carriers and manufacturers a competitive edge in world markets while encouraging conservation and American energy independence. Overall, taking on such a monumental challenge was appropriate for the Federal Government through NASA due to the technical risks involved and the nonproprietary role the Agency would play in disseminating the program's results to all of industry.[9]

The ACEE was what has been termed a "focused" research and development program. In other words, NASA allocated large amounts of funding and engineering resources to mature fundamental technology successes already underway to create full-scale demonstration technology. Overall, NASA intended the program as a partnership with industry with a clear path of technology transfer originating from the Agency.[10] The project also facilitated parallel and intertwined research into noise and emissions since needed improvements in one sphere affected the others. As a result, the fuel-savings-oriented ACEE program incorporated NASA's most comprehensive noise and emissions effort to date.[11]

The industry and Government partners in the ACEE program addressed six major areas divided into three groups. The aerodynamics effort included the Energy Efficient Transport and laminar flow control, while the structures portion addressed composite materials.[12] Lewis Research Center, under the leadership of program manager Donald Nored, managed the propulsion component, which represented three levels of development toward reducing fuel consumption. For the short term, the Engine Component Improvement (ECI) Program targeted a 5-percent reduction in existing engines, specifically the Pratt & Whitney JT8D and JT9D turbofans and the GE CF6 that powered the majority of airliner fleets around the world. The Energy Efficient Engine (E^3, also EEE, and pronounced "E-cubed") turbofan program was intended to decrease fuel consumption by 12 percent in the new engines of the late 1980s. NASA leaders believed that new fuel-conserving technology exhibiting increased thermodynamic and propulsive efficiencies would appear in derivative or entirely new commercial engines as early as 1983. The Advanced Turboprop Project (ATP) offered the most fuel savings, upwards of 15 percent, for its intended introduction in the early 1990s. NASA believed in the possibility of efficient, economic, and acceptable commercial turboprop airliners capable of cruising at Mach 0.8 and at altitudes above 30,000 feet.[13]

The energy crises of the 1970s and high fuel costs made all three propulsion-related ACEE programs, and their required expenses, palatable, but in different ways. The ECI and E^3 dovetailed with the conservative nature of the commercial aviation industry by offering refinement and the advancement of the current state of the art, the turbofan engine. The ATP, on the other hand, was a radical departure from the norm, with the potential of the propeller's return to the forefront of commercial aviation. Donald Nored remarked, "The climate made people do things that normally they'd be too conservative to do."[14]

ACEE 1: Engine Component Improvement

Beginning in 1975, the ECI Program consisted of two parts that focused on improving the performance and fuel efficiency of new production engines and three specific turbofan engines already in service with the introduction of advanced components. The Performance Improvement (PI) effort led by John E. McAulay at Lewis addressed the GE CF6 and the Pratt & Whitney JT8D and JT9D. Those engines—the first generation of American turbofans introduced in the United States—were instantly successful, and the aviation industry anticipated their long-term use into the early 21st century. The Engine Diagnostics effort addressed the reduction of performance deterioration by 1 percent after the CF6 and JT9D engines had gone into service by focusing on revised maintenance procedures.[15]

The PI effort began with a feasibility study that included extensive industry cooperation. Eastern Airlines and Pan American worked in a direct advisory capacity with NASA. Both GE and Pratt & Whitney collaborated directly with Boeing, Douglas, United Airlines, and American Airlines in advisory capacities. Pratt & Whitney formed a partnership with Trans World Airlines to create a cost/benefit methodology. Regardless of the particular details of the partnerships, the interaction provided the appropriate modeling to simulate airline usage and provided the basis for estimating the needed level of engine modification to improve efficiency.[16]

A derivative of the TF39 turbofan that powered the Air Force's Lockheed C-5A Galaxy, the CF6, was GE's first major high-bypass turbofan engine for the commercial aviation market upon its introduction on American Airlines' McDonnell Douglas DC-10 aircraft in 1971. The 40,000-pound-thrust engine was ideal for the ECI program. The modular design that facilitated the incorporation of easily removable and interchangeable components for better maintenance made it easier to modify for increased performance. Part one of the project addressed reducing the fuel usage of the CF6. The initial goal was to achieve a 5-percent reduction by 1982. The research team identified an

The Quest for Propulsive Efficiency, 1976–1989

Figure 4-1. Three General Electric CF6 engines powered the McDonnell Douglas DC-10. (Boeing)

improved-efficiency fan blade, a short core exhaust system, and an improved high-pressure turbine as promising methods.[17]

The CF6 turned out to be the long-lived engine the researchers of the ECI program anticipated. Besides McDonnell Douglas, both Boeing and Airbus incorporated the CF6 into their wide-body and long-distance airliners, including the 747 and A300, as the major airlines used them on their long-distance commercial routes all over the world. By 2012, the CF6 had become the most-produced high-bypass-ratio turbofan to date, with 7,000 units delivered to more than 250 operators in 87 countries. Overall, they accumulated 367 million flight hours. Choosing a successful engine that would have operational longevity was a key to the success of the ECI and a stunning example of the ability of NASA to target a problem and work to make it better.[18]

Pratt & Whitney also stood to benefit greatly from the ECI program. The 14,000-pound-thrust JT8D low-bypass turbofan debuted on the Boeing 727 in 1964. It quickly became one of the most successful commercial jet engines ever built, with over 673 million flying hours. Pratt & Whitney produced more than 14,750 for use on the Boeing 737 and the McDonnell Douglas DC-9 and MD-80.[19]

The 45,000-pound-thrust JT9D was the first high-bypass turbofan from the Connecticut manufacturer and the first to power a wide-body airliner

The Power for Flight

Figure 4-2. The Boeing 727 featured three Pratt & Whitney JT8D turbofans. (Boeing)

after its introduction on Pan Am's Boeing 747s in January 1970. Besides the 747, JT9Ds powered the Boeing 767, the McDonnell Douglas DC-10, and the Airbus A310.[20]

For Pratt & Whitney, the ECI advanced the state of the art in its engine designs. The program demonstrated the advantages of updated cooling, sealing, and aerodynamic design in the JT8D high-pressure turbine and compressor. For the JT9D, improved component technologies included thermal barrier coatings, ceramic seal systems, advanced turbine clearance control, and single-shroud fan design. Even though the work was Government-funded, the results were proprietary. Pratt & Whitney directly applied the knowledge learned to its new 38,000-pound-thrust PW2000 turbofan designed for the new Boeing 757 twin-jet airliner. Nevertheless, Pratt & Whitney shared the data with NASA, which transferred the technology to its E^3 program.[21]

The 16 ECI program technology improvements are listed on the following page by engine, along with the consequent percentage of reduced Specific Fuel Consumption (SFC) values at cruise achieved as of 1980.[22]

The Quest for Propulsive Efficiency, 1976–1989

Figure 4-3. The high-bypass Pratt & Whitney JT9D powered a new generation of airliners, including the Boeing 747 jumbo jet. (NASA)

Concept	SFC at Cruise (% Reduction)
Pratt & Whitney JT8D Engine	
High Pressure Turbine (HPT) Improved Outer Seal	0.6
DC-9 Nacelle Drag Reduction	1.2
HPT Root Discharge Blade	0.9
Trenched High Pressure Compressor Blade Tip	0.9
Pratt & Whitney JT9D Engine	
3.8-Aspect-Ratio Fan	1.3
HPT Active Clearance Control	0.65
HPT Vane Thermal Barrier Coating	0.2
HPT Ceramic Outer Air Seal	0.4
General Electric CF6 Engine	
CF6 Improved Fan	1.7
CF6 Short Core Exhaust Nozzle	0.9
CF6 New Front Mount	0.1
CF6 Improved HPT	1.3–1.6
DC-10 Reduced Engine Bleed	0.7
HPT Roundness/Clearance	0.4–0.8
HPT Active Clearance Control	0.3–0.6
Low Pressure Turbine Active Clearance Control	0.3

109

Overall, the ECI technological improvements included the reduction of clearances between rotating parts, a decrease in the amount of cooling air needed, and the aerodynamic refinement required to raise component efficiencies. GE and Pratt & Whitney incorporated most of the technologies into their production engines.[23] Overall, the ECI was a success for three reasons. First, there was the short time in which the manufacturers incorporated the improvements into their engines. Turbofans incorporating ECI technology powered the new airliners of the 1980s, namely the Boeing 767 and McDonnell Douglas MD-80. Second, the enthusiasm on the part of the manufacturers for the results further cemented a strong relationship with Lewis, especially for GE. Finally, the ECI maintained the American aviation industry's edge over that of the rest of the world as those improved engines and the airliners they powered continued to dominate a large share of the commercial aviation market.[24]

ACEE 2: Energy Efficient Engine Turbofan

NASA's interest in a fuel-conserving engine began in the early 1970s. With the creation of the ACEE program, the program became the Energy Efficient Engine (E^3) project. This second-tier effort worked toward the development of a new generation of advanced fuel-efficient, high-bypass turbofans with operational introduction in 1984. These new engines were not modified turbojets or squeezed into compact nacelles. They were dedicated turbofans that incorporated large fan sections to increase the amount of air that bypassed the core engine. The efficiency of an engine was an expression of configuration, fuel consumption, cruise speed, and bypass ratio. In the mid-1970s, a turbofan like the GE CF6 with a high 5:1 bypass ratio cruised at Mach 0.8 and consumed the most fuel. The legacy QCSEE project revealed that a turbofan with an even higher 12:1 bypass ratio consumed 10 percent less fuel at Mach 0.7.[25]

The specific goals of the E^3 program targeted the reduction of fuel usage by at least 12 percent and the rate of performance deterioration by at least 50 percent while improving direct operating costs by 5 percent. Additionally, any new technology had to meet future FAA FAR 36 noise regulations and EPA exhaust emission standards. The E^3 program was not intended to create a complete engine, but to create technology for use in future engines. It was up to the discretion of the manufacturers to decide when a new or modified advanced fuel-conserving engine that integrated E^3 technology was ready and commercially viable in the aviation market.[26]

Carl C. Ciepluch managed the project at Lewis and supervised the individual $90 million E^3 contracts awarded to GE and Pratt & Whitney. Each company, which was responsible for obligating $10 million of their own funds, faced three major challenges. First, they were to design a Flight Propulsion System (FPS) that served as the engine platform for determining and evaluating the component configuration and new technology advances. Second, they investigated those new innovations through full-scale design, fabrication, and testing. Finally, the engine makers were to integrate the advanced components into an engine system for evaluation in an operational environment.[27] Achieving the E³ goals required aerodynamic, mechanical, and system technologies that were well in advance of current production engines and required successful demonstration in component rigs, a core engine, and a turbofan ground-test engine.

A NASA-led team consisting of the two engine makers and including Boeing, Douglas, and Lockheed defined the engine configurations based on airplane/mission definition and engine/airframe integration. They identified the GE CF6-50C and Pratt & Whitney JT9D-7A as the points of departure. The new 36,000-pound-thrust engines would exhibit higher turbine inlet temperatures for better fuel economy and higher pressure ratios of 50 percent, which was a 20-percent increase from the original goal.[28]

E³ allowed the team members to go beyond the simple refinement of existing turbofan designs that had their origins in the 1950s. With NASA's support, they pushed into new territories focusing on components, nacelles, exhaust-gas mixers, control systems, and accessories. The estimated fuel savings increased to 15 percent. Improved components accounted for half of that, with the remainder coming from refined engine cycles and mixer nozzles.[29] The basic E³ engine featured a single-stage fan, a 4-stage low-pressure compressor, a 9- or 10-stage high-pressure compressor, a 2-zone combustor, a single-stage high-pressure turbine, a 4-stage low-pressure turbine, and a mixer. Despite the similarities, GE and Pratt & Whitney went about developing new and different engine configurations and component technologies from 1976 to 1984.[30]

GE integrated its E³ components into an actual engine. A crucial focus was the annular combustors. GE's E³ combustors were a continuation of the two-stage combustor studies initiated by the earlier Experimental Clean Combustor Program (ECCP). That work contributed to a broader knowledge of how to reduce emission levels before the challenging work on engine performance and operation began.[31] At the end of the program in 1985, Donald Y. Davis, the GE E³ program manager, reported that it met all goals for efficiency, environmental considerations, and economic payoff. The FPS exhibited 16.9 percent lower specific fuel consumption than a contemporary CF6 engine while cruising at Mach 0.8 at 35,000 feet. In terms of direct operating costs, the FPS offered reductions of 8.6 percent for a short-haul domestic transport and 16.2 percent for an international long-distance transport. Moreover, GE's design met the noise- and emissions-compliance goals.[32]

The Power for Flight

Pratt & Whitney focused more on the development of individual technologies, with an emphasis on the fan and core components rather than on a complete engine. While some of the evaluated components exceeded the efficiency goals, others did not. As a result, the engineers in Hartford estimated that the overall reduction in specific fuel consumption in a flight engine would be 15 percent, with direct operating costs at 5 percent. With the exception of NO_x levels, the components failed to meet the EPA's emissions requirement.[33]

Carl Ciepluch of Lewis, Donald Davis of GE, and David Gray of Pratt & Whitney proclaimed the E^3 a success at the $200 million program's completion in 1985. They were confident that the technology developed during its course was, and would continue to be, effectively employed in both current and future advanced transport aircraft engine designs.[34] Later estimates claimed that the performance of the E^3 demonstration engines exceeded the expectations of the program even more, with overall fuel use reduced 18 percent and operating costs lowered between 5 and 10 percent.[35]

The experience and generated knowledge of the E^3 led to the introduction of new engines from GE and Pratt & Whitney. One of the program's legacies was its role in the great "engine war" of the early 1990s that surrounded the anticipated introduction of the revolutionary Boeing 777 airliner in 1996. Boeing bet its own future as a manufacturer on a twin-engine aircraft. Two engines were cheaper to maintain, but required certification for Extended Range Twin Operations (ETOPS) over water. The three leading engine manufacturers—Pratt & Whitney, Rolls-Royce, and GE—held 20, 19, and 14 percent of the world commercial airliner market, respectively, in 1994. The engines they developed for the 777 were the largest and most powerful ever produced, with ratings in the area of 100,000 pounds of thrust. Their fan sections were approximately 10 feet in diameter, with an overall size nearly as wide as the fuselage of a Boeing 737. Pratt & Whitney invested $500 million into the PW4084, which was less fuel-efficient but easier to maintain. Rolls-Royce spent approximately $1 billion on the Trent series. GE spent $1.5 billion to develop the GE90 family, which generated the most thrust. The comparable size and performance of each placed more emphasis on the financial negotiations and airline partnerships. The payoff was huge. At $10 million or more per engine, potential sales of $60 billion over 20 years were possible. As of April 1994, Pratt & Whitney led with 64 orders, while Rolls-Royce followed with 56 and GE with 54.[36]

The replacement for the highly successful and long-lived JT9D, the Pratt & Whitney PW4000 engine series, took advantage of E^3 technology. It featured single-crystal materials, powdered metal disks, low-NO_x combustor and turbine technology, and improved full-authority digital electronic control (FADEC). The 86,760-pound-thrust PW4084 was the launch engine for the

Figure 4-4. Pictured is a Boeing 767 with Pratt & Whitney PW4000 turbofans in the mid-1990s. (Delta Air Lines)

777, which entered service in 1995. The follow-on PW4098, certified in 1998 for the 777-200ER and 777-300, was the first engine to operate with approval for 207-minute ETOPS. Overall, several airlines used the PW4000 series on many of their wide-body aircraft, including the Airbus A310, A300-600, and A330; the Boeing 767; and the MD-11.[37]

The GE90 also reflected the legacy of GE's involvement in the E³ program with its 10-stage high-pressure compressor that developed an impressive 23:1 pressure ratio. After entering service in 1995, GE90s powered variants of the Boeing 777, including the long-range 777-300ER model. One of the latter completed an unprecedented five-and-a-half-hour ETOPS flight in October 2003. The GE90 was regarded as the most fuel-efficient, silent, and environmentally friendly engine in commercial service in the early 21st century. It is also the largest in size and output, with the GE90-115B capable of 115,000 pounds of thrust. The *Guinness Book of World Records* honored the GE90 as the "World's Most Powerful Commercial Jet Engine" in 2001.[38]

ACEE 3: Advanced Turboprop Project

On August 20, 1986, a Boeing 727 airliner took to the skies from the civil flight-test center at Mojave Airport, CA, on the first flight-test exploration of GE's new GE36 Unducted Fan (UDF) ultra-high-bypass (UHB) turboprop engine. The new engine, instantly recognizable by its two rows of fan blades on the outside of the nacelle, offered the potential of a 25- to 45-percent increase in efficiency over existing turbofan engines. An astute industry observer, Craig Schmittman, noticed that parked on ramps and the desert, off the runway, were the "dinosaurs" of the early civil aviation Jet Age, the "fast and thirsty" airliners of the 1950s and 1960s that were no longer economical in a world beset by high fuel prices and increased concern for the environment. Manufacturers proposed new aircraft like the McDonnell Douglas MD-91, a UDF-powered 100- to 150-passenger airliner that offered reduced fuel consumption and increased revenues for the airlines of the 1990s. For Schmittman, a "new era in jet aviation" had begun, and it was not a question of if, but when, the airlines would embrace UHB technology.[39] The UDF was one of two new and advanced turboprop engines that reflected the central role of NASA's ATP project from 1976 to 1987. The ATP, which garnered the prestigious Collier

Figure 4-5. This image shows the GE36 UDF installed on the Boeing 727 demonstrator for flight tests. (National Air and Space Museum, Smithsonian Institution, NASM-9A11738)

Trophy for 1987, amounted to a reinvention of the turboprop and a possible revolution in aeronautics.

The reemergence of the propeller as a viable high-performance propulsion technology in the late 20th century resulted from NASA's search for more fuel savings. In 1974, Lewis engineer Daniel Mikkelson met with Carl Rohrbach, who worked for the last major propeller manufacturer in the United States—Hamilton Standard of Windsor Locks, CT—to discuss an advanced turboprop concept with multiple highly loaded, swept blades called a "propfan." A turboprop was efficient in terms of thrust and fuel economy, and it generated less noise than both piston and turbojet engines. There was an additional two decades of experience in the use of computational fluid dynamics and structural mechanics in the design of supersonic wings, helicopter rotors, and fan blades. The application of that knowledge to the propeller was the key to increased fuel savings. As part of the ACEE, the ATP aimed to address the technical issues inherent in turboprop engines that, if overcome, would encourage the engines' increased use by manufacturers and commercial air carriers. As with the E^3 project, the new system needed to be safe, efficient, and clean overall.[40]

The ATP was a large-scale, multimillion-dollar collaborative effort between NASA, industry, and academia. On the NASA side, each of the four aeronautics Centers—Lewis, Langley, Dryden, and Ames—played major roles in providing research expertise, test and evaluation equipment facilities, and management of over 40 industrial contracts and 15 university grants. The industrial partners included Hamilton Standard, GE, Lockheed, Allison, Pratt & Whitney, Rohr Industries, Gulfstream, McDonnell Douglas, and Boeing, who also brought expertise, their own development facilities, and specialist and airframe perspective to turboprop development. University researchers also played a major role.[41]

To achieve those goals, the ATP went through four major phases. The preliminary phase from 1976 to 1978 involved proving the initial concept that a propeller could maintain efficiency at higher Mach numbers. Lewis contracted with Hamilton Standard in April 1976 for the design, construction, and testing of 2-foot-diameter single-rotation (SR) propfan models. As those tests took place, Lewis and Hamilton Standard engineers became increasingly excited about what an ATP would offer aviation. In contrast, they faced considerable reluctance on the part of a larger aviation industry firmly committed to the turbofan engine, a fact that the Kramer Commission made apparent in 1975.[42] Nevertheless, the Hamilton Standard tests generated the necessary efficiency data to influence the formal establishment of the ATP in 1978.[43]

The next phase of program, from 1978 to 1980, addressed four important enabling technologies necessary for the design of a new ATP. First, researchers used a new tool, the computer, to analyze blade sweep, twist, and thickness to

The Power for Flight

increase cruise efficiency with new design codes. Second, a byproduct of that work, which included extensive in-flight acoustic testing of propfan models on NASA's Lockheed Jetstar multipurpose test bed aircraft maintained by Dryden Flight Research Center, was the ability to make the blades quieter in operation to alleviate internal and external noise. The perfection of an aerodynamically efficient turboprop installation was the third component, which revealed the ideal mounting in a nacelle on top of the wing for the best results. Finally, there was the mechanical design of the drive system, especially the pitch-change and gearbox mechanisms, that were both durable and economical to maintain. This led to a proposal to modify an Agency Gulfstream with a full-size propfan.[44]

As the work on enabling technology continued, Government, industry, and military studies initiated by NASA investigated the increased application of advanced turboprop aircraft over turbofan-powered aircraft. Ames contracted with McDonnell Douglas in 1979 to explore the possibility of integrating a propfan system into a DC-9/MD-80–type airliner. The resultant report concluded that a propfan installed on the aft portion of the fuselage was an important alternative to wing mounting. Lockheed Aircraft, under contract to Langley, initiated an Advanced Cargo Aircraft Study that revealed that the use of propfans reduced fuel consumption by 20 percent, noise by 15 percent, and the required runway length by 25 percent. Lewis awarded study contracts to McDonnell Douglas and Beech Aircraft for business aircraft, and those studies yielded similar results. (In its analysis, however, Beech indicated that the investment required for the new technology was an economic pitfall that would outweigh any significant financial returns.) With technical direction from Lewis, the U.S. Navy awarded contracts to Boeing, Grumman, and Lockheed

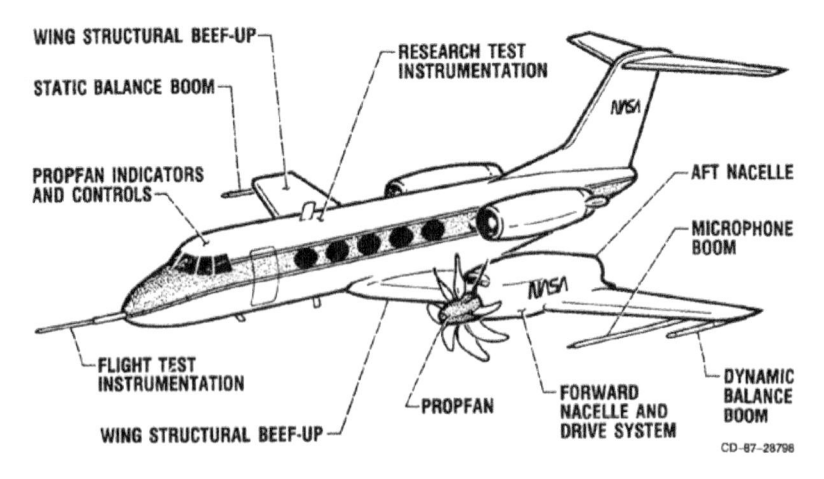

Figure 4-6. This drawing shows a proposed propfan installation on a modified NASA Grumman Gulfstream test bed. (NASA)

to evaluate concepts for subsonic, multipurpose, carrier-based aircraft. They determined that the use of propfans helped increase mission duration and loiter time, which was desirable from the perspective of operators.[45] Those studies worked to overcome the initial industry and operator resistance to the ATP.

The next phase of the ATP, large-scale integration of the propfan technology, took place from 1981 to 1987. Up to that point, the model tests and manufacturer studies were promising, but it was another matter to incorporate those data into full-scale experimental propfans. NASA initiated the Large-Scale Advanced Propfan (LAP) project in 1981. Hamilton Standard received the contract for the design, fabrication, and ground testing of a LAP with a pitch-change mechanism and the production of additional propfans for flight testing. The propeller maker finished the SR-7L in September 1985. Initial tests indicated that the blades vibrated in excess of their design threshold, which placed the program's major emphasis on structural integrity over blade efficiency and noise.[46]

After constructing the first full-scale propfan assembly, the SR-7L, Hamilton Standard sent it to the Propeller Laboratory of the Wright-Patterson Air Force Base for static testing of aerodynamic performance and blade stability. Since 1928, the Air Force and its predecessor organizations had operated three large whirl test rigs for the purpose of establishing the strength and endurance of specific propeller designs. While thrust, torque, centrifugal force, and gyrostatic moments could be calculated mathematically, physically whirling a propeller in excess of a propeller's operating regime provided concrete results, especially for determining overall safety. Air Force engineers tested the SR-7L at 1,900 rpm at 6,000 shaft horsepower, which generated 9,000 pounds of thrust. Tests confirmed that the design did not present any unanticipated structural problems in the blades or weaknesses in the pitch-change mechanism. NASA engineers added new tools to the Air Force process. They used the Air Force's optical system and a laser measurement apparatus developed by their fellow researchers to measure blade deflection caused by aerodynamic and centrifugal loads. They also made one of the first applications of another innovation, Laser Doppler Velocimetry (LDV), to measure flow velocities.[47]

The next step was high-speed rotor tests in the S1MA Continuous-Flow Atmospheric Wind Tunnel in Modane, France. Capable of Mach 0.5 to 1 performance, the tunnel was part of the extensive Office National d'Etudes et Recherches Aérospatiales (ONERA) in eastern France. Built from a nearly complete Nazi wind tunnel found at war's end in Austria and brought across the border into France, the tunnel was capable of evaluating the 9-foot-diameter SR-7L at Mach 0.8 at conditions of 12,000 feet of altitude. Two series of tests with the SR-7L assembly in two-, four-, and eight-blade configurations took place from 1986 to 1987. The full loading tests in Modane were limited to two blades due to the power limitations of the drive rig. The four- and eight-blade tests were conducted at lower power ratings. Despite those facility limitations, the tests led to the verification of the aerodynamic analyses and model data.[48]

Evaluation of the SR-7L propfan continued within the context of a complete turboprop system as part of the Propfan Test Assessment (PTA) project. As the contractor, Lockheed-Georgia had the task of verifying the structural integrity of the blades and the acoustic properties of a large-scale propfan at cruise conditions. The PTA consisted of five elements: (1) combining a large-scale advanced propfan with a drive system and nacelle; (2) proof-testing the system at Rohr Industries' Brown Field facility near San Diego; (3) conducting a series of model tests to confirm aircraft stability and control, handling, performance, and flutter characteristics; (4) modifying a Gulfstream II aircraft; and (5) flight-testing the propfan installed on the left wing of the modified aircraft.[49]

The joint NASA-industry team set about the necessary wind tunnel tests of the supporting technology and the modification of the engine hardware. The concept development and enabling technology phases of the ATP determined that a top-mounted, single-scoop inlet with a boundary layer diverter between the inlet and the top of the nacelle was the best method of directing air to the gas turbine engine. Further investigations exposed the inadequacy of that configuration. The propfan system required an S-duct diffuser that allowed the measurement of pressure recovery and flow distortion all the way from the inlet to the compressor face in the 570 engine. Lockheed-Georgia tests completed in October 1984 yielded 99 percent pressure recovery and acceptable flow distortion levels. In fact, the design of the inlet increased airflow to the compressor, creating a veritable supercharger effect, by 4 percent. Engineers selected and modified an Allison Model 570 engine and T56 gearbox as the basis for the SR-7L propfan drive system. Allison tested the components beginning in September 1985, and the assembly proved more than adequate.[50]

NASA conducted tests of ⅑-scale models of the PTA aircraft in its wind tunnels from 1985 to 1987. The configuration of the PTA was unorthodox. The platform was a model Gulfstream II business jet with two turbofans mounted on the rear fuselage. The addition of the propfan system on the left wing and a static boom on the right introduced considerable design and flying challenges in terms of aeroelastics, stability and control, performance, handling, and flow-field characteristics. Technicians in the 16-Foot Transonic Dynamics Freon Tunnel at Langley investigated the aeroelastic instability to determine overall structural integrity. Personnel in the 16-Foot Transonic Tunnel and the 4-Meter Tunnel conducted high- and low-speed tests respectively to evaluate overall stability and control of the Gulfstream II propfan configuration. Researchers took that model, split it in half into a semispan model, and conducted a flow survey test of the propfan and plan configuration in the Lewis Transonic Wind Tunnel in January 1987. Speeds ranged from Mach 0.4 to 0.86, and the angle, or tilt, of the nacelle from the wing ranged from −3 to 2 degrees.[51] Each of these tests proved successful upon completion.

PTA ground static tests of the complete propfan, engine, gearbox, and nacelle took place in May and June 1986. Rohr Industries, a leading manufacturer of aircraft engine nacelles, performed the tests at its Chula Vista, CA, facility south of San Diego. They confirmed that the propfan system's fuel consumption, operability, structural integrity, and acoustic characteristics were all within the specified limits.[52]

In July 1986, the propfan system arrived at Lockheed-Georgia's facility in Savannah for installation on the Gulfstream II, which led to extensive modification of the business jet. Internal additions included the required lines for the fuel, hydraulic, electric, compressor bleed air, and instrumentation for the propfan system and test instrumentation, which included monitoring consoles and over 600 sensors. New external structures included strengthened wings and flaps and four booms. The static and dynamic balance booms on each wing aided controllability with the propfan installed, and the microphone boom on the left wing measured free-field noise. The flight-test boom in the nose measured aircraft speed, angle of attack, and yaw. Lockheed technicians finished the installation and modifications in February 1987. The NASA Airworthiness Committee approved the PTA aircraft for flight test the following March.[53]

NASA and Lockheed-Georgia conducted flight tests of the modified Gulfstream II from April to November 1987. The program concentrated on systems evaluation, operability, structural integrity, efficiency, and noise. The initial flights involved air starts at altitudes of 5,000, 6,000, and 10,000 feet and moved on to high-altitude research between 5,000 and 35,000 feet at speeds ranging between Mach 0.4 and 0.85. The PTA team traveled to NASA Wallops Flight Facility in Virginia for low-altitude noise testing at speeds of 190 knots at

Figure 4-7. The propfan on the Gulfstream II. (NASA)

altitudes between 850 and 1,600 feet. High-altitude, en route noise data collection was conducted in cooperation with the FAA, including by a NASA Learjet mapping the noise pattern below the PTA in flight. Overall, the flight tests confirmed the initial estimates made in the early 1970s by NASA that the propfan offered a 20- to 30-percent reduction in fuel costs over existing turbofan propulsion systems.[54]

The single-rotation propfan was one configuration explored by NASA and its industry partners during the ATP. There was an inherent problem with that system. As a conventional propeller rotates and produces thrusts, it leaves a swirl of air behind it that degrades both propulsive and aerodynamic efficiency. A design utilizing two propellers, one in front of the other and rotating in different directions, offered three advantages that would contribute to the overall goals of the ATP. First, counter-rotating propellers removed swirl and increased fuel efficiency by 5 percent. Second, they offered twice the power of a single-rotation system of the same diameter. Finally, they facilitated a compact design mounted on the rear fuselage of an airliner, which helped alleviate interior cabin noise and resulted in a "clean" wing with improved lift-to-drag characteristics. The development of an advanced counter-rotating turboprop offered a solution to problems related to aerodynamic interaction between the blade rows, aeromechanical stability (the relationship between air flow and structural integrity), and acoustics.[55]

NASA recognized that investigation into counter-rotating advanced turboprops, or propfans, required the use of test rigs in wind tunnels. The process of designing and fabricating rigs to test 2-foot-diameter counter-rotating assemblies began in 1983; these assemblies enabled the testing of both tractor and pusher configurations in wing or fuselage installations. NASA contracted with Hamilton Standard and GE for the design and fabrication of several models. Those models, mounted on NASA's test rigs, underwent evaluation in wind tunnels and acoustic facilities operated by NASA and the primary contractors as well as by Boeing and United Technologies.[56]

Hamilton Standard's dual-rotation CRP-X1 design represented a tractor propfan configuration. In tests conducted in the United Technologies Research Center (UTRC) 8- by 8-Foot Wind Tunnel from April 1985 to March 1986, engineers evaluated the design's aerodynamic performance, structural integrity, and aeromechanical stability. The aerodynamic efficiency was measured at 86 percent at Mach 0.75, which was 8 percentage points higher than the efficiency of the company's most successful single-rotation propfan. Hamilton Standard researchers tested both tractor and pusher configurations of the CRP-X1 in the UTRC Low-Speed Acoustic Research Tunnel from April to June 1986. They learned that varying the angle of the propfan overall brought noise levels down. To better understand blade efficiencies, Hamilton

Standard developed with Lewis engineers a flow visualization method based on a three-dimensional Euler solution and a high-resolution grid. Their charting of leading-edge vortices and flow streamlines paralleled physical tests using the flow of oil during low-speed tests at the UTRC.[57] While this program yielded impressive results and was a glowing example of industry-Government collaboration, Hamilton Standard's CRP design never reached the full-size test and evaluation stage.

Independent of NASA, GE began investigating the feasibility of a 25,000-pound-thrust commercial counter-rotating turboprop engine for a 150-passenger aircraft in 1983. Company engineers ventured from conventional turboprop design by choosing not to use a complicated gearbox. In previous designs, a system of gears ensured that the propeller and the gas turbine turned at their individually ideal rpms. GE elected to remove the entire challenge of designing a gearbox capable of 20,000 shaft horsepower and, instead, to drive the two rows of blades with turbine stages powered by the core section from an F404 military turbofan. In other words, the power from the exhaust of the core engine was transmitted to the blades directly, without a gearbox; although unorthodox, this approach was simpler, lighter, and more efficient at higher horsepower ranges.[58] Performance estimates indicated a 32:1 bypass ratio, a 4:1 thrust-to-weight ratio, and a specific fuel consumption of 0.52 at Mach 0.8 and 35,000 feet. Estimates put fuel consumption at 30 percent less than that of most contemporary turbofan engines and 50 percent lower than that of engines used on 150-passenger airliners. GE called their new proprietary UHB concept an "unducted fan," or UDF, with the company designation GE36.[59]

NASA supported GE through contracts administered by Lewis starting in early 1984. The work focused on the initial design and ground tests, which alleviated GE's startup risks and accelerated the overall development time for the UDF. The shared objectives centered on the successful demonstration of the gearless propfan concept as a viable alternative to turbofan engines. The partnership consisted of scale-model, full-scale fan blade, and static engine tests, as well as the design of the nacelle and specific engine components.[60]

Eager to pursue the new idea, GE engineers designed and fabricated three counter-rotating model rigs for the aerodynamic, acoustic, and aeroelastic testing of various blade designs, speeds, and blade row spacing. One model went to Boeing for low- and high-speed testing in the company's 9- by 9-Foot Low Speed Wind Tunnel and 8- by 12-Foot Transonic Wind Tunnel in May 1984. The second went to GE's Cell 41 Vertical Anechoic Chamber in November 1984. The third, through a cost-sharing contract with NASA, went to Lewis for high- and low-speed testing in the 8- by 6-Foot Transonic Wind Tunnel and 9- by 15-Foot Anechoic Low Speed Wind Tunnel in July 1985. The collaborative

The Power for Flight

tests revealed that the spacing between, and the diameters of, the blade rows were instrumental in reducing overall noise. Throughout the entire process, GE used NASA data to design or refine the unique UDF components. Rig testing of the counter-rotating turbine at Lewis revealed that the F7-A7 blade set had the highest efficiency—82.5 percent, at Mach 0.72.[61]

With the concept in hand, GE invited NASA to partner in the development of a gearless propfan demonstrator. Beginning in late August 1985, GE began ground tests of a complete UDF engine at its Peebles, OH, test site. After a series of structural failures that appeared between October 1985 and February 1986, GE realized the blades needed strengthening. With the help of NASA, GE devised better and stronger ways to construct the blades for the UDF. At the conclusion of testing in July, the UDF had completed 100 hours of testing. Half of those hours consisted of concentrated endurance testing over a 2-week period during that last month. The UDF generated 25,000 pounds of thrust at sea level, with specific fuel consumption of 0.24, which was approximately 20 percent better than the fuel consumption of contemporary turbofans. GE engineers operated the UDF through the full range of flight conditions, including a demonstration of reverse-thrust capability, which was a common feature used on turbofan-powered airliners for slowing the aircraft down at landing.[62]

The next step was for the UDF to take to the air. GE and NASA began planning flight tests of the UDF in early 1985. The purpose was to confirm existing test results and to operate the engine at the speeds and altitudes flown with turbofans, primarily Mach 0.8 at 35,000 feet, to determine its suitability as a replacement propulsion system. GE and NASA joined with Boeing for the first round of collaborative tests on a 727 airliner during the summer of 1986. GE bore the responsibility of modifying the 727, installing the UDF in place of the right-side JT8D turbofan, and conducting the test program. NASA facilitated the use of Government-owned hardware, including the Agency's Learjet, and cleared the experimental airliner for flight. Both parties shared all data. Flights began on August 20 at GE's facility in Mojave, CA, with the normal airline operations profile being achieved by the following December. Beginning in January 1987, the GE-NASA team measured the outside acoustic properties and experimented with interior modifications to alleviate cabin noise. The program concluded the following February with a total of 41 hours of flight time. The UDF exhibited 30 percent lower fuel consumption than the JT8D and noise levels that were elevated but were a promising beginning for refinement toward commercial implementation.[63]

GE and McDonnell Douglas partnered in 1986 for further tests of a UDF engine installed in place of the left JT8D turbofan on an MD-80 airliner in anticipation of placing the system on the market. The primary goals of the McDonnell Douglas UHB Demonstrator program were to further reduce noise

The Quest for Propulsive Efficiency, 1976–1989

by continuing experimentation with the number of blades used in the two fan stages and to design a quieter passenger cabin. To achieve that, they compared the noise produced by an 8- by 8-blade engine versus a 10- by 8-blade configuration during a test program that ran for most of 1987. The latter engine produced a lower primary tone, which, unfortunately, did not fully meet FAR 36 noise requirements. Carbon fiber fan blades were crucial to the design. In September 1988, the UHB demonstrator flew across the Atlantic Ocean to go on display at the world aviation industry's Farnborough Air Show in England.[64] GE moved from that point to continue work on a commercial UDF engine with an improved actuation system and refined aerodynamic, mechanical, and acoustic design, with an anticipated introduction in 1992. McDonnell Douglas fully intended to offer the UDF configuration as an option for its customers in the early 1990s, whether it was as a retrofit for existing airliners or brand-new aircraft.[65]

In 1986, Hamilton Standard, Pratt & Whitney, and Allison began work on their 578-DX demonstrator engine, which incorporated the knowledge generated by the joint NASA-industry ATP studies. They started with the power section from an Allison 571 industrial gas turbine and a FADEC system derived from the Pratt & Whitney PW2037 turbofan. Unlike the GE36 UDF, the 578-DX's transmission system consisted of a complex reduction gearbox between the low-pressure turbine and the propfan blade. The Pratt & Whitney–Allison

Figure 4-8. The McDonnell Douglas MD-80 UHB Demonstrator is shown with the 578-DX engine. (NASA)

team believed that the gearbox, developed in collaboration with NASA, offered important advantages and was an overall improvement over earlier designs used in turboprops. The 578-DX featured a lighter turbine, a smaller-diameter nacelle, and a more efficient match of turbine and propeller rotational speed with an estimated 4- to 6-percent increase in efficiency over that of the UDF. Flight testing of the 578-DX installed on a McDonnell Douglas MD-80 began at Mojave, CA, in April 1989, with an anticipated commercial introduction in 1992.[66]

There was much enthusiasm for these new propeller-driven propulsion systems over the course of the 1980s. The March 1985 issue of *Popular Science* ran a cover story that asked the seemingly controversial question, "So Long, Jets?," as the aviation industry pondered the possibility that "propellers may be on the way back."[67] Despite the obvious economic benefit due to lower fuel consumption, NASA and the manufacturers could not simply say that these were new propellers. The head of GE Aircraft Engines, Brian Rowe, recognized that the public believed that fans found in jet engines were "modern" and that the "old technology was propellers"; this knowledge shaped how the developers presented the new advances. NASA and its industry partners coined "prop-fan," while GE emphatically called their design a fan.[68] McDonnell Douglas engineers, considering the installation of advanced turboprop engines on the MD-80 series of airliners, called them "propulsors."[69]

Despite the potential fuel savings and a marketing campaign that attempted to overcome the public's resistance, propfan-driven and UDF-powered aircraft did not appear in the 1990s. Issues of technical and economic risk, reliability, maintenance, purchase price, and ride quality required further exploration as the programs went on permanent hiatus. Michael A. Dornheim, engineering editor of *Aviation Week & Space Technology*, believed that UHB airliners had to offer a 10-percent reduction in overall operating costs before airlines would seriously consider implementation. With fuel between 50 and 65 cents a gallon, there was no incentive to pursue propfan technology. Even if the price of fuel rose to $1 a gallon, he felt that the technology was not "quite there" anyway and required more development.[70] In the end, a drop in oil prices negated the need for manufacturers and airlines to reequip with advanced turboprop aircraft that were estimated to cost between $3 and $10 billion to develop. They continued to use existing turbofan-powered aircraft.[71]

The technological achievement of reinventing the turboprop, however, did not go unnoticed. The National Aeronautic Association (NAA) awarded NASA and its industry partners the 1987 Collier Trophy for the development of an advanced turboprop propulsion technology for new, fuel-efficient, subsonic aircraft propulsion systems. NASA alone invested approximately $200 million in the project. Lewis's ATP represented the most significant fuel savings and was the most revolutionary of all the technologies explored in the ACEE program.[72]

ACEE in the Big Picture

During the 1970s and 1980s, NASA strove to be the voice that would shape the future of aircraft propulsion through its efforts to alleviate fuel consumption, noise, and emissions. The three propulsion programs of the ACEE, born out of the chaos of the energy crisis, represented a balance between the near, intermediate, and long terms. The ambition and the sheer scope and size of the ACEE led one observer to christen it the "Apollo of Aeronautics" for NASA and the American aviation industry.[73] NASA believed that the programs of the ACEE stimulated the industry with an estimated 5-year "jump in technology."[74]

NASA funded the ECI and E^3 programs to develop technologies suitable for energy-efficient turbofans. In both the near and intermediate terms, their work at Lewis constituted the most significant contributions to improving fuel efficiency for turbofans and commercial aircraft overall.[75] The most beneficial ACEE program for industry was E^3 since it reinforced and improved the existing turbofan paradigm rather than creating an entirely new technology like the ATP, which faced considerably more hurdles, not all of them technical. E^3 allowed engine manufacturers to invest in new innovations at a lower cost. Since engine development was a high-risk proposition, Government funding and research support made it possible. Both GE and Pratt & Whitney incorporated E^3 high-efficiency turbofan technology into their pioneering GE90 and PW4000 engines designed for the groundbreaking Boeing 777 airliner in 1995.[76] Meyer J. Benzakein, the chair of the aerospace engineering department at Ohio State University and a former GE engineer, assessed NASA's impact on jet engine technology before the House Subcommittee on Space and Aeronautics in March 2005. To Benzakein, without NASA's E^3 and the earlier QEP, GE would not "have had the composite fan blades, the high pressure-ratio core, or the low emission double annular combustor that put [the company] in a leading position in the industry."[77]

The technical legacy of ATP was significant. A new generation of "commuter propellers" driving turboprop airliners emerged in the 1990s. With NASA data in hand, Hamilton Standard and Hartzell Propeller of Piqua, OH, introduced further refinements to the variable-pitch propeller. Advanced electronic control allowed for immediate response in flight and on the ground. Echoing trends in fuselage construction and duplicating the work on the SR propellers, propeller makers engineered blades made from composite materials—which included carbon fiber, Kevlar, fiberglass, and foam—that resulted in a 50-percent reduction in weight. These new blades also featured innovative airfoil profiles that benefited, once again, from the pioneering aerodynamic research by NASA in the United States. To handle the increased power of turboprop engines, like the Pratt & Whitney Canada PW100 series, without increasing noise or the individual diameter or blade width of a propeller, designers chose five- and six-blade configurations.[78] As for the propfan and UDF, that knowledge and technology are "on the shelf" waiting for the next fuel crisis that may potentially push for their justified expense and implementation in the future.[79]

Perhaps the greatest achievement of the ATP was the collaboration between NASA, industry, and academia that advanced the state of the art in turboprop technology. Overall, the ATP required the technical expertise of the three NASA research Centers and the involvement of both industry and academia. Of the 40 contracts to the aviation industry, the primary contractors were GE for the UDF, Hamilton Standard for the LAP, and Lockheed-Georgia for the PTA. Over 15 grants went out to universities across the United States.[80]

Improving General and Business Aviation

General aviation constitutes any kind of flying other than scheduled commercial airlines and military aviation. The category represents a myriad of aircraft types and airborne activities, including aerial demonstration and sport, agricultural dusting, business travel, cargo transport, firefighting, flight training, recreation, and utility operations.

During the 1950s, the NACA's aeronautical research program focused almost entirely on meeting the challenges of high-speed flight during the second aeronautical revolution. It was not until the early 1960s that NASA began to devote limited attention to general aviation on a sporadic basis. NASA researchers held a series of meetings with general aviation manufacturers in 1967. As a result, NASA searched its research database of over 10,000 technical documents for new avenues of assistance for that particular sector of American aviation. By 1970, NASA had initiated studies investigating aerodynamic characteristics, control, handling, avionics, and propulsion.[81]

CARD and Its Impact on NASA-FAA General Aviation Research
The impetus to help general aviation grew stronger in the early 1970s. The CARD Policy Study released in 1971 identified general aviation safety as an area for Federal Government involvement. The study identified noise and emission problems caused by larger general aviation aircraft, but it did not anticipate the public concern over the environmental impact of the general aviation fleet.[82]

The mainstream awareness of the environmental impact of flight carried over into general aviation. NASA entered into a joint program with the FAA and industry to reduce noise and exhaust emissions to meet current and proposed standards. The first step was the investigation of minor engine modifications and the extent of possible emissions reduction. Another program experimented with hydrogen injection. It was believed that the introduction of small amounts of gaseous hydrogen into the fuel-air mixture permitted cleaner engine operation and reduced fuel consumption. A parallel program begun in 1975 existed for small turbofan engines found on business jets and was to lead to new and cleaner engine designs.[83]

Echoing the impetus to make jet engines cleaner in the 1970s, NASA directed considerable effort toward reducing emissions from the aircraft piston engines that dominated general aviation. With the endorsement of the EPA, NASA partnered with the FAA in supporting studies of general aviation piston engine emissions and potential ways to reduce them in 1973. They awarded grants to engine makers Avco-Lycoming and Teledyne Continental to conduct a three-phase program. Phase I evaluated five different engine types to determine the effects of variations in fuel-air ratio on emission levels and other operating characteristics such as cooling, misfiring, roughness, power, and acceleration. Conceiving minor design modifications to those engines to reduce emission levels without degrading desirable operating characteristics constituted Phase II. Phase III involved the testing of those modifications. The staff of the FAA's National Aviation Facilities Experimental Center (NAFEC) near Atlantic City, NJ, performed independent checks on Phase I engines, while NASA developed new testing equipment at Lewis for basic engine technology studies over the long term. Lewis worked closely with the FAA as the two organizations expanded the emissions studies. Through additional grants in October 1975, Lycoming and Continental explored advanced emission-reduction technology concepts that included unusual engine configurations and cycles that offered the potential of greater fuel economy and lower weight, cost, and maintenance.[84]

NASA's involvement in the piston engine emissions studies was an example of the Agency's cooperative style. NASA maintained contact with both the FAA and the EPA through two channels. At the researcher level, representatives from each organization interacted at the respective facilities in Cleveland, Atlantic City, and the EPA's National Vehicle and Fuel Emissions Laboratory (NVFEL) in Ann Arbor, MI. The respective program managers and administrators for each also met in Washington, DC, which enabled Lewis to plan NASA's longer-range activities while complementing the short-term goals of the overall project.[85]

NASA researchers also focused on the propeller to increase noise reduction and efficiency. Influenced by the groundbreaking aerodynamic knowledge of Richard Whitcomb, they incorporated supercritical airfoil shapes into

The Power for Flight

propeller blades. Researchers evaluated a shrouded propeller in Langley's Full-Scale Tunnel in 1974, which led to evaluations of a variable-pitch ducted fan in 1975. Increased testing resulted in the collection of noise and thrust data on two-, three-, and five-blade propellers, which in turn led to greater knowledge of free, unshrouded propellers and ducted fans.[86]

Work on general aviation propellers continued on into the late 1970s and 1980s. NASA's general aviation technology program addressed the refinement of propeller technology for small aircraft. The Agency conducted numerous wind tunnel and flight tests of general aviation propellers to investigate thrust efficiency and acoustic characteristics. A joint project between NASA, the EPA, Ohio State University, and MIT resulted in a new, quiet general aviation propeller design. Tests revealed a flyover noise reduction of 5 decibels, while climb performance improved at slower speeds with a tradeoff loss of cruise speed performance. Researchers in the Lewis 10- by 10-Foot Supersonic Wind Tunnel evaluated a propeller with the DR Incorporated Pro Wake Survey Probe installed. Researchers flight-tested a new design on a Cessna 206 flight demonstrator.[87]

QCGAT: Toward the Quiet and Clean Turbofan Engine

The first jet designed specifically for business aviation, the Learjet, first flew in October 1963. A new family of "bizjets," powered by small turbojet and turbofan engines, emerged. By the end of the 1970s, business aircraft accounted for

Figure 4-9. The Learjet represented a new departure for general aviation, the "bizjet." (NASA)

38 percent of intercity air passenger traffic in the United States.[88] With business aviation becoming a growing force in American air travel, NASA searched for ways to make a contribution. The Quiet, Clean, General Aviation Turbofan (QCGAT) engine program initiated by Lewis in 1976 worked to apply NASA's advances in large turbofan noise and emission reduction to the design and development of turbofan engines with thrust levels below 5,000 pounds for general aviation aircraft without compromising performance. The QCGAT program was NASA's first foray into the area of aeronautical propulsion for general aviation.

QCGAT project manager G. Keith Sievers remarked that the program "should not be a major constraint on the future growth of turbofan-powered aircraft in general aviation." The program aimed to reduce flyover noise levels by between 10 and 14 percent, which amounted to a reduction in perceived noisiness between 50 and 60 percent. Overall, Sievers believed the program was able to reduce the noise "footprint" of a business jet by 90 percent. Compared to a commercial airliner, a business jet was quiet, but NASA wanted to see if it could maximize noise reduction even further without degrading overall performance. Once the aircraft application was selected, the effective noise reduction goals ended up being 15–20 PNdB below the FAA's FAR 36 Stage 3 standard.[89]

Lewis awarded contracts to the Garrett AiResearch Manufacturing Company of Phoenix, AZ, and Avco-Lycoming of Stratford, CT, to develop the candidate engines. To achieve a reduction in noise and pollution while decreasing or maintaining fuel consumption levels, the joint NASA-industry team introduced the ideas pioneered in the Quiet Engine and Refan programs found only on the largest turbofans. The team reduced the velocity of the engine exhaust; redesigned the interior parts of engine to reflect advances in acoustics, which included sound-absorbing materials to reduce the noise produced by the fan, compressor, and turbine; added internal exhaust mixers; and eliminated fan inlet guide vanes.[90]

By 1979, testing by the manufacturers and at Lewis achieved the primary goals of the QCGAT program. There was a significant reduction of engine noise and pollutant emissions. The Avco-Lycoming engine exceeded NASA goals outright, while the Garrett AiResearch engine achieved them cumulatively.[91] The engine noise profile proved to be 10 to 14 decibels lower than the quietest business jet engine at the time, corresponding to a 50- to 60-percent reduction in perceived noise. The program also demonstrated a 54-percent reduction in carbon monoxide, a 76-percent reduction in unburned hydrocarbons, and significant reductions in nitrogen oxides.[92]

The theories, techniques, and concepts developed for large turbofan engines could be successfully applied to their smaller and less-powerful counterparts. Garrett AiResearch went on to apply the advanced acoustic technology of the QCGAT program technology into the fan, low-pressure turbine, exhaust nozzle, and nacelle of its latest TFE731 turbofan engine. The compound mixer nozzle mixed the bypass and core thrust together before it left the engine, which improved thrust while reducing noise and smoke emissions. The improved TFE731 entered service in 1983 and became the engine of choice for modernizing existing business jets like the Dassault Falcon and the British Aerospace BAe 125 series.[93]

As part of NASA's program to advance small gas turbine engine technology, Lewis opened the Low-Speed Centrifugal Compressor Facility in April 1988. Operational compressors in turboprop and turboshaft engines measured a small 8 inches in diameter and rotated at high rpm. In a reversal of the tradition

Figure 4-10. The Low-Speed Centrifugal Compressor Facility at Lewis. (NASA)

of building models in anticipation of full-scale tests, Lewis researchers installed an oversized, 60-inch steel model compressor that they rotated at a much slower 1,920 rpm. The large diameter facilitated the mounting of advanced instrumentation, while windows in the side enabled researchers to use the laser Doppler velocimeter system to visualize the airflow along the compressor channels. The testing verified advanced computer codes and enabled the creation of detailed models that reduced the time required to design future generations of fuel-efficient engines intended for helicopters and general aviation and commuter aircraft. As a result, the work conducted there provided a more thorough understanding of airflow in the geometrically complex channels of a centrifugal compressor.[94]

Aircraft Propulsion Research in the 1980s: In Sum

At the end of the 1980s, the United States continued to lead the world in aeropropulsion technology for military, commercial, and general aviation aircraft. NASA was a significant part of that achievement through the demonstration and introduction of new innovations through the ACEE and general aviation program. Lewis continued to serve as NASA's center of research in aeropropulsion technology. The Center hosted approximately 500 high-level representatives of aeropropulsion, airframe, and related industries; numerous smaller companies; several Government agencies; and the academic world at the "Aeropropulsion '87" conference in November. The purpose of the 3-day meeting was to present an unclassified and comprehensive summary of the aeropropulsion research accomplished at Lewis during the decade.[95]

NASA Deputy Associate Administrator Robert Rosen opened the meeting with a talk addressing the themes of change and challenge that the Agency faced. There had been great achievement in the development of innovative propulsion technologies, but he wanted the attendees to take note of what lay ahead. Increasing competition with foreign manufacturers, the exponential growth of air travel, and dwindling Government budgets threatened America's technical ascendancy. The goal of NASA, and of Lewis, was to make sure that the Agency's aircraft propulsion program accomplished three things: The research had to be fundamental and at the leading edge. A successful transfer of that knowledge had to be made to industry to reap the benefits. Above all else, the cooperation between the Government and the rest of the aeropropulsion community, which included industrial, Government, military, and academic partners, had to be as good as it possibly could be. That was the only way the "tremendous capability" of Lewis could make aircraft propulsion technology better.[96]

Endnotes

1. James C. Tanners, "Fueling Inflation: Sharp Increases Seen in Prices of Gasoline and Most Other Fuels," *Wall Street Journal* (October 31, 1973): 1.
2. Richard Witkin, "Government Control of Jet Fuel May Lead to 10 Percent Cut in Flights," *New York Times* (October 13, 1973): 36.
3. Roy D. Hager and Deborah Vrabel, *Advanced Turboprop Project* (Washington, DC: NASA SP-495, 1988), p. v.
4. John D. Anderson, Jr., *Introduction to Flight*, 7th ed. (New York: McGraw-Hill, 2012), pp. 46–49. For John Anderson, "farther," or distance, is a result of design requirements for specific aircraft and is not as fundamental a performance parameter as "higher" (altitude) and "faster" (speed).
5. Harrington, "Leaner and Greener: Fuel Efficiency Takes Flight," p. 817.
6. United States Senate, *Committee on Aeronautical and Space Sciences, United States Senate, 1958–1976* (Washington, DC: Government Printing Office, 1977), pp. 9, 56–57, 85–86.
7. "Halving of Fuel Needs for Jetliners Is Seen," *Washington Post* (February 7, 1975): A9.
8. NASA Office of Aeronautics and Space Technology, "Aircraft Fuel Conservation Technology Task Force Report," NASA Technical Memorandum No. X-74295 (September 10, 1975), pp. 18–53. Predecessors to the ACEE program included the Advanced Transport Technology (ATT) and the Energy Trends and Alternate Fuels (ETAF) programs of the mid-1970s. ATT addressed cruise aerodynamics, the area-rule fuselage, supercritical airfoils, composite materials, active controls, and advanced propulsion technology. ETAF investigated laminar flow control and winglets and supported propulsion studies focusing on engine cycle, refined and new component technology, and the construction of advanced turbofan demonstrators. GE and Pratt & Whitney both received contracts to study fuel efficiency and explore advanced unconventional engine concepts. Ethell, *Fuel Economy in Aviation*, pp. 6–7.
9. Mark D. Bowles, *The "Apollo" of Aeronautics: NASA's Aircraft Energy Efficiency Program, 1973–1987* (Washington, DC: NASA SP-2009-574, 2010), pp. 16, 23–24.
10. Bowles, *"Apollo" of Aeronautics*, pp. xiii, xv.
11. St. Peter, *History of Aircraft Gas Turbine Engine Development*, p. 398.
12. Those three ACEE programs, composite structures, the Energy Efficient Transport, and laminar flow control fall outside the boundaries of this

study but were equally important components of the push toward aircraft efficiency.

13. D.L. Nored, "Propulsion," *Astronautics and Aeronautics* 16 (July–August 1978): 47; J.M. Klineberg, "Technology for Aircraft Energy Efficiency," Paper A79-14126 03-03 in *Proceedings of the International Air Transportation Conference, Washington, D.C., April 4–6, 1977* (New York: American Society of Civil Engineers, 1977), pp. 127–171; Mark D. Bowles and Virginia P. Dawson, "The Advanced Turboprop Project: Radical Innovation in a Conservative Environment," in *From Engineering Science to Big Science: The NACA and NASA Collier Trophy Research Project Winners*, ed. Pamela E. Mack (Washington, DC: NASA, 1998), p. 324.
14. Donald Nored, quoted in Bowles, *"Apollo" of Aeronautics*, p. xiii.
15. John E. McAulay, "Engine Component Improvement Program: Performance Improvement," AIAA-80-0223 (paper presented at the 12th AIAA Aerospace Sciences Meeting, January 14–16, 1980), p. 1.
16. McAulay, "Engine Component Improvement Program: Performance Improvement," pp. 1, 5.
17. A.J. Albright, D.J. Lennard, and J.A. Ziemanski, "NASA/General Electric Engine Component Improvement Program" (paper presented at the 14th AIAA/SAE Joint Propulsion Conference, Las Vegas, NV, July 25–27, 1978), p. 1, available online at *http://ntrs.nasa.gov/search.jsp?R=19780061189&hterms=propulsion+lewis+1979&qs=Ntx%3Dmode%2520matchallpartial%26Ntk%3DAll%26Ns%3DPublication-Date%7C0%26N%3D0%26No%3D40%26Ntt%3Dpropulsion%2520lewis%25201979* (accessed August 18, 2013).
18. GE Aviation, "The CF6 Engine Family," 2012, at *http://www.geaviation.com/engines/commercial/cf6/* (accessed August 18, 2013).
19. Pratt & Whitney Media Relations, "JT8D Engine Family: The Low-Cost Performer," October 2012, at *http://www.pw.utc.com/Content/JT8D_Engine/pdf/B-1-7_commercial_jt8d.pdf* (accessed August 18, 2013).
20. Pratt & Whitney Media Relations, "Pratt & Whitney's JT9D Engine Family," October 2012, *http://www.pw.utc.com/Content/JT9D_Engine/pdf/B-1-8_commercial_jt9d.pdf* (accessed August 18, 2013).
21. W.O. Gaffin, "NASA ECI Programs: Benefits to Pratt & Whitney Engines" (paper presented at the 27th ASME International Gas Turbine Conference and Exhibit, London, England, April 18–22, 1982), available online at *http://ntrs.nasa.gov/search.jsp?R=19820051913* (accessed August 18, 2013).

22. McAulay, "Engine Component Improvement Program: Performance Improvement," pp. 5, 9. Specific fuel consumption was the weight of the fuel consumed per pound of thrust per hour.
23. Louis J. Williams, *Small Transport Aircraft Technology* (Washington, DC: NASA, 1983; repr., Honolulu: University Press of the Pacific, 2001), pp. 37–38.
24. Bowles, *"Apollo" of Aeronautics*, pp. 77–78.
25. "Fuel Per Passenger-Mile," NASA HQ RA76-336, August 13, 1975, NASA HRC, file 011151.
26. Ethell, *Fuel Economy in Aviation*, pp. 29–30.
27. Ibid., p. 30; Bowles, *"Apollo" of Aeronautics*, p. 79.
28. Ethell, *Fuel Economy in Aviation*, p. 31.
29. Ibid., p. 31.
30. Carl C. Ciepluch, Donald Y. Davis, and David E. Gray, "Results of NASA's Energy Efficient Engine Program," *Journal of Propulsion and Power* 3 (November–December 1987): 562–567; St. Peter, *History of Aircraft Gas Turbine Engine Development*, p. 399.
31. Ciepluch, Davis, and Gray, "Results of NASA's Energy Efficient Engine Program," p. 561; St. Peter, *History of Aircraft Gas Turbine Engine Development*, p. 402; Daniel E. Sokolowski and John E. Rohde, "The E^3 Combustors: Status and Challenges," NASA TM-82684, July 1981; D.L. Burrus, C.A. Chahrour, H.L. Foltz, P.E. Sabla, S.P. Seto, and J.R. Taylor, "Combustion System Component Technology Performance Report," NASA CR 168274, July 1984.
32. D.Y. Davis and E.M. Stearns, "Energy Efficient Engine: Flight Propulsion System Final Design and Analysis [GE Design]," NASA CR-168219, August 1985, pp. 1–8, 147.
33. Ciepluch, Davis, and Gray, "Results of NASA's Energy Efficient Engine Program," pp. 565–567.
34. Ibid., p. 567.
35. U.S. Congress, Office of Technology Assessment, *Federal Research and Technology for Aviation*, OTA-ETI-610 (Washington, DC: Government Printing Office, 1994), p. 78.
36. David Field, "Engine Makers in Three-Sided 'War' To Provide Power for Boeing 777," *Washington Times* (May 17, 1994): B7, B12; St. Peter, *History of Aircraft Gas Turbine Engine Development*, p. 398.
37. Smithsonian National Air and Space Museum, "Pratt & Whitney PW 4098 Turbofan Engine," 2007, Registrar's File 20070002000.
38. "Boeing 777-300ER Performs 330-Minute ETOPS Flight," October 15, 2003, *http://www.prnewswire.com/news-releases/boeing-777-300er-performs-330-minute-etops-flight-72530852.html* (accessed January 15,

2016); GE Aviation, "The GE90 Engine Family," 2012, *http://www.geaviation.com/engines/commercial/ge90/* (accessed June 30, 2013).
39. Craig Schmittman, "Ultra High Bypass (UHB)," *Aerospace-Defense News* (1989), available online at *http://www.youtube.com/watch?v=zxVAaIsfPIY* (accessed July 5, 2013).
40. Bowles, *"Apollo" of Aeronautics*, pp. 122–123; Bowles and Dawson, "The Advanced Turboprop Project," pp. 321, 323; Hager and Vrabel, *Advanced Turboprop Project*, p. 2.
41. Bowles, *"Apollo" of Aeronautics*, p. 122.
42. "Aircraft Fuel Conservation Technology Task Force Report," p. 44.
43. Bowles, *"Apollo" of Aeronautics*, pp. 122, 125–126; Hager and Vrabel, *Advanced Turboprop Project*, pp. 6–10.
44. Bowles, *"Apollo" of Aeronautics*, pp. 127–128; Hager and Vrabel, *Advanced Turboprop Project*, pp. 11–42.
45. Bowles, *"Apollo" of Aeronautics*, pp. 127–128; Hager and Vrabel, *Advanced Turboprop Project*, pp. 43–48.
46. Bowles, *"Apollo" of Aeronautics*, p. 128; Hager and Vrabel, *Advanced Turboprop Project*, pp. 52, 55.
47. Hager and Vrabel, *Advanced Turboprop Project*, pp. 56–57.
48. Ibid., p. 57. See also ONERA, "S1MA-Continuous-Flow Wind Tunnel," 2009, *http://windtunnel.onera.fr/s1ma-continuous-flow-wind-tunnel-atmospheric-mach-005-mach-1*.
49. Hager and Vrabel, *Advanced Turboprop Project*, p. 59.
50. Ibid., pp. 60, 62.
51. Ibid., pp. 63, 65.
52. Bowles, *"Apollo" of Aeronautics*, p. 128; Hager and Vrabel, *Advanced Turboprop Project*, p. 67.
53. Hager and Vrabel, *Advanced Turboprop Project*, pp. 69, 71.
54. Bowles, *"Apollo" of Aeronautics*, p. 133; Hager and Vrabel, *Advanced Turboprop Project*, pp. 71–74.
55. Hager and Vrabel, *Advanced Turboprop Project*, p. 75.
56. Ibid., p. 76.
57. Ibid., pp. 77–79.
58. Philip Schultz, quoted in Craig Schmittman, "Ultra High Bypass (UHB)."
59. Hager and Vrabel, *Advanced Turboprop Project*, pp. 84, 86.
60. Ibid., pp. 86–87.
61. Ibid., pp. 80–82, 88–90.
62. Ibid., pp. 91–93.
63. Hager and Vrabel, *Advanced Turboprop Project*, pp. 93–97.
64. GE Aviation, "Aircraft Engine History and Technology," 2009, *http://www.youtube.com/watch?v=4lip8lPWFLo* (accessed July 2, 2013).

65. Hager and Vrabel, *Advanced Turboprop Project*, pp. 98–101.
66. Hager and Vrabel, *Advanced Turboprop Project*, p. 101; "Whatever Happened to Propfans?" *Flightglobal*, June 12, 2007, *http://www.flightglobal.com/news/articles/whatever-happened-to-propfans-214520/* (accessed July 2, 2013).
67. Jim Schefter, "New Blades Make Prop Liners as Fast as Jets," *Popular Science* 226 (March 1985): 66.
68. Martha M. Hamilton, "Firms Give Propellers a New Spin," *Washington Post* (February 8, 1987): H1.
69. Greg Johnson, "Something New for Airliners: Propellers," *Los Angeles Times* (June 16, 1986): SD-C1.
70. Michael A. Dornheim, quoted in Craig Schmittman, "Ultra High Bypass (UHB)."
71. "Whatever Happened to Propfans?"
72. Bowles and Dawson, "The Advanced Turboprop Project," p. 342; Johnson, "Something New for Airliners: Propellers"; Hager and Vrabel, *Advanced Turboprop Project*, pp. vi–vii.
73. Bowles, *"Apollo" of Aeronautics*, p. xx.
74. Ethell, *Fuel Economy in Aviation*, p. 30.
75. Bowles, *"Apollo" of Aeronautics*, p. 60.
76. GE Aviation, "Aircraft Engine History and Technology"; Reddy, "Seventy Years of Aeropropulsion Research at NASA Glenn Research Center," p. 202.
77. Dr. Mike J. Benzakein, "The Future of Aeronautics at NASA," statement before the Subcommittee on Space and Aeronautics, Committee on Science, House of Representatives, in Cong. Rec., 109th Cong., 1st sess., serial no. 109-8, March 16, 2005, available at *http://commdocs.house.gov/committees/science/hsy20007.000/hsy20007_0.HTM* (accessed September 12, 2014); Miller, "NASA," p. H12545.
78. Patrick Hassell, "A History of the Development of the Variable Pitch Propeller" (paper presented before the Royal Aeronautical Society, Hamburg Branch, Hamburg, Germany, April 26, 2012); George Rosen, *Thrusting Forward: A History of the Propeller* (Windsor Locks, CT: United Technologies Corporation, 1984), pp. 84–90.
79. Bowles and Dawson, "The Advanced Turboprop Project," p. 342.
80. Hager and Vrabel, *Advanced Turboprop Project*, pp. iii, 105.
81. "General Aviation Technology Program," NASA News Release 75-65, March 1975, available at *http://ntrs.nasa.gov/archive/nasa/casi.ntrs.nasa.gov/19770074219_1977074219.pdf* (accessed July 1, 2012).
82. Ibid.
83. Ibid.

84. Joseph P. Allen to Charles F. Lombard, March 22, 1976, NASA HRC, file 012336.
85. Ibid.
86. "General Aviation Technology Program," NASA News Release 75-65.
87. Williams, *Small Transport Aircraft Technology*, pp. 40–41.
88. "Test Engines Promise Less Noise from Small Aircraft," NASA News Release 79-24, March 2, 1979, NASA HRC, file 012336.
89. "Quieter Engines for Small Aircraft Possible, Tests Show," *Aviation Week & Space Technology* 16 (March 16, 1979): 1; R.W. Koenig and G.K. Sievers, "Preliminary QCGAT Program Test Results," NASA TM-79013, 1979; Huff, "NASA Glenn's Contributions to Aircraft Noise Research," p. 250.
90. "Test Engines Promise Less Noise from Small Aircraft."
91. R.W. Heldenbrand and W.M. Norgren, AiResearch QCGAT Program, NASA-CR-159758, 1979, p. 3; J. German, P. Fogel, and C. Wilson, "Design and Evaluation of an Integrated Quiet Clean General Aviation Turbofan (QCGAT) Engine and Aircraft Propulsion System," NASA CR-165185, 1980.
92. Mary Fitzpatrick, "Highlights of 1979 Activities: Year of the Planets," NASA News Release 79-179, December 27, 1979, p. 18.
93. Richard A. Leyes II and William A. Fleming, *The History of North American Small Gas Turbine Aircraft Engines* (Reston, VA: AIAA, 1999), pp. 683–684, 723–724.
94. Jerry R. Wood, Paul W. Adam, and Alvin E. Buggele, "NASA Low-Speed Centrifugal Compressor for Fundamental Research," NASA TM-83398, June 1983, p. 1; photo 88-H-123, April 14, 1988, NASA HRC, file 011151.
95. *Aeropropulsion '87: Proceedings of a Conference Held at NASA Lewis Research Center, Cleveland, Ohio, November 17–19, 1987*, NASA CP-3049, 1990, pp. iii, 13.
96. Robert Rosen, "The 1987 Aeropropulsion Conference: Change and Challenge," in *Aeropropulsion '87*, pp. 1–12.

Dryden researchers used a NASA F-15 to test the Digital Electronic Engine Control (DEEC) system from 1981 to 1983. (NASA)

CHAPTER 5
Propulsion Control Enters the Computer Era, 1976–1998

Through the first half of the 20th century, engine diagnostic systems primarily consisted of a pilot physically monitoring gauges indicating oil pressure, temperature, and fuel capacity on the instrument panel. That basic human-machine interface evolved into the development of automated onboard diagnostic systems. Those new systems assessed engine health and recorded data for postflight troubleshooting and maintenance through the use of mathematical models.

Engine Controls from the Piston Engine to the Afterburning Turbofan

The configuration of piston engine control systems up to the early 1930s reflected a direct and simple approach that was, at its core, similar to that of other internal combustion engine systems. The pilot manipulated the throttle to set the amount of power needed. Carburetors or fuel injectors metered the appropriate amount of fuel to the engine's combustion chambers. As airflow for combustion and flight conditions changed, the system maintained the power at the desired level.[1]

The introduction of the jet engine in the 1940s brought new challenges. The parallel rise of aviation electronics created new avenues that enabled growing sophistication, increased capability, and better control of the new technology. That development represented four phases during the latter half of the 20th century. The initial phase, from 1942 to 1949, witnessed the first steps toward new control systems and highlighted the limitations of existing technology. The growth phase marked the rise of practical applications of military and commercial jet aircraft from 1950 to 1969. The electronic phase, from 1970 to 1989, witnessed the pioneering introduction of new and revolutionary engine control systems, while the period 1990 to 2002 marked their increased integration.[2]

The first U.S. jet engine, the British-derivative centrifugal-flow GE I-A, featured a hydro-mechanical governor upon its introduction in 1942. The governor metered the fuel flow going into the engine to be proportional to

The Power for Flight

the difference between the speed set by the pilot and the actual speed of the turbine. A minimum-flow stop in the fuel-metering valve prevented "flame-out," or the extinguishing of ignition in the combustion chamber, caused by a pilot's incorrect throttle setting or overzealous acceleration. A maximum flow schedule prevented the engine from going too fast or overheating. This system possessed the basic components for controlling a single-spool turbojet engine during the early days of the Jet Age.[3]

In 1948, GE introduced its new J47 turbojet. It became the first axial-flow engine approved for commercial use in the United States, as well as a mainstay engine for the American military through the early Jet Age into the late 1960s. The most notable applications included the Boeing B-47 Stratojet, the first operational jet bomber for the Strategic Air Command, and the MiG-mastering North American F-86 Sabre of Korean War fame. The engine featured a standard and reliable hydro-mechanical fuel control for its combustion chambers. In a variant used on the F-86D, K, and L interceptors, the J47 incorporated a thrust-increasing afterburner, one of the earliest engines to do so. The afterburner's electronic fuel control relied upon fragile vacuum tubes for operation, which proved unsatisfactory in the harsh environment of the turbojet, characterized by high heat and vibration. In collaboration with GE, NACA researchers at Lewis conducted testing in the Altitude Wing Tunnel (AWT) that revealed a solution to the problem. GE first utilized frequency

Figure 5-1. Pictured is the North American F-86D Sabre. (United States Air Force via National Air and Space Museum, Smithsonian Institution, NASM 74-3342)

response techniques to control the J47 afterburner. NACA testing indicated that noise in the speed sensor, coupled with the high gain of the speed governor, limited the operation of the engine. GE and NACA engineers used the time-domain step response analysis method to fix the problem by reducing the control gain at altitude. The industry-Government cooperation established a knowledge base for the design of newer and better engine controls.[4]

Edward W. Otto and Burt L. Taylor III at Lewis studied the behavior of single-shaft turbojet engines in 1948. They focused on shaft speed because it was nearly equivalent to thrust and could be easily and more accurately measured. Otto and Taylor learned that the transfer function from fuel flow to engine speed, regarded as the dynamic characteristic of a turbojet engine, could be represented by a first-order lag linear system with a time constant.[5]

Engine capabilities increased in the 1950s with new twin-spool turbojets like the Collier Trophy–winning Pratt & Whitney J57 and high-compression-ratio, bypass-flow turbofans. Engine control technologies followed suit. The state of engine control technology matured to incorporate variable-geometry controls at the compressor stator, intake, and nozzle. The increased complication posed challenges to effective evaluation by testing alone. A development that took place outside of aviation, the introduction of increasingly sophisticated computer technology, facilitated the use of computer-based real-time dynamic simulations for engine control design and analysis. Researchers James R. Ketchum and R.T. Craig at Lewis initiated this work with electronic analog computers in the early 1950s. They simulated the response of a turbojet engine to a step change in fuel flow and validated the results, which proved applicable to different types of gas turbine engines.[6] That first step toward the accurate simulation of complex engine dynamic behavior led to further advancements that kept the cost and time involved in control design development and validation in line with those of other technologies.[7] In other words, efficient engine control would not be a bottleneck in the development of new aircraft. By the late 1960s, researchers were contemplating moving beyond traditional control systems and architectures to new ones making use of emergent powerful, lightweight airborne computers.

The Advent of Digital Electronic Engine Control

The advent of stability augmentation had stimulated controls research that evolved into electronic flight control. Research in the 1950s had led to ever-more-heavily "augmented" aircraft that preceded the genuine "fly-by-wire" aircraft of the 1970s–1980s pioneered by NASA.[8] First applied to aircraft flight controls, electronic control then migrated to propulsion controls as well. The introduction of more powerful and sophisticated military and commercial turbofans in the 1970s and 1980s taxed the capabilities of decades-old hydro-mechanical engine control systems. To continue with them meant increasing their size, weight, and expense, which introduced ramifications regarding the design and performance of the aircraft overall.[9]

Integrated Propulsion Control, Highly Integrated Digital Electronic Control, and Adaptive Engine Control

At the request of the U.S. Air Force, NASA researchers at the Dryden Flight Research Center at Edwards Air Force Base in the high desert of California and at Lewis had assisted in the development and flight test of the first digital integrated propulsion control system (IPCS) to be flown in an aircraft in 1976. The team, which included Boeing, Pratt & Whitney, and Honeywell, selected an F-111 as the platform for two primary reasons. First, the aircraft featured a variable-geometry inlet and two afterburning turbofans, which allowed experimentation with the left-side engine while the other remained unaltered to ensure flight safety. Second, the IPCS proved capable of duplicating the standard hydro-mechanical inlet and engine controls that manipulated the inlet spike and expanding the cone, fuel supply, compressor bleed, and nozzle area. Flight tests in 1976 exhibited faster throttle response, increased thrust, extended range at Mach 1.8, and—perhaps most importantly for the F-111—stall-free operation. The success of the IPCS in enhancing propulsion systems in terms of efficiency, operability, reliability, and maintenance led to the widespread use of digital inlet-engine controls technology in both military and commercial aircraft.[10]

As well, NASA used its F-15 to investigate and demonstrate a new method of obtaining optimum aircraft performance with computer-controlled engines called Highly Integrated Digital Electronic Control (HIDEC).[11] The experimental program explored new innovations to improve engine diagnostics, control, and efficiency. The major components of HIDEC were a Digital Electronic Flight Control System (DEFCS), a Digital Electronic Engine Control (DEEC), an onboard general-purpose computer, and an integrated architecture that allowed all of the components to interact with each other.[12]

A full-authority DEEC regulated the operation of the F-15's PW1128 turbofan. The DEEC scheduled and maintained the engine operating point through the use of two main control loops. The first used the main burner fuel flow to regulate the low rotor speed. The second controlled engine pressure with actuation of the nozzle throat area. The DEEC also controlled the front and rear compressor variable vanes. The system monitored seven individual parameters: fan and high-pressure compressor speed, engine face and fan turbine inlet temperature, and engine face, burner, and augmenter pressure. An RS-422 universal asynchronous receiver/transmitter (UART) bus transmitted the data

Figure 5-2. NASA, the Air Force, Boeing, Honeywell, and Pratt & Whitney used the prototype F-111E for integrated propulsion control system (IPCS) flights from 1975 to 1976. (NASA)

to onboard computers for processing by a priority-selected cache algorithm.[13] In addition to the mechanical and analog electronic flight control system found on operational U.S. Air Force F-15s, the F-15 HIDEC also had a dual-channel, fail-safe digital flight control system. Engineers could program it using any of the major computer languages, including Pascal, Ada, and FORTRAN. H009 and Military Standard 1553B data buses linked all of the electronic systems together.[14] The DEEC program was a major step in increased computer control of key engine functions. The Air Force specified them for the F100 engines powering the F-15 and General Dynamics F-16 Fighting Falcon fighters, and Pratt & Whitney incorporated them into the new PW2037 turbofans that powered the Boeing 757 airliner.[15]

A critical problem to increased jet engine performance was compressor stall. NASA began to address the problem in 1983 with the development of the Adaptive Engine Control System (ADECS). The integrated and computerized flight and engine control systems monitored the engine stall margin—the amount that engine operating pressures needed to be reduced to provide a margin of safety—based on the flight profile and real-time performance needs. That information allowed the ADECS to maximize engine performance that would otherwise be held in reserve to meet the stall margin requirement. In essence, the ADECS exchanged excess engine stall margin for improved

The Power for Flight

Figure 5-3. The F-15 HIDEC is shown in flight over the Mojave Desert. (NASA)

performance. As a result, the ADECS increased thrust, reduced fuel usage, and lowered engine operating temperatures.[16]

ADECS research and demonstration flights began at Dryden in 1986. The F-15 displayed increased engine thrust from 8 to 10.5 percent at 10,000 and 30,000 feet, respectively, and up to 16 percent lower fuel consumption at 30,000 feet. The increased engine thrust improved the rate of climb 14 percent at 40,000 feet and reduced time to climb from 10,000 feet to 40,000 feet by 13 percent. Increases of 5 to 24 percent in acceleration were also experienced at intermediate and maximum power settings, depending upon altitude. Overall, engine performance improvements in terms of rate of climb and specific excess power were in the range of 10 to 25 percent at maximum afterburning power. The research pilots tried to induce stalls to validate the ADECS methodology, but no amount of aggressive maneuvering could cause one.[17]

Performance Seeking Control: Progressing Beyond HIDEC and ADECS

The integration phase of the history of engine control systems from 1990 to 2002 saw dual-channel FADEC systems become the standard for jet engines.[18] Another NASA F-15 HIDEC flight research program that worked to optimize overall engine operation was the Performance Seeking Control (PSC) project, which began during the summer of 1990. Previous control modes used on the HIDEC aircraft utilized stored schedules of optimum engine pressure ratios for an average engine on a normal day. PSC used highly advanced techniques that identified the condition of the engine components and optimized the overall system for best efficiency based on the actual engine and flight conditions encountered on a given day. Specifically, the new system employed integrated control laws to use the digital flight, inlet, and engine control systems to ensure the availability of peak engine and maneuvering performance at all times. The overall result was that PSC reduced fuel usage at cruise conditions, maximized excess thrust during accelerations and climbs, and extended engine life by reducing the fan turbine inlet temperature. A byproduct was the capability to monitor the degradation of engine components. When combined with regularly scheduled preventative maintenance, the PSC enabled greater operational efficiencies and longevity for high-performance aircraft.[19]

The PSC system could be applied to a wide variety of aircraft but was especially suited to high-performance military aircraft. Pratt & Whitney used the self-tuning onboard model in its advanced engine controllers, including those on the F119-PW-100 engine used on the Lockheed Martin F-22 Raptor aircraft. The manufacturer applied other aspects of HIDEC technology in the improved F100-PW-229, the most widely used fighter engine in the world, to increase performance and operational longevity. The flight demonstration and evaluation performed at NASA Dryden in the F-15 HIDEC contributed to the rapid transition of the technology into operational use.[20]

Response to Tragedy: Toward Propulsion-Controlled Aircraft

A series of aircraft accidents through the 1970s and 1980s illustrated the need for better methods of flight control. One of the surprising outcomes was the

Figure 5-4. Shown is the Lockheed Martin F-22A Raptor. (U.S. Air Force)

demonstration of how engine throttle manipulation could alleviate the problem as a "last resort" flight control system.

In April 1975, a U.S. Air Force Lockheed C-5 Galaxy transport evacuating 300 orphans from Saigon, Vietnam, as part of Operation Babylift, lost all flight controls in the tail after the rear bulkhead failed. Left with only roll control, the pilots used their throttles to regain limited pitch (up and down) control authority. The giant transport entered into a motion called a phugoid, a roller coaster–like oscillation of pitching up and slow climbing followed by pitching down and rapidly descending. Despite the crew's best efforts, the lumbering transport crashed on approach to Tan Son Nhut Air Force Base, causing the loss of 139 on board; tragically, many of the dead were young children and infants being evacuated.

A decade later, in August 1985, Japan Airlines Flight 123 suffered an explosive decompression in the rear fuselage after taking off from Tokyo International Airport. The decompression blew most of the vertical stabilizer away and disabled all hydraulic control. The crew flew the uncontrollable 747 with the throttles and electrically actuated flaps for half an hour before the plane disastrously crashed into Mount Takamagahara, resulting in the loss of all 520 people on board. Overall, more than 1,100 crew and passengers died following the failure or destruction of hydraulic control systems by 1996. Other flights encountered close calls that were just as terrifying. In one case, in April 1977, after a horizontal stabilizer jammed, the crew of a Delta Air Lines Lockheed L-1011 avoided a stall by using their throttle controls to change the aircraft's pitch, managing to land safely.[21]

A major wake-up call came during the summer of 1989. At 2:09 p.m. on the afternoon of July 19, United Airlines (UAL) Flight 232, a McDonnell Douglas DC-10, rose swiftly from Denver's Stapleton International Airport, bound for Chicago. The McDonnell Douglas DC-10 was a tri-jet with an engine on each wing and in a nacelle integrated into the vertical fin of the aircraft. At first all went well, and the big jetliner climbed to its cruising altitude of 37,000 feet. Then, at 3:16 p.m., a loud bang followed by vibration and shuddering alerted the crew that the plane had just experienced a catastrophic engine failure. The fan disk in the center GE CF6 engine had disintegrated because of an undetected fatigue crack, scattering shrapnel that disabled all hydraulic systems used for aircraft control. Consequently, the plane's control columns and rudder pedals were useless. It seemed certain that the plane would shortly plunge 7 miles to Earth, killing all aboard in a horrific crash.[22]

The determined and courageous flight crew—Captain Alfred C. Haynes, First Officer William R. Records, Second Officer Dudley J. Dvorak, and Training Check Airman Captain Dennis E. Fitch—were not about to give up, and for not quite 45 minutes, they ingeniously controlled the DC-10 as best they could by manipulating the thrust of the two remaining engines on the wings. They overcame the challenge of compensating for oscillations in pitch and roll by manipulating the engine throttles as they tried to make their way to Sioux City, IA, for an emergency landing. But despite their best efforts, the ailing DC-10 could not be controlled with any degree of precision. Thus, at 4:00 p.m., as the crippled airliner approached to land, its right wingtip touched the ground first, followed by the right landing gear. The DC-10 then skidded, rolled over, burst into flame, and cartwheeled across the runway, coming to a rest, blazing furiously. Of the 296 people on board, 111 were killed and a further 172 injured, 47 seriously (one of whom died a month later).[23] The National Transportation Safety Board (NTSB) determined that no amount of training could prepare other aircrew to cope successfully in a similar situation with existing equipment and encouraged the research and development of a backup flight control system. "Under the circumstances," the Safety Board concluded, "the UAL flightcrew performance was highly commendable and greatly exceeded reasonable expectations."[24]

The Sioux City crash was a tragedy, and in its accident report, the NTSB recommended "research and development of backup flight control systems for newly certificated wide-body airplanes that utilize an alternative source of motive power separate from that source used for the conventional control system."[25] The response from NASA was typical of the Agency's search for solutions to common challenges and problems facing the operation of aircraft. From the 1970s until the end of the 20th century, NASA flight research conducted at Dryden contributed to the development and demonstration of advanced integrated flight and propulsion control system technologies that contributed to the maneuverability, fuel efficiency, and safety of new generations of aircraft. Once again, the research airplane became just as important a tool for NASA's work in propulsion as the computer and wind tunnel.[26]

During a commercial flight to St. Louis shortly after the Sioux City disaster in 1989, Frank W. "Bill" Burcham, the Chief Propulsion Engineer at Dryden, started to ponder whether there was a solution: a backup landing technique—one that relied solely upon the thrust of its engines—for an aircraft that had lost its flight controls. The key was using digital engine control computers found in contemporary airliners. Burcham sketched on a cocktail napkin the basic system that became the propulsion-controlled aircraft (PCA) concept. Beginning with the control stick, the system went to a DEFCS computer, then individually to the right and left engines, which then routed to the F-15 HIDEC. A connection between the DEFCS and HIDEC kept the system integrated. By the end of the flight, Burcham and his fellow traveler, NASA project manager Jim Stewart, had a test program outlined.[27]

The Power for Flight

Figure 5-5. Bill Burcham's sketch of the PCA concept. (NASA)

Burcham's goal was to create a backup landing system—based solely on the thrust of an aircraft's engines—for aircraft that had lost their flight controls. The key was using technology already found in the latest aircraft: digital flight and engine control computers. The question was whether that equipment could be utilized for that purpose. PCA amounted to a reconfiguration program where Burcham and Stewart's team replaced traditional means of control with a new system based on engine thrust.[28]

The first step in the program was to ascertain if a pilot could alter the course of an airplane through the use of engine throttles, or Throttles Only Control (TOC). By manipulating the throttles, a pilot could maneuver the airplane with two forms of thrust. Collective thrust controlled flightpath, or lateral control, while differential thrust controlled bank angle.[29] Starting in the simulator, Burcham had the F-15's controls locked in place. A pilot himself, he maneuvered the aircraft using only the engine throttles. He advanced one throttle and retarded the other to roll the aircraft. Pushing both throttles forward pitched the nose up. Pushing them back dipped the nose down. After a few crashes, Burcham was able to safely land the F-15 in the simulator.[30]

The software model of an aircraft and its flying environment found in a simulation offered flight experience without the risks of actually being in the air. Besides the airplane's various systems, including the propulsion system, the simulator included an airport and weather, which all amounted to "an elaborate video game" for the research team. "Flying" the Dryden Boeing 720 simulator revealed that a PCA aircraft suffered from a lethargic response of the engines to control inputs during TOC.[31]

PCA firmly straddled both worlds of flight control and propulsion research. In response, both communities were lukewarm to the idea initially, and it played to very mixed reviews. One initially skeptical engineer remembered, "PCA wasn't intuitively obvious," while another labeled it as "hare-brained!" NASA Headquarters feared premature regulatory action by Federal agencies centered on safety that would curtail manufacturer interest and development before the idea could be fully explored. They advised Dryden Director Ken Szalai to discontinue work on PCA.[32]

Nevertheless, the work continued, and it fortified the reasons for the program. The addition of well-known research pilot and former Space Shuttle astronaut C. Gordon Fullerton to the team instantly added credibility. He experienced the same problems with the lethargic response of the controls. If pilots continued to use the control stick, they would expect the same quick response as from traditional flight controls, which could lead to fatal errors from over- or under-compensation. Fullerton suggested a new control system based on the twin thumbwheels used in autopilot systems. One wheel controlled lateral movement while the other offered longitudinal movement. To answer the concern over what would happen to a PCA-controlled aircraft in bad weather, Dryden engineer Joe Conley designed and incorporated an instrument landing system (ILS) component.[33]

PCA in Flight Test: The F-15 and MD-11 Experience

With an influx of funds from the Air Force, Burcham went in search of a research aircraft for the project. NASA Dryden's extensively instrumented McDonnell Douglas F-15 could be employed for the tests, and the F-15 simulator at McDonnell Douglas in St. Louis would be available for setup and preparation of the flight-test program. Earlier projects had left the aircraft loaded with test instrumentation from programs like HIDEC, as well as two computer systems, a digital flight control computer (FCC) and DEEC, which were programmable and capable of communicating with each other in flight. Researchers incorporated interim control-system software to produce a slower engine response at low power settings to emulate the characteristics of high-bypass turbofan engines. But there was a disadvantage with the F-15 as well, in that the close proximity—just 1 foot—between the two Pratt & Whitney

The Power for Flight

engines was not ideal for controlling and evaluating the differential thrust between the left and right power plants.[34]

The first dedicated TOC flight took place on July 2, 1991. Nearly 2 years later, the program realized a significant achievement when Fullerton landed the F-15 using only engine power to turn, climb, and descend on April 21, 1993. Fullerton descended in the F-15 in a shallow approach to approximately 20 feet above the runway. The rate of descent increased dramatically, but the veteran test pilot brought the aircraft down safely and effectively proved the capability of the PCA system. Burcham remarked that while the technology was proven for incorporation into future aircraft designs, he hoped "it never has to be used."[35] A series of guest pilots from industry and the military went on to fly the F-15 in a series of trial approaches and go-arounds, and their enthusiasm for PCA was unanimous.[36]

The next step in the program concerned expanding the flight research program to include an actual multi-engine airliner, the type of aircraft Burcham and the rest of the team originally envisioned for PCA. That required the expansion of institutional involvement beyond Dryden. During December 1992, industry and airline executives, Government administrators, and NASA Center Directors met in Washington, DC. Dwain Deets, acting Director of Dryden Research Engineering, and Burcham presented the case for PCA. They faced resistance within and outside NASA. They had to navigate internal NASA resistance to PCA, which came mostly from Langley. The Center in Tidewater

Figure 5-6. Gordon Fullerton uses only engine power to land the NASA F-15 at Dryden on April 21, 1993. (NASA)

Virginia was traditionally the facility that investigated problems in subsonic aeronautics. Second, they needed the endorsement of industry. Boeing especially had been unconvinced of the potential of PCA. Bob Whitehead, Director of Subsonic Transportation in the Office of Aeronautics and Space Technology (OAST), who convened the meeting, believed that industry had to support the project for it to receive the appropriate funding and move forward. He introduced the topic and set in motion a roundtable discussion of the program. At the end of the meeting, everyone was in agreement that the PCA project should move forward. They created a full-fledged Government-industry collaborative project with $2.5 million in funding.[37]

McDonnell Douglas became the primary contractor. The company built the latest-generation airliners, which were a perfect example of the type of aircraft that needed PCA. The MD-11 was a three-engine wide-body airliner derived from the earlier DC-10, but with the latest computerized full flight control system and digital engine controls. The company started with simulations, to which the initial reactions were very good.[38]

With the help of Drew Pappas, McDonnell Douglas project manager, the project received an MD-11 for flight testing. The first PCA flight over the manufacturer's Yuma, AZ, facility occurred on August 27, 1995. The pilots climbed to 10,000 feet and turned on the PCA system. It held the wings level and on the desired flightpath with minor deviations in direction and altitude. Two days later, Gordon Fullerton flew the MD-11 during its first landing

Figure 5-7. In August 1995, Gordon Fullerton brings the MD-11 in for the first PCA landing of an airliner. (NASA)

under engine power alone at Edwards Air Force Base, which offered longer, wider, and safer runways. After successive approaches that went as low as 100, 50, and 10 feet, Burcham proceeded to land at a sink rate of 4 feet per second. The video recording of the landing revealed that the control surfaces remained motionless as the engines guided the airliner down to the runway.[39]

To spread the word, NASA invited two dozen guest pilots to fly the PCA-equipped MD-11 on a flightpath that included an approach within 100 feet of the runway during November 29–30, 1995. They represented Government agencies including NASA, the FAA, the Air Force, and the Navy; manufacturers McDonnell Douglas, Boeing, Airbus, and Honeywell; airlines such as American, Delta, Japan Air, Royal Flight of Saudi Arabia, and Swissair; and the aviation press, *Aviation Week* and *Flight International*. The addition of the ILS software and an autoflare brought improved control and proved the potential for hands-free emergency touchdowns. One of the guest pilots commented that PCA changed "what had been a very challenging, if not impossible, situation into what could be considered a textbook lesson with no exceptional pilot skills required."[40]

The next step in the program was to fly the MD-11 with its three hydraulic systems completely disabled. Burcham's original intention was to directly address the tragedy of Flight 232, which was a standard scenario in all the aircraft simulations from the beginning of the project. In reality, PCA needed to show that in the event of the loss of the hydraulic system, control was possible. Also, unlike the pioneering August 29, 1995, flight, the flight surfaces in the event of control failure would not be straight and neutral; they would be stuck in or would float to different positions, which would affect stable control. Initially, test pilot Dana Purifoy flew the airliner over the Pacific in September in the "Whiskey" test area, where there was uninterrupted room to conduct further flight tests without endangering anyone on the ground. Fullerton flew the next flight over the Mojave Desert in November. The final test, on November 28, presented exactly the kind of scenario that Burcham envisioned. With all hydraulics disabled, Fullerton and the flight-test crew flew the MD-11 under the control of PCA.[41] The test by the NASA-industry team clearly demonstrated the capability of PCA.

While the November 1995 MD-11 flights proved the capability of PCA to increase airliner safety, the potential of implementation was another matter. The system only benefited airliners equipped with FADEC—a minority amongst then-current commercial fleets. A full two-thirds of airliners did not have FADEC, and their service-life projections were for decades—into the 21st century. In other words, PCA was a solution for an airline fleet that had yet to exist. The FAA was not going to implement a safety standard that could not yet be applied. Nor was a cost-conscious industry going to make the necessary investments to facilitate the incorporation of PCA.[42]

All along, the Dryden researchers worked together with their contemporaries at Ames Research Center on devising real-world-scenario simulations for PCA. They began with a full PCA system for the Boeing 747. One flight simulation saw the airliner lose its hydraulic system at 35,000 feet and roll upside down. The use of the PCA system righted the airplane, leveled the wings, and brought the behemoth down for a safe landing. Another simulation demonstrated how, with the loss of flight controls and one engine, pilots could use PCA to transfer fuel from one tank to another to counterbalance an airliner's center of gravity toward the operating engine. An industry-NASA team led by John Bull expanded its focus on other aircraft, including the Boeing 757, the McDonnell Douglas F/A-18 Hornet multirole fighter, and the advanced McDonnell Douglas C-17 Globemaster III transport. These simulations addressed a variety of in-air emergencies and how PCA could help avoid tragedies like Flight 232.[43]

In response, Burcham set out to work to develop a simpler and cheaper version of PCA to facilitate near-term implementation in existing designs in May 1995. After a discussion with a Delta Air Lines pilot, he sketched a new concept that eliminated the need for changing engine control software. "PCA Lite," as it was called by Ken Szalai, utilized systems most aircraft already had installed. The autothrottle and the digital thrust trim system provided pitch and lateral control respectively. For PCA Lite, John Bull and Ames researchers demonstrated effective simulations for a number of aircraft ranging from military fighters to jumbo jets.[44]

At the lowest tier of airliners in commercial aviation were aircraft with no digital engine controls, meaning they had neither autothrottle nor an engine thrust trim system. "PCA Ultralite" was a method wherein a pilot operated the throttles for lateral control manually. To improve the process, the Dryden-Ames partnership added a flight director needle in the cockpit that indicated cues that the pilot used to manipulate the throttles. Evaluations of the system at Ames in 1998 substantiated the results.[45]

PCA was an inexpensive technology that required only software modification and aircrew training to achieve widespread implementation. There were factors that stymied progress in that direction. There was the question of the rigorous process of FAA certification. It also remained for the airline industry and the manufacturers to want to incorporate the technology. A Honeywell software engineer involved in the project, Jeff Kahler, estimated optimistically that PCA would be 100-percent effective if commercial and military aviation adopted it for everyday use.[46]

In the absence of installed PCA systems becoming part of new aircraft, NASA continued to push TOC and PCA. Burcham and Fullerton continued to advocate TOC as a means of safely flying and landing a crippled airplane. They suggested techniques for flying with throttles only and making a survivable landing using the principles of TOC.[47]

The lingering challenge for TOC/PCA was fast engine response. That parameter was a major issue for the system's latest incarnation, Integrated Resilient Aircraft Control (IRAC), which was a program sponsored by NASA's Aviation Safety Program in 2009. Researchers believed that through the use of a Commercial Modular Aero-Propulsion System Simulation (C-MAPSS) developed at Glenn, an engine controller could be modified for faster response and overthrust operation, defined as speed in excess of throttle setting, for more power in emergency situations. The modification did have its drawbacks, primarily increased wear on the engine with the possibility of catastrophic failure. The biggest challenge was creating a universal system capable of being adapted to the specific characteristics of individual engines.[48]

Thrust Vectoring for Propulsion Control

While PCA offered increased safety in the operation of a crippled airplane, another form of propulsion control, thrust vectoring, enhanced the maneuverability of high-performance military aircraft. In the wake of the American air combat experience during Vietnam, aircraft manufacturers introduced a new generation of fighters capable of dogfighting. In these new aircraft, if a fighter pilot pulled sharply back on the control stick, the nose would pitch up while the fighter continued in its original direction. Engineers called the angle of the aircraft's body and wings in relation to its flightpath the alpha, or angle of attack. The problem with high-alpha maneuvers was that airflow disturbances resulted in loss of wing lift, which degraded control and overall performance. NASA initiated the High-Angle-of-Attack Technology Program (HATP) in 1987 to address that problem in partnership with the Department of Defense, industry, and academia. Besides state-of-the-art fighter jets, potential applications of HATP research included hypersonic vehicles and high-performance civilian aircraft.[49]

The primary objectives of HATP were to provide flight-validated aircraft design tools and to improve the maneuverability of aircraft at high alpha. The program placed particular emphasis on aerodynamics, propulsion, control-law research, and handling qualities, which required participation from all four research Centers. Langley managed the program and, in partnership with Ames, conducted wind tunnel testing and calculations using advanced control laws and CFD. Lewis's and Dryden's responsibilities centered on inlet and engine integration and flight research respectively. NASA received an early-model McDonnell Douglas F/A-18 Hornet, an aircraft known for its high maneuverability, as surplus from the U.S. Navy and modified it for the HATP mission with the designation F/A-18 High Alpha Research Vehicle (HARV).[50]

The HARV flight research program consisted of three phases. The first and third flight-test phases were not directly related to propulsion modifications.[51] The second phase, 193 flights conducted between July 1991 and June 1994, explored the use of vectored thrust to enhance maneuverability and control at high angles of attack. Primary contractor McDonnell Douglas designed a multi-axis thrust-vectoring system for installation on the exhaust nozzles of the F/A-18's two GE F404 turbofan engines. The system consisted of a research flight control system that directed three paddle-like vanes, one set for each engine, made from the heat-resistant alloy Inconel, to deflect engine thrust. Dryden project manager Donald H. Gatlin described it charitably as "crude" and never intended for an operational aircraft.[52] Nevertheless, the system worked when the conventional aileron, rudder, and stabilator (a slab-surfaced combined horizontal stabilizer and elevator) aerodynamic controls were ineffective. The thrust-vectoring system increased the high-alpha capability of the F/A-18 by a third, up to 70 degrees. Additional modifications

Figure 5-8. The use of thrust vectoring in the HARV program was an important precedent for later programs. (NASA)

The Power for Flight

to the HARV included a sophisticated engine inlet pressure measurements system between the inlet entrance and the engine face. They measured pressure fluctuations of up to 250 Hertz at over 2,000 samples per second, which contributed to a broader understanding of what happened to engine airflow under extreme maneuvering.[53]

While the HARV thrust-vectoring nozzles permitted high-alpha investigations, the technology offered only a temporary means of evaluation for the program. NASA's Advanced Control Technology for Integrated Vehicles (ACTIVE) program specifically investigated thrust vectoring for application in future subsonic and supersonic commercial and military aircraft. Fighters benefited from enhanced maneuverability, while airliners like the much-anticipated 300-passenger Mach 2 High-Speed Civil Transport would allow drag and noise reduction through smaller control surfaces supplemented by thrust vectoring.[54] They also wanted to build upon previous thrust-vectoring programs while introducing the new element of safety.[55] NASA and the Air Force were the Government partners; McDonnell Douglas and Pratt & Whitney served as the industrial partners. Project manager Don Gatlin remarked that ACTIVE was "an example of government and industry cooperating to bring an important technology to maturity."[56]

NASA began the preparation of a two-seat F-15B transferred from the U.S. Air Force for that purpose in 1993 at Dryden.[57] ACTIVE relied upon an Inner Loop Thrust Vectoring (ILTV) system for control. It integrated the F-15's standard aerodynamic flight control surfaces, ailerons, stabilators, and rudders with thrust vectoring so that the pilot controlled the aircraft with both the control stick and the rudder pedals. Pratt & Whitney's Large Military Engines Division developed the axisymmetric system, which featured a pair of pitch and yaw (left and right) balanced beam nozzles for each of the two new F100 turbofan engines. They were capable of redirecting engine exhaust flow up to 20 degrees in any direction. The system was much lighter than previous thrust-vectoring designs, and Pratt & Whitney designed it for easy retrofitting to existing aircraft as well as direct installation in future aircraft. An advanced Improved Digital Electronic Engine Controller (IDEEC) system, strengthened duct cases able to withstand the vectored thrust, and improved engine mounts and rear fuselage construction completed the fighter's modification to the ACTIVE configuration. Ground testing of the nozzles began in November 1995 at the Air Force Flight Test Center's universal horizontal thrust stand.[58]

ACTIVE flight testing began in March 1996. The NASA goal was to fly up to 100 hours at speeds of up to Mach 1.85 and at angles of attack of up to 30 degrees.[59] A series of four flights between October 31 and November 1, 1996, witnessed the first time thrust vectoring was accomplished at speeds approaching Mach 2.[60] In regard to the flight program, which ran through

Figure 5-9. The Pratt & Whitney pitch-yaw balance beam nozzle system enabled vectoring horizontally (yaw) and vertically (pitch). The ACTIVE program achieved the first supersonic yaw-vectoring flight on April 24, 1996. (NASA)

1998, Jim Smolka, the ACTIVE project pilot, enthusiastically commented on the "exceptional handling" he and his fellow NASA research pilots experienced flying the F-15. For those early flights, the integrated system operated only at higher altitudes for the purposes of pilot familiarization. ACTIVE chief engineer Gerard Schkolnik remarked that one major accomplishment was the use of the pitch-vectoring control. The pilots trimmed the stabilators to steady the F-15 in flight and pitched the aircraft up and down with its exhaust. The next phase was to use the system throughout the entire flight from takeoff to landing.[61]

During the summer of 1997, the ACTIVE team incorporated Lewis's High Stability Engine Control (HISTEC) project into its flight program. A joint effort by Lewis, Pratt & Whitney, the Boeing Phantom Works (formerly McDonnell Douglas at St. Louis), and the Air Force's research laboratories at Wright-Patterson Air Force Base, HISTEC's goal was to improve engine operating stability to encourage the development of higher-performance military aircraft and more fuel-efficient commercial airliners. In combat, fighter pilots employed dogfighting maneuvers such as high angles of attack (up to 25 degrees), full-rudder sideslips, windup turns, and split-S descents, which created turbulent flow at the engine inlet.[62]

To avoid sudden in-flight compressor stalls and engine failures, the NASA-industry team created a computerized system called Distortion Tolerant Control, which sensed inlet airflow distortion at the front of the engine and made the necessary trim changes to accommodate changing distortion conditions in real time. The end result was a higher rate of engine stability in adverse airflow conditions.[63] The two engines facilitated the installation of the high-speed processor and control instrumentation and equipment on only one of the F100-PW-229 engines for increased safety. The flight program consisted of two phases. The first, flown during July and early August 1997, gathered the needed baseline date. The second took place during the remainder of August and used those data, stored in the Stability Management Control, in the F-15's electronic engine control, which inputted commands into the right engine to accommodate airflow distortion.[64] The primary benefit of Distortion Tolerant Control was its ability to set the stability margin requirement online and in real time. That allowed reduction of the built-in stall margin, thus maximizing propulsive performance. The result, as expressed by John DeLaat, NASA Lewis research engineer, would be "higher-performance military aircraft and more fuel-efficient commercial airliners."[65]

The successful completion of the HARV and ACTIVE flight research programs resulted in a better understanding of aerodynamics, the effectiveness of flight controls, and airflow phenomena at high angles of attack. Armed with that experience and that of the Enhanced Fighter Maneuverability X-31 and F-16 Multi-Axis Thrust Vectoring (MATV) programs that ended in 1995, the American military aircraft industry moved on to incorporate high-angle-of-attack technology into new aircraft.[66]

The Lockheed Martin F-22 Raptor, the U.S. Air Force's advanced air superiority fighter of the early 21st century, is a case in point. Its dual afterburning Pratt & Whitney F119-PW-100 turbofans generated approximately 35,000 pounds of maximum thrust per engine. The propulsion system incorporated pitch-axis thrust vectoring with a range of plus or minus 20 degrees, which made the fighter extremely agile at both supersonic and subsonic speeds. With thrust vectoring, a pilot could fly the F-22 through high-angle-of-attack maneuvers like the Herbst J-Turn, the Kulbit, and Pugachev's Cobra.[67] The Raptor's top speed is Mach 2.25, or 1,500 mph, and it is capable of supercruise, or extended supersonic flight, without the use of afterburners, thus consuming less fuel while racing through the sky at Mach 1.82, or 1,220 mph. The flat, two-dimensional shape of the nozzles also reduces infrared emissions from the engines and the chance of detection from heat-seeking missiles.

NASA played an important role in the development of the multidimensional inlet. All jet engines are circular in configuration; and, by extension, their exhaust nozzles follow that pattern for optimum efficiency. In the 1960s, new,

Propulsion Control Enters the Computer Era, 1976–1998

Figure 5-10. Shown are the F/A-18 HARV, the X-31, and the F-16 MATV, all thrust-vectored research aircraft. (NASA)

advanced, multirole twin-engine fighters like the F-111 and the F-15 suffered from drag problems related to the airflow at the back of the airplane where the angular boxy fuselage interacted with the round exhaust nozzles. To reduce that area of drag, the U.S. Air Force Flight Dynamics Laboratory worked closely with Bill Henderson of Langley to develop a series of two-dimensional nozzles beginning in the 1980s.[68]

Their work coincided with the Air Force's new requirement for an Advanced Tactical Fighter (ATF) that incorporated composite materials, lightweight alloys, advanced flight control systems, more powerful propulsion systems, and stealth technology. After a challenging design competition starting in 1986, the team consisting of Lockheed, Boeing, and General Dynamics won with its YF-22 in 1991.

The F-22 was the first aircraft to utilize the two-dimensional engine exhaust nozzle for both of its F119 engines. Lockheed became aware of the two-dimensional nozzle program and recognized its value to the overall mission of the fighter. Thrust vectoring provided more stability at high angles of attack during close-in maneuvering, or dogfighting, which accentuated the fighter's long-range standoff missile capability. The faceted shape of the exhaust nozzles also enhanced the stealth characteristics of the F-22 because they generated less radar return from the back of the aircraft.[69]

Endnotes

1. Link C. Jaw and Sanjay Garg, "Propulsion Control Technology Development in the United States: A Historical Perspective," NASA TM-2005-213978, October 2005, p. 2.
2. Ibid., p. 3.
3. Ibid., p. 4.
4. Ibid., p. 4.
5. Edward W. Otto and Burt L. Taylor III, "Dynamics of a Turbojet Engine Considered as a Quasi-Static System," NACA TR 1011, 1950; Jaw and Garg, "Propulsion Control Technology Development in the United States," p. 4.
6. James R. Ketchum and R.T. Craig, "Simulation of Linearized Dynamics of Gas-Turbine Engines," NACA TN-2826, November 1952, pp. 2, 11.
7. Sanjay Garg, "Aircraft Turbine Engine Control Research at NASA Glenn Research Center," NASA TM-2013-217821, April 2013, p. 9.
8. See James E. Tomayko, *Computers Take Flight: A History of NASA's Pioneering Digital Fly-By-Wire Project* (Washington, DC: NASA SP-4224, 2000).
9. Jaw and Garg, "Propulsion Control Technology Development in the United States," p. 6.
10. Frank W. Burcham, Jr., and Peter G. Batterton, "Flight Experience with a Digital Integrated Propulsion Control System on an F-111E Airplane" (AIAA Paper 76-653, presented at the 12th Joint AIAA/SAE Propulsion Conference, Palo Alto, CA, July 26–29, 1976), p. 1; Burcham et al., "Propulsion Flight Research at NASA Dryden from 1967 to 1997," pp. 2, 6, 21.
11. NASA obtained the F-15 from the U.S. Air Force in January 1976 and modified it into a one-of-a-kind aircraft that featured extensive instrumentation and an integrated DEFCS to carry out complex and sophisticated research projects, including the ADECS, PSC, and PCA programs. Besides propulsion control, researchers used the former fighter in advanced research projects involving aerodynamics, control integration, flight-test techniques, human factors, instrumentation

development, and performance. Dryden Flight Research Center, "F-15 Flight Research Facility," fact sheet, December 3, 2009, available at *http://www.nasa.gov/centers/dryden/news/FactSheets/FS-022-DFRC.html* (accessed February 19, 2012).
12. James F. Stewart, "Integrated Flight Propulsion Control Research Results Using the NASA F-15 HIDEC Flight Research Facility," NASA Technical Memorandum 4394, June 1992, p. 9.
13. Mark Bushman and Steven G. Nobbs, "F-15 Propulsion System: PW1128 Engine and DEEC," NASA Dryden Flight Research Center Document N95-33012, 1995, pp. 37–38.
14. Dryden Flight Research Center, "F-15 Flight Research Facility."
15. Banke, "Advancing Propulsive Technology," p. 756.
16. Dryden Flight Research Center, "F-15 Flight Research Facility."
17. Ibid.
18. Jaw and Garg, "Propulsion Control Technology Development in the United States," pp. 17–18.
19. Stewart, "Integrated Flight Propulsion Control Research," pp. 3, 7; Dryden Flight Research Center, "F-15 Flight Research Facility."
20. Dryden Flight Research Center, "F-15 Flight Research Facility"; Pratt & Whitney, "F100 Engine," n.d. [2013], available at *http://www.pw.utc.com/F100_Engine* (accessed August 22, 2013).
21. Frank W. Burcham, Jr., John J. Burken, Trindel A. Maine, and C. Gordon Fullerton, "Development and Flight Test of an Emergency Flight Control System Using Only Engine Thrust on an MD-11 Transport Airplane," NASA Technical Paper 97-203217, October 1997, pp. 3–5; Frank W. Burcham, Jr., C. Gordon Fullerton, and Trindel A. Maine, "Manual Manipulation of Engine Throttles for Emergency Flight Control," NASA TM-2004-212045, January 2004, pp. 63–66; Tom Tucker, *Touchdown: The Development of Propulsion Controlled Aircraft at NASA Dryden*, Monographs in Aerospace History, no. 16 (Washington, DC: NASA, 1999), p. 12.
22. NTSB, *Aircraft Accident Report: United Airlines Flight 232, McDonnell Douglas DC-10-10, Sioux Gateway Airport, Sioux City, Iowa, July 19, 1989*, NTSB/AAR-90/06 (Washington, DC: NTSB, 1990), *http://www.airdisaster.com/reports/ntsb/AAR90-06.pdf* (accessed January 30, 2014).
23. Ibid., pp. 1–5, 72–73, 102. See also David Young and George Papajohn, "3 Pilots Struggled to Keep Jet Aloft," *Chicago Tribune* (July 22, 1989): 4.
24. NTSB, *Aircraft Accident Report: United Airlines Flight 232*, p. 76.
25. Ibid., p. 102.

26. Frank W. Burcham, Jr., Ronald J. Ray, Timothy R. Conners, and Kevin R. Walsh, "Propulsion Flight Research at NASA Dryden from 1967 to 1997," NASA TP-1998-206554, July 1998, p. 1.
27. Tucker, *Touchdown*, p. 1.
28. Ibid., pp. 1, 4.
29. Frank W. Burcham, Jr., and C. Gordon Fullerton, "Controlling Crippled Aircraft—With Throttles," NASA TM-104238, September 1991, pp. 1, 3; Burcham et al., "Manual Manipulation of Engine Throttles for Emergency Flight Control," p. 11.
30. Tucker, *Touchdown*, pp. 6–7.
31. Ibid., pp. 8, 10.
32. Ibid., pp. 9–10.
33. Ibid., pp. 10–11, 12.
34. Frank W. Burcham, Jr., Trindel Maine, and Thomas Wolf, "Flight Testing and Simulation of an F-15 Airplane Using Throttles for Flight Control," NASA Technical Memorandum 104255, August 1992, pp. 1–2; Dryden Flight Research Center, "F-15 Flight Research Facility"; Tucker, *Touchdown*, p. 14.
35. Burcham, quoted in "NASA F-15 Makes First Engine-Controlled Touchdown," NASA News Release 93-75, April 22, 1993, NASA HRC, file 011664.
36. Tucker, *Touchdown*, pp. 15, 16, 22. For detailed excerpts from the PCA F-15 guest pilot program, see appendix D, Tucker, *Touchdown*, pp. 40–45.
37. Ibid., pp. 23–24.
38. Ibid., p. 24.
39. Frank W. Burcham, Jr., John J. Burken, Trindel A. Maine, and C. Gordon Fullerton, "Development and Flight Test of an Emergency Flight Control System Using Only Engine Thrust on an MD-11 Transport Airplane," NASA TP-97-206217, October 1997, p. 34; Tucker, *Touchdown*, pp. 25–26.
40. Burcham et al., "Development and Flight Test of an Emergency Flight Control System Using Only Engine Thrust on an MD-11 Transport Airplane," pp. 35, 75; Tucker, *Touchdown*, pp. 27, 47.
41. Tucker, *Touchdown*, pp. 28–30.
42. Ibid., p. 30.
43. Tucker, *Touchdown*, p. 31; John Bull, Robert Mah, Gloria Davis, Joe Conley, Gordon Hardy, Jim Gibson, Matthew Blake, Don Bryant, and Diane Williams, "Piloted Simulation Tests of Propulsion Control as a Backup to Loss of Primary Flight Controls for a Mid-Size Jet Aircraft," NASA TM-110374, December 1995, p. 1.

44. Frank W. Burcham, Jr., Trindel A. Maine, John J. Burken, and John Bull, "Using Engine Thrust for Emergency Flight Control: MD-11 and B-747 Results," NASA TM-1998-206552, May 1998, pp. 1, 8; Tucker, *Touchdown*, pp. 30–31.
45. Tucker, *Touchdown*, p. 31.
46. Ibid., p. 31.
47. Burcham et al., "Manual Manipulation of Engine Throttles for Emergency Flight Control," pp. 28–39.
48. Jonathan S. Litt, Dean K. Frederick, and Ten-Huei Guo, "The Case for Intelligent Propulsion Control for Fast Engine Response," NASA TM-2009-215668/AIAA 2009-1876, p. 15.
49. Jim Matthews, "Thrust Vectoring," *Air & Space Smithsonian* (July 2008), available at *http://www.airspacemag.com/flight-today/Thrust_Vectoring.html* (accessed September 4, 2012).
50. Albion H. Bowers, Joseph W. Pahle, R. Joseph Wilson, Bradley C. Flick, and Richard L. Rood, "An Overview of the NASA F/A-18 High Alpha Research Vehicle," NASA TM-4772, 1996, pp. 2–4; Albion Bowers and Joseph W. Pahle, "Thrust Vectoring on the NASA F/A-18 High Alpha Research Vehicle," NASA TM-4771, 1996, pp. 3–4.
51. The first phase generated experience in the aerodynamic measurement of high angles of attack and provided the baseline evaluation methodology for the F/A-18 from April 1987 through 1989. The third investigated the concept of using deployable winglike surface strakes in the nose for increased yaw control from March 1995 to September 1996. Dryden Flight Research Center, "F/A-18 High Angle-of-Attack (Alpha) Research Vehicle," fact sheet, December 3, 2009, available at *http://www.nasa.gov/centers/dryden/news/FactSheets/FS-002-DFRC.html#.UiNUsjaTiHM* (accessed September 1, 2012).
52. Breck W. Henderson, "Dryden Completes First Flights of F/A-18 HARV with Thrust Vectoring," *Aviation Week & Space Technology* (July 29, 1991): 25.
53. Bowers et al., "An Overview of the NASA F/A-18 High Alpha Research Vehicle," pp. 24–28; Dryden Flight Research Center, "F/A-18 High Angle-of-Attack (Alpha) Research Vehicle."
54. Dryden Flight Research Center, "F-15 ACTIVE Nozzles," November 13, 1995, NASA HRC, file 011664.
55. "F-15 ACTIVE Achieves First-Ever Mach 2 Thrust-Vectoring," NASA Dryden News Release 96-62, November 8, 1996, available at *http://www.nasa.gov/centers/dryden/news/NewsReleases/1996/96-62_pf.html* (accessed September 4, 2012).

56. "NASA Tests New Nozzle To Improve Performance," NASA News Release 96-59, March 27, 1996, NASA HRC, file 011664.
57. "NASA F-15 Being Readied for Advanced Maneuvering Flight," NASA News Release 93-115, June 16, 1993, NASA HRC, file 011664.
58. James W. Smolka, Laurence A. Walker, Gregory H. Johnson, Gerard S. Schkolnik, Curtis W. Berger, Timothy R. Conners, John S. Orme, Karla S. Shy, and C. Bruce Wood, "F-15 ACTIVE Flight Research Program," in *1996 Report to the Aerospace Profession: Fortieth Symposium Proceedings of the Society of Experimental Test Pilots, September 1996*, available at *https://www.nasa.gov/centers/dryden/pdf/89247main_setp_d6.pdf* (accessed September 4, 2012); "NASA F-15 Being Readied for Advanced Maneuvering Flight," NASA News Release 93-115, June 16, 1993; Dryden Flight Research Center, "F-15 ACTIVE Nozzles," November 13, 1995; "Flight Research Reveals Outstanding Flying Qualities Result from Integrating Thrust Vectoring with Flight Controls," NASA Dryden Flight Research Center News Release, January 4, 1999, NASA HRC, file 011664.
59. "NASA Tests New Nozzle To Improve Performance," NASA News Release 96-59, March 27, 1996, NASA HRC, file 011664.
60. "F-15 ACTIVE Achieves First-Ever Mach 2 Thrust-Vectoring."
61. "Flight Research Reveals Outstanding Flying Qualities Result from Integrating Thrust Vectoring with Flight Controls," NASA Dryden Flight Research Center News Release, January 4, 1999, NASA HRC, file 011664.
62. "NASA Researching Engine Airflow Controls To Improve Performance, Fuel Efficiency," NASA News Release 97-183, August 27, 1997, NASA HRC, file 011664; Burcham et al., "Propulsion Flight Research at Dryden," pp. 17–18.
63. "ACTIVE F-15," *X-Press* (October 17, 1997): 4, NASA HRC, file 011664.
64. "NASA Researching Engine Airflow Controls To Improve Performance, Fuel Efficiency."
65. Lori Rachul, "HISTEC To Boost Performance, Fuel Efficiency," *Lewis News* (November 1997): 5, NASA HRC, file 011664.
66. Dryden Flight Research Center, "F/A-18 High Angle-of-Attack (Alpha) Research Vehicle."
67. The Herbst J-Turn is a quick reversal of direction using a combination of high-angle-of-attack and rolling maneuvers. The Kulbit is an extremely small-diameter loop. Pugachev's Cobra involves a drastic vertical rising of the nose of the aircraft that tips it slightly backward before quickly going back down to resume normal forward flight.

68. G. Keith Richey, interview by Squire Brown, August 31, 2006, interview no. 2, transcript, Cold War Aerospace Technology Project, Wright State University Libraries Special Collections and Archives, p. 9.
69. Ibid., p. 10.

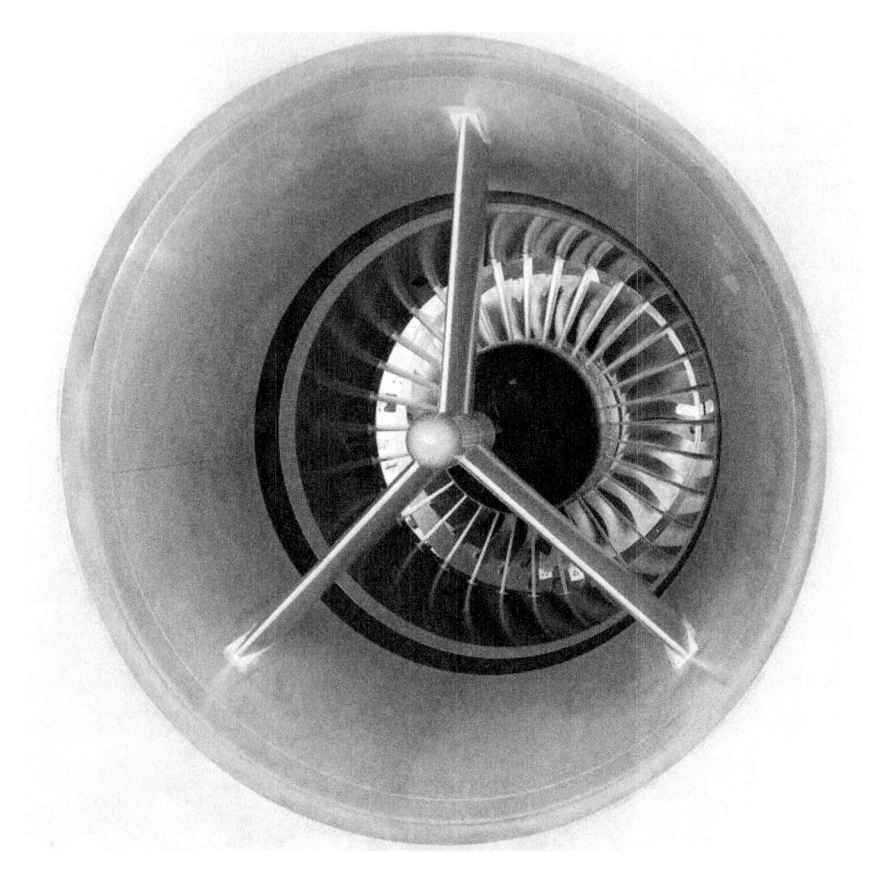

Shown above is an inlet view of a proof-of-concept two-stage compressor evaluated during the Ultra-Efficient Engine Technology Program at Glenn Research Center in 2005. (NASA)

CHAPTER 6
Transiting to a New Century, 1990–2008

Through the decade of the 1990s and into the first of the 21st century, NASA's propulsion specialists continued their work on several important projects that had their origins in the 1960s. The refining of jet engines for increased efficiency, emissions and noise reduction, investigations into high-speed flight, and participation in large-scale joint propulsion projects cemented NASA's role as not a competitor, but a collaborator, in both long- and short-term projects. Over those years, NASA pursued a diverse range of propulsion initiatives and projects.

Materials Research for Improved Propulsion Efficiency and Safety

While NASA researchers worked on aircraft-centered programs from the 1960s through the 1980s, their colleagues produced results from more basic research. One of the areas that had long-term applications in aircraft propulsion was advanced materials. Innovations in the use of superalloys, polymer-matrix composites, thermal barrier coatings, structural ceramics, and ceramic matrix composites facilitated the refinement of gas turbine engine components for long life, reliability, and higher performance. The joint aircraft engine development programs, as well as the individual manufacturers, increasingly applied this technology through the 1980s and 1990s.

New Advances in Materials

The turbine disks and blades undergo the harshest of conditions within a gas turbine engine, with temperatures ranging from 1,200 to 2,100 degrees Fahrenheit. NASA worked toward the introduction of Oxide Dispersion Strengthened (ODS) superalloys in the 1960s, especially in areas of formulating the required thermomechanical processes and failure prediction to increase durability, strength, and temperature resistance. In a collaborative effort with the International Nickel Company and several universities, NASA facilitated the introduction of two popular alloys in sheet and bar form: iron-based MA956 and nickel-based MA754. They were contributions to the problem of sigma phase instability and the creation of alloys containing tantalum for turbine blade applications.[1] NASA's work to strengthen alloys by adding refractory metals became the industry standard by the late 20th century and could be found in the majority of all new engines being produced. Pratt & Whitney and GE went on to incorporate more advanced single-crystal orientation and gamma prime "rafting" behaviors in their latest-generation turbine blade alloys.[2]

New materials also facilitated lightweight solutions to aircraft engines. To reduce fuel consumption, weight, and emissions while increasing passenger and payload capability, NASA in the mid-1970s introduced a new family of high-temperature polymers, called Polymerization of Monomer Reactants (PMR).[3] Manufacturers embraced the new materials for both military and commercial applications. The principal polyimide, PMR-15, offered 10,000 hours of use at temperatures reaching 550 degrees Fahrenheit and quickly became the state-of-the-art material for engine bypass ducts, nozzle flaps, bushings, and bearings. GE used them for the F404 outer bypass duct and the GE90 center vent tube, while Pratt & Whitney incorporated PMRs into the F-100-229 exit flaps. Later PMR formulations could withstand higher temperatures—up to 650 degrees Fahrenheit—which opened up the range of applications to include engine aft fairings and a compressor case for a U.S. Air Force–U.S. Navy Joint Technology Demonstrator Engine (JTDE) program.[4]

Other materials advances NASA researchers introduced were thermal barrier coatings and structural ceramics. In 1976, Lewis researchers Curt H. Liebert and Francis S. Stepka discovered that the application of an insulating ceramic layer over metal components in the "hot" section of an engine, such as turbine blades, reduced temperature and the amount of coolant flow required, which permitted the use of cheaper and simpler materials.[5] Their work extended component life and enabled a broader understanding of metallic coatings.[6] The success with ceramic coatings led to new work in the late 1970s on monolithic ceramics, such as reaction-bonded silicon nitride (RBSN), that were capable of resisting 3,000 degrees Fahrenheit. That work led to NASA's collaborative work with DOE in the automotive-oriented Advanced Gas Turbine (AGT) and Advanced Turbine Technology Applications (ATTAP) programs in the 1980s. The Agency also investigated ceramic matrix composites in the 1980s. The work led to the use of the material as the combustor liner design for two Government programs to reduce cooling flows and increase engine efficiency: the Enabling Propulsion Materials (EPM) effort within the High-Speed Research (HSR) program and the Air Force's Advanced Turbine Engine Gas Generator (ATEGG) engine as part of the Integrated High Performance Turbine Engine Technology (IHPTET) program. Both the HSR and IHPTET programs are discussed subsequently.[7]

HITEMP: Advancing Materials and Structures

To further influence the development of UHB turbofans for the 21st century, NASA initiated the Advanced High Temperature Engine Materials Technology Program (HITEMP) in 1988 at Lewis. HITEMP endeavored to create advanced materials and structures, along with the necessary analytical and evaluation frameworks to increase fuel economy, reliability, and service life while reducing operating costs. The program placed the primary focus on developing advanced high-temperature composite materials for fan, compressor, and turbine rotor blades; stator vanes, disks, and shafts; thrust bearings, gearbox bearings, and linings; combustor cases and linings; nacelles; and thrust-reversers. Those new materials included polymer-matrix, metal-matrix, intermetallic-matrix, and ceramic-matrix composites.[8]

HITEMP placed a specific focus on materials and structures, but it was also an integrated program that reflected NASA's fundamental research goals and supported the work of component development programs. It accomplished its work through direct NASA research as well as grants and contracts to academia and industry. NASA dispersed the information from the project through annual conferences and the annual publications *HITEMP Review* and *Research & Technology*, both published by Lewis Research Center. HITEMP research investigations involved coordination with other NASA and Government programs, primarily EPM, IHPTET, the Aerospace Industry Technology Program, and the Advanced Subsonic Technology (AST—to be discussed later in this chapter) program. Keeping communication open and sharing research with other programs ensured that HITEMP innovations could be applied to those efforts as well.[9]

The research generated during HITEMP made two valuable contributions to industry. The first was a minimally intrusive, high-temperature, thin-film strain gauge able to measure both dynamic and static stress and was adopted by GE, AlliedSignal, and the Ford Motor Company. The other was a new ultrasonic imaging method utilizing a single transducer. The R&D 100 Awards, long considered an indicator of excellence in technology innovation, recognized HITEMP for those two developments. Through release agreements, the aviation and software industries adopted HITEMP developmental codes, which included the ceramic matrix composite analyzer (CEMCAN) computer code. Lewis researchers used HITEMP analytical models to make recommendations on tooling and processing that enabled Textron Specialty Materials to produce defect-free titanium-matrix composite rings used in reinforcing engine components. A cooperative program between AlliedSignal, Lincoln Composites, and Lewis led to a collaborative investigation into the feasibility of using the Lewis-created Vehicle Charging And Potential (V-CAP) polyimide resin matrix for high-temperature jet engine applications.[10] The contributions of HITEMP were many, but they were beneath the surface as industry adopted them and advanced the state of the art as they refined materials, design, and the processes needed to manufacture the new technology.

Two instances of research and development originated during HITEMP, and their direct application to the new UHB turbofans of the 21st century concern the GE GEnx engine. HITEMP researchers aimed to reduce the weight of low-pressure turbine blades. They identified titanium aluminide (TiAl) as an ideal solution due to the material's low density and high-temperature properties that offered to reduce weight by 40 percent, but it did exhibit limited ductility. Lewis researchers Bradley Lerch, Susan Draper, J. Michael Pereira, Michael Nathal, and Curt Austin designed a laboratory impact test that simulated potential blade damage resulting from the ingestion of foreign objects. They revealed that TiAl alloys could withstand considerable impact damage without catastrophic failure.[11] As a result, the designers of the GEnx utilized TiAl in the low-pressure turbine stages, specifically in the thick leading edges. GE materials and process engineering general manager Robert Schafrik remarked that after "years of research," the use of TiAl alloys was a "key breakthrough" in minimizing the weight of the GEnx.[12]

Another HITEMP contribution to the GEnx involved NASA researchers in Cleveland and researchers with GE, who maintained a longstanding collaboration centered on developing a use for nickel aluminide (NiAl) alloys as structural materials. A team consisting of Ronald Noebe, Robert Miller, Anita Garg, and Ivan Locci of the University of Toledo redirected the focus toward using NiAl as a bond coat for high-pressure turbine blades. A bond coat served to promote adhesion between the blade structure itself and the thermal barrier coatings applied to increase high-temperature performance. HITEMP sponsored tests of pure and customized NiAl bond-coat alloys. Those evaluations led to various patents, including one jointly held by NASA and GE, and their application to the GEnx in the mid-2000s.[13]

EPM: Exploiting Materials Research for Lower Weight and Safety

The catastrophic failure of the fan on the center engine of the United Airlines Flight 232 DC-10 airliner and the subsequent loss of all hydraulic power in 1989 was a chilling reminder of what happened when fan blades were not contained. Engine makers used metal alloys in their fan casings to deflect broken blades and contain them within the engine nacelle. Unfortunately, those casings were also very heavy due to the required high margins of safety, which translated into poor fuel efficiency, shorter flights, and decreased cargo capacity. They also required expensive and time-consuming physical testing. In order to advance the next generation of commercial aircraft, new solutions needed to provide a balance between safety and decreased weight to improve efficiency.[14]

NASA's EPM program, begun in 1994, utilized ballistics research to investigate the use of composite casings for fan blade containment in 1994. Researchers first used flat composite material panels to assess their performance compared to that of aluminum in simulated blade-out events. They fired projectiles at the panels and analyzed the level of penetration for each material. The results revealed that composites experienced lower stress levels and were a promising replacement for heavy aluminum fan casings. Those initial experiments encouraged NASA researchers to investigate more precise and cheaper methods of simulating the use of composites in fan blade containment.[15]

The turbofan is the standard engine for the latest commercial airliners such as the 777. The crucial element of those systems is the fan, which contributes to the high efficiency, high thrust at low speeds, low fuel consumption, and reasonable noise levels. NASA researchers realized that the fan enclosure contributed extra weight and decreased fuel efficiency. The fan enclosure protected the engine and airframe in case a malfunction or unexpected obstruction led to one of the fan blades breaking off at very high speed, which is called blade-out. Engine manufacturers used heavy-metal alloys to build the fan casing so that it would be robust enough to absorb the blades. Additional stress to the fan casing caused by rotor imbalance occurred at engine shutdown. NASA recognized that decreasing that extra weight while retaining adequate levels of safety would contribute to a new generation of advanced commercial aircraft.[16]

NASA recognized that the large turbofan engines used on commercial airliners constituted a significant portion of the overall weight of the airplane. A single engine weighed approximately 10,000 to 15,000 pounds; thus, a four-engine aircraft could have 60,000 pounds in engine weight alone. By targeting the fan casing, used to contain failed blades and the largest component in these engines, NASA believed that further weight reductions could be made.

There were attempts to strike that balance. One concept, called the "hard wall" approach, minimized weight through the use of lighter metals, such as high-grade aluminum alloys, in the place of heavier and stronger materials like steel. The thick aluminum walls deflected stray fan blades and keep them within the engine. Another approach, first used on the GE CF34 turbofan engine during the early 1990s, relied upon a "soft wall" design that featured a thick, high-strength fabric wrapped around a thinner aluminum fan casing. The fabric absorbed the damaged blades until they could be removed by maintenance personnel. Commercial engine manufacturers incorporated both types of reinforced damage-tolerant fan casings through the 1990s and 2000s.[17]

Researchers at Glenn Research Center (formerly Lewis) investigated carbon fiber/polymer matrix composite materials, fiber architectures, and design concepts for use in the manufacture of lighter-weight fan casings starting in 1999. Their work was part of the Ultra Safe Propulsion Project, which was part of

The Power for Flight

NASA's larger Propulsion and Power Base Research and Technology Program. Advanced composites offered high strength, increased safety, low weight, and cheaper operational costs, and they were readily available to the aerospace industry. Glenn staff conducted comprehensive research along with extensive ballistic impact testing for concept validation that identified a promising approach for developing the all-composite fan case. Early on, NASA worked to overcome the challenges of structural strength and safety to bring the commercial aviation industry one step closer to improved fuel efficiency, increased payload, and greater aircraft range.[18]

Glenn and its industry and academic partners in the Jet Engine Containment Concepts and Blade-Out Simulation Team announced the results of 4 years of research in 2003. They introduced TEEK, a low-density, lightweight, flame-resistant polyimide foam that provided high-performance structural support while serving as an excellent thermal and acoustic insulation material. The team also developed the first advanced composite jet engine fan blade containment system concept and explored new ways of manufacturing composite fan casings. Finally, team members also created new tools, primarily simulation software that could be used to study and predict the dynamics of blade-out occurrences without the use of real engines.[19]

NASA's Aviation Safety Program pushed toward the next step, the actual development of composite fan casings for use on turbofan engines, with funding for Glenn scientists to investigate the possibility. As part of the process, Glenn issued a Small Business Innovation Research (SBIR) grant to A&P Technology, Inc., of Cincinnati, OH, to ensure industry involvement. A&P Technology was the ideal choice since it was a leader in the manufacture of braided composites and already had experience working with engine makers Williams International and Honeywell International. The grant from Glenn funded A&P Technology's development of a new generation of carbon fiber–reinforced polymer (CFRP) suitable for use in advanced lightweight fan casings. The new material provided increased strength and durability due to the use of T-700 12K carbon fiber and EPON 862 bisphenol F–based epoxy resin. Additionally, the triaxial braided fiber construction greatly reinforced the material's structural integrity and increased its resistance to the formation of cracks.[20]

A&P Technology and Glenn researchers faced the challenge of automating the production process to make the technology efficient, reliable, and affordable. Their manufacturing of composite structures involved two main processes. Workers first laid out the pre-formed dry fibers, a difficult process when the structure was as large and complex as a cylindrical fan casing. The next step involved the impregnation of the fibers with resin using a transfer molding process. NASA desired a system that was adaptable to different engine designs. A&P Technology created a robust system that braided the fiber directly around a capstan shaped to the profile of the particular containment case without any warping. The work facilitated the invention in 1997 of the A&P Technology Megabraider, the largest braiding machine in the world, with 800 individual carriers, in 1997.[21]

Glenn collaborated with researchers at the University of Akron to develop the software capable of replacing physical ballistics testing of fan casings. The work required an in-depth analysis of multiple blade-loss scenarios and the individual roles of angular acceleration, mass, orientation, and speed. With a better understanding of those dynamics, the research team used LS-DYNA, a unique analytical code developed by Livermore Software Technology Corporation, to program computers to simulate blade-out scenarios with different fan-case materials. The new software generated new insights while maintaining significantly lower costs than those of previous impact testing.[22]

Despite the promised low costs of computer simulation, the state of composites research reached a point where material testing fell behind. As a result, the lack of adequate material property data and validated material models limited the overall success of computer simulations. To push the research forward, Glenn worked with researchers from the University of Akron and ATK Space Systems to initiate impact testing of new tribraided composites. They started with small, flat panels and worked their way to full-scale fan-case models. The physical testing confirmed that A&P Technology's composite materials were more than capable of resisting fan blade impacts. More importantly, the impact tests revealed that the new composite structures were stronger than traditional metal alloy fan casings.[23]

After an errant fan blade struck a fan casing, there was an additional, potentially destructive scenario for the engine. The fan casing endured secondary loads during the spool-down stage of the engine after the loss of the blade. Specifically, impact debris and the out-of-balance fan assembly could lead to the creation of cracks in the casing, which became a serious safety issue. Due to the tribraided fiber construction, composites exhibited a strong resistance to crack formation. That discovery, along with the successful impact testing of sample panels, encouraged the move toward full-scale testing.[24]

The certification of new composite fan casings began with full-scale engine blade-out tests. They confirmed that the new case safely contained the stray blade and retained its structural stability during the large dynamic loads generated during the engine's spool-down stages as the rotation of the fan slowed down. The successful completion of those tests ensured that manufacturers would use composite fan casings in their new and lighter turbofan engines.[25]

A constant in the modern aviation industry is the quest to improve fuel efficiency to save money while continuing to provide the same service. Composite fan casings offered to reduce engine weight by up to 40 percent, which directly

The Power for Flight

translated into longer flight distances, greater cargo capacity, improved fuel burn, and increased safety for new commercial aircraft. A decade of NASA investment and industry collaboration made that possible.[26]

GE recognized the benefits of composite fan casings and selected them for the revolutionary GEnx high-bypass turbofan engine, the first whose fan case and fan blades were made completely of composite materials. GKN Aerospace developed and manufactured the front fan containment case, which allowed for a weight reduction of up to 800 pounds for a two-engine aircraft. Final GEnx testing occurred in 2006, with certification in 2007. Boeing and Airbus used the new engines on their highly anticipated 787 Dreamliner and A350 aircraft.[27]

The search for new and better materials to reduce the weight of aircraft has been a constant since 1903. Since their introduction in the 1960s, turbofan engines represented a significant portion of the weight of a commercial airliner. The increasing usability of advanced composite materials and the development of effective ballistics testing methods in the early 2000s contributed to new developments in composites manufacturing for large aerospace structures. NASA has recognized the benefit of replacing existing metal fan casings with

Figure 6-1. The advanced fan blades and composite fan casing of the GE GEnx-2B engine reflect NASA's pioneering work. (General Electric)

safer and stronger composite structures to reduce the weight of commercial aircraft engines. The improved safety, reduced fuel burn, increased aircraft range, and expanded cargo capabilities facilitated by composite fan-casing technology benefited both airlines and passengers.[28]

NASA's award of an SBIR to A&P Technology indicated its commitment to work with industry to develop the materials and manufacturing techniques to make composite fan-casing technology a reality for commercial aviation. The NASA-industry-academic team responsible for the containment concepts and blade-out simulations received the NASA Turning Goals into Reality Award for its dedicated research in July 2004. With incorporation into the GEnx being the first step, composite fan-casing technology stood poised to benefit aviation for decades to come.[29]

Shape Memory Alloy Research

NASA and industry achieved their past improvements in gas turbine engine efficiency—performance, noise, and emissions—through a focus on combinations of new component designs and new and lighter materials capable of withstanding higher temperatures. NASA researchers believed that the key to increased performance in the future involved the removal of various static and heavy structures such as electric, hydraulic, or pneumatic actuators from the airplane. In their place, adaptive, or reconfigurable, components utilizing advanced shape memory alloys (SMAs) would permit lighter and more dynamic inlets, nozzles, flaps, variable-geometry chevrons, and blades. An SMA facilitated two configurations within a single component. At ambient temperature, the component was one shape; with the application of heat, it changed into another. Recognizing the potential for what shape-shifting components could do, NASA initiated a 5-year development effort on SMAs in 2003.[30]

High-temperature SMAs also have proven crucial to another new innovation, active flow control. Advanced sensors detect the onset of incipient stall in the compressor, which would then make small adjustments to the flow geometry that would achieve both improved efficiency and tolerance against stall conditions. In 2008, NASA announced that it had successfully demonstrated a design that utilized a high-temperature SMA wire to actuate a control rod to change airflow.[31]

Advancing Gas Turbine Technology

Integrated High-Performance Turbine Engine Technology

DOD established the Integrated High Performance Turbine Engine Technology (IHPTET) program in November 1987 to stimulate the development of 21st-century high-performance military turbine engines. The program grew out of an earlier military-industry project sponsored by the Air Force's Aero Propulsion Laboratory at Wright-Patterson Air Force Base near Dayton, OH. DOD was so impressed that it expanded the program to include the Army, the Navy, the Defense Advanced Research Projects Agency (DARPA), and NASA. The program possessed a broad charter to study all engine components and work toward technological maturity through testing and demonstration, with an overall goal to double the power of military jet engines by 2005. Moreover, the IHPTET engines were to be robust and affordable and to exhibit high performance under all conditions.[32]

IHPTET was a coordinated, three-phase Government and industry initiative that served as the framework for practically all Government- and industry-sponsored research and development on military turbine engines. Specifically, a joint DOD-NASA steering committee coordinated eight separate plans for the U.S. Government and individual industry participants, which included Pratt & Whitney, General Electric, Allison, Williams International, Teledyne Ryan Aeronautical, AlliedSignal Aerospace, and Textron Lycoming. There were component technology panels that addressed the following systems: compression, combustion, turbine, exhaust, control, mechanical, and demonstrator engines.[33]

The increasing costs and diminishing market share for aircraft engines in the late 20th century resulted in a push for advanced universal, or dual-use, technology—primarily materials, computational fluid dynamics (CFD) design codes, engine controls and logic, turbine cooling concepts, bearings, and structures.[34] Still, there were major differences between military and commercial engines and in what direction innovations took them. Low-bypass-ratio military engines required reduced numbers of stages for lighter weight and increased reliability, thrust vectoring/reversing capability, stealth and low observable signatures, and expendable engines. Commercial high-bypass turbofans needed to be quiet and clean, and there was considerable interest in novel regenerative cycles and universal fuels. Addressing those solutions required different pathways in combustor design, operating temperatures, emissions, operational durability, and exhaust design.[35]

Within IHPTET, NASA performed three specific roles that were fundamental to development regardless of their final application. First, the Agency provided its unique testing and evaluation facilities as its researchers worked to innovate essential high-temperature, high-strength materials and devised new, advanced design analysis capabilities. NASA worked to gain a broader understanding of durability, modeling, high-temperature materials and structures, increased fuel economy, and the reduction of emissions and noise. NASA's Small Engine Component Test Facility (SECTF), located in Glenn's Engine Research Building (ERB), which opened in 1992, was the main test facility for the IHPTET. The SECTF consisted of two individual cells dedicated to compressors and turbines. Both replicated the operating conditions of an actual engine over a broad range of speeds and temperatures.

Second, NASA utilized its long history in the development of advanced materials, like polymer matrix composites, superalloys, structural ceramics, ceramic matrix composites, and thermal barrier coatings, that led to lighter, stronger, and better engine components capable of operating in high temperatures. Those NASA-innovated materials easily crossed over into the new technology going into military jet engines.[36]

Finally, Glenn's expertise in CFD code design led to the Center's assuming the leadership of IHPTET's CFD technology development panel. The use of computer codes led to NASA's taking the lead in innovating new and effective tools for the component design, evaluation, and operational analysis of high-performance jet engines.[37] Researchers used the NASA Average-Passage turbomachinery flow analysis code called "APNASA" to optimize the performance of a composite, forward-swept, shrouded fan. Created at Glenn by John Adamczyk in 1985, APNASA enabled the prediction of the interaction between stationary and rotating parts of multistage components such as the fan, compressor, and turbine.[38] They also used the 3D Combustor Simulation Code to model liquid spray droplet fuel injection for improved combustor designs.[39]

Glenn's hallmark Numerical Propulsion System Simulation (NPSS) allowed the complete modeling of a running engine throughout a simulated flight. Initiated in 1994, NPSS served as a "virtual wind tunnel" that enabled engineers to explore multiple design options simultaneously in terms of performance, affordability, stability, operational life, and certification requirements. Those options centered on fluid mechanics, heat transfer, combustion, structural mechanics, materials, controls, manufacturing, and overall economics. The core of the system contained three main elements: engineering application models and two kinds of system software, for the simulation and high-performance computing environments respectively. With NPSS, designers did not have to resort to costly and time-consuming physical construction and tests of jet engines. The NASA/Industry Cooperative Effort agreement partnered Glenn with the Air Force's aerospace engineering organizations and universities with GE, Pratt & Whitney, Boeing, Honeywell, Rolls-Royce, Williams International, and Teledyne Continental to develop NPSS for both military and commercial engines.[40]

177

The Power for Flight

NPSS offered to cut the development time and cost of a high-performance jet engine in half, from 10 years and $2 billion to 5 years and $1 billion. It became an attractive tool for the virtual design of other technologies, including airframes, rocket engines, fuel cells, and ground-based power systems. There was also the possibility that the software could support nuclear power, water treatment, biomedicine, chemical processing, and marine propulsion. The result was significant recognition for the program by 2001. The NPSS team received the NASA Office of Aerospace Technology Turning Goals into Reality Award and the Agency's overall Software of the Year Award. The program itself was named a Top 16 Government Software Project and a finalist for the *Journal of Defense Software Engineering*'s Top 5 Projects.[41]

The three phases of IHPTET worked to maximize technology transition to both military and commercial users. Phase I research and development demonstrated a 30-percent increase in propulsion capability. Engines that benefited from that work included improved versions of Pratt & Whitney's F100 and GE's F101 for the F-15 and F-16, GE's F414 for the F/A-18E/F, and Pratt & Whitney's F119 engine for the F-22. Phase II targeted a 60-percent increase in propulsion capacity that facilitated the introduction of the

Figure 6-2. This image shows the Pratt & Whitney F119 Engine for the F-22. (Pratt & Whitney)

supersonic Lockheed Martin F-35 Lightning II Short Takeoff and Vertical Landing (STOVL) aircraft. Phase III looked toward the future with a goal of 100 percent maximized propulsion capability for larger and faster air superiority and STOVL aircraft, a large helicopter, and an intercontinental air-launched cruise missile (ALCM). IHPTET concluded in 2005 as a successful program, although it did not fully achieve all of its goals.[42]

The Versatile Affordable Advanced Turbine Engine (VAATE) program succeeded IHPTET and retained the organization structure. The Air Force continued its direction of the program as it worked with its partners to define goals and offered competitive bids to industry for the Army's Advanced Affordable Turbine Engine (AATE) for helicopters and the Air Force's Adaptive Versatile Engine Technology (ADVENT) program. The partnership allowed it to maximize evaluation and funding for specific tasks while relying on NASA's fundamental research expertise to provide a sound technical foundation for the innovative work.[43]

The Ultra-Efficient Engine Technology Program, 1999–2003

Glenn continued to work toward its goal of developing and transferring enabling technologies to industry through the Ultra-Efficient Engine Technology (UEET) Program beginning in October 1999. Under the management of Robert Shaw at Glenn, the 6-year, nearly $300 million program included participation from Ames, Langley, and Goddard Space Flight Center; engine manufacturers GE, Pratt & Whitney, Honeywell, Allison/Rolls-Royce, and Williams International; and aircraft builders Boeing and Lockheed Martin.

UEET continued the legacy of ECI, E^3, QCSEE, ECCP, and ATP through its seven main component areas: low emissions, highly loaded turbomachinery for increased efficiency, high-temperature materials and structures, intelligent controls, propulsion-airframe integration, integrated component technology demonstrations, and the integration and assessment of the overall technology relevant to the program in general. Those projects led to the identification of areas that contributed to the two interrelated goals of reducing fuel consumption by 15 percent and carbon dioxide (CO_2) and NO_x emissions by 70 percent. They included advanced compressor, combustor, and turbine design and new alloy and ceramic materials and coatings. The program also included a management component that integrated and assessed the individual technologies and brought them together in workable systems.[44]

One of those legacy projects, turbine disks made from high-temperature materials, facilitated NASA's contribution to the advanced turbofan engines of the early 21st century. Turbine disks are critical to safety, efficiency, and overall engine performance. Development of the ME3 turbine disk alloy began in 1993 under the auspices of the EPM project by a team consisting of members from

Glenn, GE, and Pratt & Whitney. Refinement during UEET and after led to a new alloy capable of withstanding temperatures of 1,300 degrees Fahrenheit, which was a 100-degree increase over the limits of operational turbine disks.[45] ME3 first appeared on GE and Pratt & Whitney's Engine Alliance GP7200 engine in 2007 and became a central element of GE's GEnx turbofan.[46]

UEET evolved into the NASA Office of Aerospace Technology's Vehicle Systems Program in 2003. It coordinated its efforts with IHPTET and VAATE, as well as similar programs sponsored by DOE, the FAA, and the EPA to avoid duplication and maximize the resources of each organization.[47]

Programs like UEET kept Lewis operating through the 1990s and early 21st century, but personnel could suffer the ups and downs of congressional funding. When the House Science Committee cut the funding, enthusiastic bipartisan and active lobbying by Representative Dennis Kucinich (D-OH) and others from the Cleveland area restored a minuscule $29.5 million of the overall $14 billion NASA budget, but they kept 3,000 jobs in Cleveland.[48] There was a balance between pushing technology to keep American aviation competitive and keeping congressional districts employed and supported with Federal funding.

The AGATE and GAP Programs

The Advanced General Aviation Transportation Experiments (AGATE) program—a consortium of NASA, the FAA, industry, universities, and nonprofit groups—worked to revitalize the U.S. general aviation industry's role in the global marketplace in the 1990s. Since its heyday in 1978, when annual airframe production reached 17,800, the industry suffered from a steady decline that bottomed out in 1993 with only 964 aircraft leaving factories. Moreover, the average general aviation aircraft flying in the early 1990s was of 1960s vintage with an outdated cockpit, airframe, and propulsion system. The impetus to improve the capability of general aviation was there. At the time, the American general aviation community served 18,000 airports, proved to be the only means of air transportations to many areas throughout the Nation and the world, and employed hundreds of thousands of people across the Nation.[49]

After NASA Administrator Daniel S. Goldin met with industry representatives at the Experimental Aircraft Association Convention in Oshkosh, WI, during the summer of 1994, the Agency convened AGATE the following spring. Before the conclusion of the program in December 2001, there were 76 members in 31 states with overall direction provided by Bruce A. Holmes of the general aviation office at Langley. The technical portion of AGATE strove to create new approaches to advanced airframe, cockpit, and propulsion technologies that could be applied to the design and manufacture of safer and more affordable small aircraft.[50] AGATE's propulsion component was the General Aviation Propulsion (GAP) program, managed by Leo Burkhardt at Lewis.

To inaugurate the GAP, Goldin invited members of Congress, Government officials, and aviation industry executives to the Agency's headquarters in Washington, DC, for a formal signing ceremony and press conference on December 16, 1996. At the event, NASA announced the selection of Teledyne Continental Motors of Mobile, AL, and Williams International of Walled Lake, MI, to develop new piston and gas turbine engine systems, respectively. Goldin stressed that the goal of the program was to develop the technology and manufacturing processes for "revolutionary, low cost, environmentally-compliant general aviation propulsion systems and test them on advanced aircraft" in the year 2000.[51] GAP subsequently involved two efforts, one to develop a diesel engine and the other to develop a small turbofan.

The GAP Diesel

Four- and six-cylinder air-cooled opposed engines, meaning the cylinders were arranged horizontally across from each other, manufactured and serviced by Continental and Lycoming, were the standard for general aviation. Originating during the late 1930s, they were a step above the water-cooled engines that were in use at the time. Unfortunately, they were noisy, caused a lot of vibration, required periodic maintenance, and were expensive to purchase and operate. NASA's GAP program promised a new generation of high-performance engines that were smooth, quiet, user-friendly, and, most importantly, affordable. Overall, the development of a new and innovative GAP system with the cockpit and airframe technologies developed under AGATE would contribute toward the establishment of a new small aircraft transportation system in the United States.[52]

The diesel aircraft engine had not been a serious possibility for American aviation since the NACA stopped its research into the system in 1940.[53] There was continual interest rooted in the simplicity and reliability of the design in the decades that followed, but they were heavy, especially for general aviation applications. Additionally, there were late-20th-century concerns over the long-term availability of aviation gasoline, or avgas. The GAP program goal for piston engines was to reduce engine prices by half while eliminating the need for leaded gasoline and to substantially improve reliability, maintainability, ease of use, and passenger comfort. To achieve this goal, Teledyne Continental Motors and an industry team that included Hartzell Propeller, propulsion control specialist Aerosance, and airframe manufacturers Cirrus and Lancair partnered with NASA Glenn to develop a highly advanced piston engine, the GAP diesel engine.[54]

The Power for Flight

The intended airframe for the GAP diesel was the archetypal single-engine, four-seat monoplane—capable of cruising at speeds of up to 200 knots—that dominated general aviation. To compete with contemporary piston aircraft engines, the engine combined the two-stroke operating cycle with an innovative and lightweight modular construction that permitted low-cost mass-production manufacturing methods and a price tag half of what conventional engines cost. For operational economy and practicality, the new engine burned readily available jet fuel at a low rate of approximately 25 percent less than other engines. Advanced-design, low-speed propellers that benefited from joint research between NASA and Hartzell offered quiet operation for both passengers and airport neighbors.[55]

The GAP diesel engine also incorporated a single-lever power control (SLPC) system, a simplified engine control compared to that in older systems. General aviation pilots had to manipulate as many as five levers to control fuel-air mixture, propeller pitch, and other parameters in flight. Introduced by team member Aerosance of Farmington, CT, in 1999, the SLPC worked in tandem with a FADEC to control the propulsion system in a manner similar to depressing the gas pedal of a car with an automatic transmission on the road. Moving the power lever automatically set the amount of fuel flow, air flow, ignition timing, and propeller pitch to maximize the power of the engine and propeller during takeoff, cruise, and landing. Overall, the use of the SLPC increased fuel efficiency, decreased the time between overhauls, and ensured the best engine and propeller performance for all flight phases.[56] The SLPC was not exclusive to the GAP diesel, and general aviation manufacturers embraced the technology for a new generation of aircraft.

NASA boasted in 2004 that the GAP diesel would provide "pilots and passengers with the same kind of quiet, easy-to-use power that we have come to expect in our automobiles."[57] Industry observer Mike Busch remarked that the GAP diesel was "what the future of piston general aviation should be."[58] In the end, however, the NASA-industry collaboration did not lead to a production engine. The manufacturer's extra costs involved in making it practical for flight, the overwhelming dominance of the four-stroke gasoline aviation engine, and the wider availability of avgas in the largest general aviation market, the United States, negated the need for an alternative power plant. Nevertheless, the demonstration of the engine at important venues like the Experimental Aircraft Association's AirVenture convention in Oshkosh, WI, acted as a bridge in the late 1990s to a new generation of innovators in the United States and Europe. They continued development and introduced aircraft diesels in the early 21st century.[59]

The GAP Small Gas Turbine Initiative

The other component of the GAP program was the development of a small gas turbine engine using low-cost manufacturing techniques. The use of turbine engines by the commercial aviation industry proved their desirability in terms of reliability, smooth operation, use of readily available jet fuel, and low noise and emissions. The limiting factor preventing the widespread use of turbine engines in the general aviation market was their high cost. Williams International of Walled Lake, MI, entered into a $37 million development program with NASA for the design, construction, and flight demonstration of a turbofan called the FJX-2. The small jet would be cheaper, lighter, and easier to manufacture than current engines in order to promote adoption by the general aviation industry. Williams and its team of industry partners shared the funding with NASA in a 60-/40-percent agreement. The GAP program endeavored to reduce the cost of small turbine engines by a factor of 10 and revolutionize the concept of personal air transportation with the introduction of a new class of general aviation aircraft that were safe, affordable, and fast.[60]

The FJX-2 was a high-bypass-ratio turbofan engine that produced 700 pounds of thrust while weighing only 100 pounds, which was approximately one-fourth the weight of a general aviation piston engine. The FJX-2 team applied many lessons learned from automotive gas turbine engines to reduce costs. Revolutionary design concepts included a shrouded fan rotor, a low-pressure fuel system, an electrically driven fuel pump, a blowdown scavenge lubrication system, a remotely mounted gearbox, and a high-speed starter/alternator. The team placed specific emphasis on simplifying the design and reducing the number of parts. Low-cost design techniques and advanced automated manufacturing methods made the FJX-2 the first turbine engine that was cost-competitive with piston engines.[61]

The core technology of the FJX-2 engine was the high-pressure compressor. It was the culmination of extensive cooperation between Glenn and Williams International aerodynamicists using the latest advancements in 3D viscous flow analysis tools. The availability of NASA's APNASA CFD code, data from compressor rig testing, and the expertise of the NASA-industry team was invaluable in the advancement of the compressor design. The collaboration resulted in a compressor that was 85 percent efficient, the most capable component of its size ever designed.[62]

Glenn provided significant support to the FJX-2 throughout the development program. Design and analysis included combustor turbomachinery modeling using CFD, structural and control system analysis, and noise predictions. The first prototype engine was ready to run in December 1998. Testing included engine altitude testing in the PSL from March to April of 2000. Overall, the FJX-2 fell well below FAA and EPA standards for noise

The Power for Flight

Figure 6-3. William Guckian of Williams and Ray Castner of NASA are pictured with the GAP FJX-2 Turbofan at Glenn Research Center in March 2000. (NASA)

and emissions.[63] John Adamczyk, a 30-year-veteran of Lewis, recalled that his involvement in the FJX-2 project was "one of the high points" of his career.[64]

An important byproduct of the program was the derivative 550-horsepower TSX-2 turboprop. It differed from its turbofan counterpart only by the removal of the fan assembly and installation of a gearbox and five-blade propeller. Williams believed that the commercial success of the FJX-2 would facilitate a low purchase price for the TSX-2.[65]

The FJX-2 was the culmination of a dream of Sam Williams, the head of Williams International, who had pioneered the concept of the small jet engine beginning in the 1950s. After a long and successful career innovating power plants for drones and cruise missiles, which garnered him the 1978 Collier Trophy, he shifted his focus toward the creation of a new generation of very light jets (VLJs) facilitated by the FJX-2. Williams contracted Burt Rutan of Scaled Composites to design and build a demonstrator aircraft called the V-Jet II. The flights of the V-Jet II at Oshkosh in 1997, powered by different engines, excited the crowds there, which included aviation entrepreneur Vern Raburn. The former Microsoft Corporation executive quickly acquired the exclusive rights to the commercial version of the FJX-2 called the EJ22 for his new

Eclipse 500 VLJ. Unfortunately, the EJ22 was not ready for FAA certification, and Eclipse quickly replaced the underpowered and temperamental engines.[66] Despite winning the 2005 Collier Trophy for its proposed production VLJ, Eclipse faced continued problems as quality-control issues and weak financing led to the dissolution of the company in 2009.[67]

The Advanced Ducted Propulsor

In June 1993, Ames Research Center began tests of a new jet engine that promised to cut fuel consumption by 12 percent as well as dramatically reduce noise. Called the Advanced Ducted Propulsor (ADP), the engine was a joint development between NASA and Pratt & Whitney. The ADP featured three innovative design elements. The approximately 10-foot (3-meter)-diameter variable-pitch fan system incorporated 18 fan blades able to adjust for the most efficient positions for takeoff, cruise, and reverse thrust at landing. The system permitted the removal of heavy, unreliable, and expensive conventional thrust reversers found on current operational engines. Finally, the 40,000-horsepower fan-drive gear system and a high-speed, low-pressure turbine facilitated a maximum forward thrust of more than 50,000 pounds.

Previously, Pratt & Whitney had tested a one-seventh-scale model of the ADP at Lewis in 1991 and at Langley in 1992. The Connecticut engine maker collaborated with MTU Aero Engines of Munich, Germany, and Fiat Avio of Turin, Italy, on design and construction of the actual demonstrator engine. Full-scale testing began at Pratt & Whitney's West Palm Beach, FL, facility during the fall of 1992; that testing confirmed that the engine was operational. The only facility capable of evaluating the ADP's variable-pitch fan at reverse thrust under simulated landing conditions was the National Full-Scale Aerodynamics Complex (NFAC) wind tunnel at Ames. Tests of the ADP, the largest engine evaluated at the NFAC, continued for 12 weeks.[68] Optimistic for the future, Project Director Clifton Horne believed that ADP-style engines would be available for use in 300- to 700-seat commercial airliners by the early 21st century.[69] NASA continued with follow-on investigations of scale-model versions of newer fan blade designs at Glenn in 1995 and 1996 that showed increased fan efficiency and low noise.

The ADP program did not lead directly to a production engine. The complex thrust-reversing fan blade mechanism and the extensive use of expensive composite materials were costly pitfalls that ended the program. Pratt & Whitney did use the experience to develop the Geared TurboFan, a new family of engines producing thrust in the range of 24,000 to 35,000 pounds, in partnership with MTU Aero Engines in Germany. The company engineers incorporated a planetary reduction gearbox that connected the core to the low-pressure system; a conventional thrust reverser mechanism; and an advanced

The Power for Flight

Figure 6-4. The Pratt & Whitney and NASA team prepare the Advanced Ducted Propulsor (ADP) for a flow visualization test in the National Full-Scale Aerodynamics Complex (NFAC) 40- by 80-foot Wind Tunnel at the Ames Research Center. (NASA)

fan with fixed, wide-chord blades. The gearbox between the fan and the low-pressure shaft allowed each to run at their optimum rotational speeds. That arrangement enabled fewer stages to be used in both the low-pressure turbine and the compressor.[70]

The latest version, designated the PurePower PW1000G, became the engine of choice for a new generation of narrow-body, medium-range jetliners manufactured by Airbus, Bombardier, Embraer, Irkut, and Mitsubishi that would go into production in 2013. Pratt & Whitney celebrated the attributes of its engines, which reflected the decades-long quest for fuel-efficient, clean, and quiet engines pursued so assiduously by NASA. The PW1000G was 16 percent more fuel-efficient, 50 percent cleaner, and 75 percent quieter than contemporary engines.[71]

Turning the Tide on Noise

As the 1990s began, there were indications of the payoff in noise-reduction efforts facilitated by NASA and the FAA. The proportion of quieter aircraft used by American airlines increased from 52 percent (2,685 aircraft) to 59 percent (3,450 aircraft) in 1992. At the regulation level, acting FAA Administrator Joseph M. Del Balzo stated, "[T]he battle is far from over, but the figures clearly show that the tide is turning." The Aviation Noise and Capacity Act enacted by Congress in 1990 directed the elimination of Stage 2 operations by the end of the decade. The requirements for Stage 3 aircraft reflected new approaches to noise suppression. Engine designers muffled noise at takeoff by reducing the speed of exhaust and installing sound-absorbing material in the large turbofan engines found on modern Stage 3 aircraft. Stage 3 technology created an improvement of up to 25 dB over first-generation Stage 1 aircraft, or an 80-percent reduction in perceived noise. Six years later, the FAA imposed Stage 4, which applied to all aircraft designed after January 2006.[72] There were also increasingly strict international noise regulations. American airliners landing at European airports came under the authority of the International Civil Aviation Organization (ICAO), the United Nations agency responsible for worldwide noise standards and its Annex 16 regulations.

The Advanced Subsonic Technology Program

By the early 21st century, the aviation industry in the United States was accounting for annual sales in excess of $36 billion and employing nearly 1 million workers. Ever-expanding globalization encouraged a travel boom that stood to increase the industry's worldwide market share. The key to success for both American aviation and the country's economy as a whole was remaining competitive regarding aircraft technology.[73] With environmental concerns increasingly taking an equal priority alongside technical and economic factors in the shaping and adoption of new technology, quieter airplanes possessed an advantage in the aviation marketplace.

The challenge of overcoming aircraft noise persisted into the late 20th century. Previous NASA and industry jet noise research focused on either subsonic mixed-flow long-duct nozzle systems or supersonic low-bypass turbojet nozzle systems. There was little emphasis on the high-bypass-ratio, non-mixed, separate-flow, short-duct exhaust systems that were part of the GE, Pratt & Whitney, and Rolls-Royce turbofan engines found on wide-body airliners like the Boeing 747. They were not the major source of noise. As aircraft weights increased, the need for higher-thrust engines resulted in higher jet velocities, temperatures, and pressure ratios that were exponentially noisier.[74] Increased air traffic and population growth into areas surrounding airports resulted in a larger percentage of communities being impacted by noise. As a result, the desire to reduce noise around airports intensified.

NASA estimated that the technology capable of reducing the "noise annoyance footprint" around an airport would not be available until 2024 at the

The Power for Flight

earliest. In response, the Agency worked to accelerate that development of the required technologies in a much shorter timeframe—by 2000. The program was under the umbrella of the Advanced Subsonic Technology (AST) program. AST originated in February 1992 as a partnership between NASA, the FAA, and the U.S. aviation industry. The program aimed to develop high-payoff technologies that enabled a safe, highly productive global air transportation system. At the core of the system was a new generation of environmentally compatible and economical aircraft and engines.[75]

NASA identified four distinct classes of airliners that permitted a thorough evaluation of noise-reduction technology over a broad range of technical parameters. They represented large four-engine aircraft like the Boeing 747, medium twin-engine airliners such as the Airbus A330, small twin-engine transports like the Boeing 737, and business jets like the Learjet 25. Aircraft engine manufacturers became involved according to their product specialties. GE and Pratt & Whitney participated in testing and analyses associated with the large, medium, and small airliners with engines that ranged in thrust from 30,000 to 90,000 pounds. Rolls-Royce and Honeywell worked primarily on their power plants for business jets whose engines produced 15,000 pounds or less.[76]

Langley, Ames, and Lewis played a role in AST, with researchers at Langley evaluating and distributing reports on the overall progress of the program. They collaborated and managed the research through contracts and subcontracts with universities and the U.S. aviation industry. The organizations and individuals participating did change over the course of the AST program. The representation from industry was comprehensive; participants included GE, Pratt & Whitney, Rolls-Royce, Honeywell, Rohr, Boeing, and Lockheed. A key element in guiding the AST was a working group composed of industry, NASA, and FAA membership. They identified technical needs, research activities, and team arrangements, and they coordinated between the myriad groups involved. Additionally, an industry steering committee provided manufacturer and airline experience to the process that facilitated a balance between partner needs, implementation, and program advocacy.[77]

AST's immediate technical focus centered on the creation of noise-reduction technology that would enable the U.S. aviation industry to feed unrestrained market growth and secure its economic position while remaining compliant with international environmental standards. NASA's answer was the cumulative reduction of subsonic aircraft noise by 30 EPNdB by the year 2000. NASA was confident that the AST noise-reduction program would accomplish that goal.[78]

The AST adopted NASA's Technology Readiness Level (TRL) methodology to measure progress in the development of individual technologies as well as comparing them with others. Researcher Stan Sadin at Headquarters in

Washington, DC, originated the first TRL scale in 1974 for space technology. The idea spread throughout Government and military research organizations and agencies through the 1980s. By 1995, a refined set of nine overlapping TRLs was in place to aid in the evaluation of the AST:

Tier	TRL	Description
System test and operations (TRLs 8–9)	9	Actual system "flight proven" in operational flight
	8	Actual system completed and "flight qualified" through test and demonstration
System/subsystem development (TRLs 6–8)	7	System prototype demonstration in a flight environment
	6	System/subsystem model or prototype demonstration in a relevant environment
Technology demonstration (TRLs 5–6)	5	Component and/or breadboard validation in relevant environment
Technology development (TRLs 3–5)	4	Component and/or breadboard test in a laboratory environment
	3	Analytical and experimental critical function, or characteristic proof of concept
Research to prove feasibility (TRLs 2–3)	2	Technology concept and/or application formulated
Basic technology research (TRLs 1–2)	1	Basic principles observed and reported

The AST program set the target TRLs for the noise-reduction technology investigated to be 5 or 6, which reflected the demonstration phase of development. TRL 6 also marked the dividing line between the Government (1–6) and industrial (7–9) roles in the development of new technologies.[79] It was clear where the responsibilities lay in terms of support, which offered clarity regarding NASA's appropriate place within American aviation.

The AST divided its effort to make quieter aircraft into specific elements: engine, nacelle, airframe integration and system evaluation, interior environment, and community noise. With its expertise in aeropropulsion and recognition as a Center of Excellence in turbomachinery, Lewis (later Glenn) managed the Engine Noise Reduction Element of the AST program. The element's goal was to develop technology that would reduce noise at the fan and exhaust by 6 EPNdB by the program's completion in 2000. That was no small feat since a 10-EPNdB noise reduction was equivalent to lessening the noise level by 50 percent. The program originally aimed to improve the soon-to-be-introduced higher-bypass engines like the GE90 but modified that goal to include current turbofans to increase the chance of an immediate result. Approximately 75 percent of the AST program's budget went to propulsion system noise reduction.[80]

Starting in 1992, Glenn and representatives from industry established a baseline centered on current technical capabilities and designs. GE, Pratt & Whitney, Rolls-Royce, and Honeywell submitted data based on the performance of their engines and the resultant noise produced. The team adopted a standard nomenclature for engine-component noise to assist in overall communication. Those abbreviations—inlet, aft fan, core, turbine, and jet—greatly facilitated a common point of departure for research. The team members then were able to investigate several concepts generated by NASA, industry, and academia to reduce overall engine noise for the four representative aircraft.[81]

The Engine Noise Reduction Element centered on six primary concepts. The development of advanced liners relied upon work in nacelle aeroacoustics that focused on refining the cowling that houses the engine below the wing. To meet the goal, researchers collaborating through the Boeing Nacelle Aeroacoustic System Technology Assessment worked on new designs that absorbed, canceled, or redirected engine noise. The avenues explored included the analytical modeling of nacelles to predict their effect on noise propagation; laboratory experiments in passive, adaptive, and active control treatment to improve duct noise; and scale-model and full-scale tests to validate the new designs.[82] NASA identified that the best arrangement for inlet noise reduction was a scarf inlet, a design that featured a protruding lower lip to redirect noise away from the ground, combined with a seamless full acoustic liner that included the inlet lip. The reduction was 2.3 and 4.0 EPNdB at landing approach and cutback respectively.[83]

Another avenue of research involved work with Herschel-Quincke (HQ) tubes. They were passive devices that suppressed tone noise through the use of tubes tuned to specific frequencies. The adjustment of the length of an HQ tube allowed maximum suppression of targeted frequencies, primarily in the range of 2,500 to 3,100 Hertz. Researchers at Virginia Polytechnic Institute and State University (Virginia Tech) tested three HQ configurations on a JT15D inlet and found that a double array of 20 and 16 tubes tuned to different, but similar, frequencies provided effective noise reduction.[84]

Two concept investigations involved modified fans and stators (stationary blades that directed airflow) to minimize the wake interaction between the fan and stator to reduce noise at both the inlet and aft fan. The first incorporated stators that were swept aft. Glenn researchers tested them with a high-speed fan in 1999 and found that they yielded suppression levels at approach of 1.4 EPNdB at the inlet. The second concept combined stators that were swept aft and leaned to the side to further minimize wake interactions with a forward-swept fan that delayed the onset of high-speed-rotor multiple pure tones. Tests in the Glenn 9- by 15-Foot wind tunnel of designs submitted by a GE-Allison team and Honeywell revealed that the latter generated the best suppression levels. The work demonstrated that a forward-swept fan combined with swept and leaned stators reduced inlet and aft-fan noise levels at both the sideline and cutback certification points by 2.5 EPNdB.[85]

Evaluation of the acoustic liners, HQ tubes, and other noise-reduction solutions was made possible through NASA's Engine Validation of Noise Reduction Concepts (EVNRC) program. Beginning in August 1997, the EVNRC provided the necessary funding for Honeywell, Pratt & Whitney, Boeing, and other contractors to validate noise-reduction concepts and technologies that evolved during the AST through engine testing.[86] The contracts supported work at industrial research and development facilities through the 2000s.

NASA also saw the need for the acoustic and aerodynamic integration of turbofan engines with the high-lift flap-and-slot systems found on commercial airliners. Researchers aimed to reduce airframe noise by 4 dB at both takeoff/climb and approach/landing below 1992 levels while maintaining performance. The solution was the introduction of new subcomponent airframe noise-prediction codes that allowed for the proper evaluation of interrelationship between the disparate structures.[87]

The remaining two research elements centered on improving the overall noise environment for passengers inside the airliner and people on the ground. The research to reduce cabin interior noise by 6 dB relative to 1992 technology integrated studies in source identification (a combination of engine noise and vibration with the aerodynamic boundary layer and the turbofan jet flow) with interior sound prediction and new noise-control concepts. Researchers also endeavored to create technology that reduced the impact of noise on communities surrounding an airport. They concentrated on achieving a 3-EPNdB reduction through the application of new aircraft technologies and operational procedures, enhanced noise impact modeling and prediction, and a better understanding of the relationship between human response and aircraft noise.[88]

The majority of the aircraft and engine noise-reduction technologies created for the AST program reached the desired TRL of 5 by the program's end. That accomplishment reflected efficient coordination between the Government, industry, and academia. The researchers evaluating the program acknowledged that the AST partnership prepared for "an orderly and effective transition" from research and development to operational use as NASA and industry carried the technology forward into the 21st century.[89]

Toward Active Noise Control

Active noise control was a new approach to reducing noise generated at the fan inlet. It was attractive as an enhancement option, or a complete alternative overall, to acoustic lining and HQ tubes due to the shorter inlet lengths found on

The Power for Flight

the newer higher-bypass engines from Pratt & Whitney, GE, and Rolls-Royce with large-diameter fans. The concept was simple. Sensors detected noise disturbances in the engine. They triggered negative-noise generators that canceled out the undesirable sound waves. The end product was no noticeable noise. The creation of active noise-control technology was a multidisciplinary effort requiring expertise in duct acoustics, controls, and actuator/sensor design.[90]

NASA overall was optimistic regarding the "potentially high payoffs" of active noise control, which would be a major contribution to the overall 6-dB noise-reduction goal of the AST program. The Agency fully intended for the validated technology to be available for use by all U.S. engine manufacturers. The result would be a new generation of economical and environmentally friendly aircraft and engines.[91]

Both Pratt & Whitney and GE investigated active noise control as part of the AST program. Central to the evaluation of their new fans was Lewis's unique testing facility, the 9- by 15-Foot Low Speed Wind Tunnel. The Active Noise Control Fan, developed at Lewis, was a 4-foot-diameter low-speed fan. Several concepts, including an arrangement of two circumferential arrays of acoustic actuators developed with GE, successfully canceled selected acoustic modes.[92] Pratt & Whitney's 22-inch ADP low-speed fan model mounted in a nacelle provided the data for the AST's analysis. Active noise control reduced inlet fan noise by 1.5 EPNdB at approach but increased noise at cutback. The work never reached the desired TRL of 5, so NASA removed active noise control from the overall evaluation process during the AST program.[93]

Nevertheless, the potential benefits to industry were obvious by the late 1990s. Rob Howes, supervisor of acoustics and structural dynamics for Cessna, remarked that while the research at Lewis was "on the cutting edge," what impressed him the most was the "timeliness" of the research and its direct applicability to the then-current market challenges.[94] The next step was to go beyond the ground-based testing facilities of models and perform full-scale validation of the noise-reduction technology on full-scale turbofan engines. The most promising concepts from the model scale testing would be selected for real-environment demonstration.[95] A concept using active noise control in an inlet of a PW4098 engine was tested at Pratt & Whitney's static engine stand in West Palm Beach, FL. Unfortunately, mechanical problems prevented the successful completion of the validation test.

Chevrons: The Deceptively Simple Solution

During the 1980s, the United States Air Force started looking for ways to reduce aircraft infrared signature by mixing the engine exhaust with free stream air. NASA later observed that the same nozzles reduced noise emissions as well. The AST Program's Steering Committee and Technical Working Group created the Separate-Flow Nozzle (SFN) Jet Noise Reduction Test Program in 1995 at Lewis. The program aimed to avoid higher aircraft noise levels without resorting to expensive and time-consuming engine and nacelle redesign. The source of the problem was the combination of the high-velocity gas flow, or jet, from the core engine (consisting of the high-pressure compressor, combustor, and high-pressure turbine); the slower air from the fan bypass duct; and the surrounding air. Some engine manufacturers, such as Rolls-Royce, resorted to heavy and expensive long fan duct mixed-flow nacelles with internal mixers to reduce jet noise. NASA's solution was the development of lightweight external noise-suppression devices that were easily incorporated into existing separate-flow exhaust nozzles with no noticeable loss in thrust. The goal was to mix the jet exhaust as it exited the engine with the free stream flow in a way to promote the suppression of the exhaust noise of the engine.[96] The benchmark was a 3-decibel reduction in jet noise as compared to 1992 technology.[97] The SFN program led to investigations into a range of new mechanical-suppression devices that would hold great promise for the future of jet noise reduction.[98]

Pratt & Whitney and GE received AST contracts to design and build scale models of separate-flow exhaust nozzles employing a range of experimental jet noise external suppression devices. Pratt & Whitney worked with data provided by two important subcontractors. Boeing's phased array tests revealed that there were two distinct jet noise sources at low- to mid-frequencies. The first, called buzz saw noise, was upstream, at the nozzle exit; the other, called shockcell noise, was downstream, beyond the exhaust. The United Technologies Research Center (UTRC) used CFD analyses to investigate the flow fields of the nozzle concepts. The engine maker delivered nine suppression devices that relied upon varying types of tab, scarf, offset, and lobed configurations to reduce jet noise. GE's submissions included vortex generator doublets and chevrons. Lewis researchers evaluated the models in the Nozzle Acoustic Test Rig (NATR) from March 20 to June 18, 1997, under static and simulated flight conditions (which included collecting far-field acoustic data) and analyzed the results. They concluded that Pratt & Whitney's inward-facing chevrons and flipper-tabs on the primary and fan exhaust nozzles reduced suppression levels nearly to the 3-EPNdB goal of the SFN. As the AST program drew to a close, the chevron element reached the TRL goal of 6.[99]

The deceptively simple designs that required no modifications to existing short-duct, separate-flow, nonmixed nozzle exhaust systems that emerged from the SFN program effectively decreased the downstream noise with only a minimum increase in upstream noise.[100] In other words, the jagged edges smoothed the mixing of both the hot air from the engine core and the cooler air blowing through the engine fan to reduce the turbulent hissing of the two flows shearing against each and creating noise.

The Power for Flight

As an outgrowth of testing derived from noise-suppressor concepts for military and civilian aircraft engines, the chevron nozzle had become a promising new concept, and research and refinement accelerated. Glenn researchers performed tests of 6 and 12 chevron nozzles on turbojet engines used on business jets in the spring of 2000. They achieved a 2-EPNdB reduction in noise over that of a standard conical nozzle. A NASA flight research team validated the results at full-scale in a Learjet 25 operating out of Estrella Sailport near Phoenix, AZ, in March 2001.[101] The computational and experimental research sponsored by the AST developed an in-depth understanding of the fluid mechanics of the chevron nozzle concept. Work began on the practical implementation of chevrons as the ideal configuration. The period 2001–2005 witnessed the ground testing in engine stands and flight evaluation in relevant environments.[102]

The Quiet Technology Demonstrator Program

NASA continued its sponsorship of the development of noise-reducing technologies through the Quiet Technology Demonstrator (QTD) program initiated in early 2000. To support research performed by the Boeing Company and Rolls-Royce, it provided the funding for new technologies that addressed the three primary sources of aircraft noise generated by the engine fan, the mixing of the engine and bypass exhaust with the surrounding air, and the airframe's wings and landing gear.[103] The QTD program built upon the work of NASA's SFN program and conducted both static model testing and in-flight validation of chevrons installed on high-bypass-ratio turbofans found on large commercial airliners. During the fall of 2001, the 3-week flight-test program of a modified Rolls-Royce Trent 800 turbofan installed on an Boeing 777 airliner on loan from American Airlines led to significant results. The researchers investigated combinations of sawtooth chevrons mounted on the primary and secondary exhaust nozzles and an expanded acoustic lining of the inlet. The best arrangement yielded reductions of 13 dB at the inlet fan and 4 dB at the exhaust. Hoping to achieve only 7 dB and 3 dB respectively, Boeing noise specialist Belur Shivashankara concluded that the "test results came out better than we expected."[104]

NASA Langley's Quiet Aircraft Technology Initiative

The Quiet Aircraft Technology (QAT) team led by NASA Langley continued to address the difficult problem of reducing noise from flying aircraft beginning in 2001. With a budget of $45 million, the project aimed to achieve a reduction of one-half in perceived community noise by 2007 and a reduction of 75 percent in 2020. They examined the persistent sources of noise addressed by the AST program. Airframe noise consisted of sound generated from wing slats and flaps and landing gear at takeoff and landing. Scarfed engine inlets, noise-absorbing treatments in the inlet, and chevron engine nozzle exit concepts addressed engine noise. Additionally, researchers scrutinized aircraft patterns around airports to determine new flight routes to lessen the noise impact on surrounding communities.[105] NASA Glenn was responsible for engine noise and focused on the fan and jet.

NASA-Funded Chevron Studies

NASA went on to encourage further study into noise-reduction technologies through support and partnership with Boeing. Researchers in 2003 conducted comparison studies to determine the ideal configuration for internally mixed nozzles. They concluded that while traditional lobe mixers were quieter, their use also resulted in a higher loss of thrust. Chevron mixers, which up to that point had been used only for separate flow nozzles, suffered no appreciable loss in thrust and were lighter and easier to manufacture. Overall, the researchers believed that chevrons were a suitable replacement for lobe mixers.[106] The following year, NASA Langley's Aeroacoustics Branch collaborated with Boeing to examine the question of whether azimuthally varying chevrons could reduce total jet-related noise radiated toward the ground. The investigation centered on taking advantage of the asymmetric flow and acoustic environment created by the pylon, the wing, and the interaction of the exhaust jet with flaps on the wing. The team concluded that T-fan chevrons with deeper scallops at the top of the nozzle were superior to chevrons with a uniform azimuth.[107]

Another barrier to the widespread adoption of chevrons was the technology's effect on overall aircraft fuel efficiency. The use of chevrons decreased noise at takeoff, which met the needs of regulators and communities in the vicinity of airports. When an airliner reached cruise altitude, those same devices degraded thrust efficiency because they protruded, or were immersed, into the jet flow. The resultant higher fuel costs would not satisfy the airline industry's profit margins. Both NASA and Boeing investigated systems that optimized chevron immersion into jet flow at takeoff and cruise. Boeing produced variable chevrons that incorporated heat-activated nickel-titanium SMA actuators. The flight crew could control the amount of immersion or rely upon autonomous variation.[108]

According to one observer, NASA's sponsorship of the QTD program and the additional industry research efforts had "led chevron technology to the brink of commercial application by 2005."[109] That same year, the first commercial engine with chevron exhaust nozzles was GE's 20,000-pound-thrust CF34, introduced in 2005. Over 60 operators purchased over 1,400 CF34s and accumulated over 13 million flight hours and 9 million cycles in service, primarily on Embraer E190 and E195 airliners.[110]

The Quiet Technology Demonstrator 2 Program

Ground and flight tests by NASA and its industry partners in 2005 and 2006 under the Quiet Technology Demonstrator 2 (QTD2) program proved that the new scalloped chevron design reduced noise levels both in the passenger cabin and on the ground.[111] The 3-week 777 flight-test program in August 2005 was a partnership between General Electric, Goodrich Corporation, Boeing, and All Nippon Airways conducted at Boeing's facility at Glasgow, MT.

The tests validated the effectiveness of a number of significant airplane noise-reduction concepts developed in computer simulations and wind tunnels. The combination of acoustic liners and chevrons created an effective noise-suppression system and removed the need for several hundred pounds of sound insulation installed in the fuselage. For the airlines, less weight translated to greater operational efficiency and higher revenues for the airlines. The team also evaluated variable-geometry chevrons made with a temperature-reactive SMA. Called "smart," or active, chevrons, they automatically warped in the jet exhaust flow to reduce noise at takeoff and landing and reverted to a streamlined position at cruise altitude. The new fan and engine core chevron exhaust configurations reduced "community noise" by 2 dB. The low-frequency rumble heard in the aft cabin by passengers at cruise altitude was reduced 4 to 6 dB. The new Goodrich "seamless" sound-absorbing liner inside the engine inlet reduced fan tones heard in front of the aircraft by up to 15 dB, to where they were "almost inaudible."[112]

While chevrons reduced noise, they still imposed a cruise performance penalty. An adaptive-geometry chevron would lessen noise at takeoff and retract during cruising flight. Complementing Boeing's work on "smart" chevrons, NASA further refined the new technology. Glenn developed a high-temperature SMA alloy that Continuum Dynamics, Inc., integrated into a new chevron design as the solution.[113] In 2007, Langley researchers investigated two types of active chevrons that differed on when the power for the SMA actuators was applied, either during takeoff (immersed) or cruise (retracted). Their tests simulated flow conditions representing the bypass exhaust of commercial jet engines. They concluded that the power-off-retracted (POR) chevron was the better configuration. It exhibited precise and rapid control that used the very aerodynamic forces acting against it to immerse and retract at a rate of deflection greater than that of the power-off-immersed (POI) chevron.[114] The challenges for the new design involved reaching the needed high actuation forces, limited volume for actuator placement, and high operating temperatures.

Chevron nozzles evolved from being a promising proof of concept in the mid-1990s to an undeniable part of a quieter future for aviation in the early 21st century. GE incorporated a sawtooth-shaped nozzle on its revolutionary GEnx engine destined for the highly anticipated Boeing 787 Dreamliner in 2006. By 2011, NASA-influenced chevron technology existed on the

GE, Pratt & Whitney, and Rolls-Royce engines powering the Airbus A321; Boeing 747-8; Boeing 787-8 and -9 Dreamliners; Embraer Lineage 1000 and 170, 175, 190, and 195 series E-Jets; and the Bombardier Regional Jet CRJ700. Many manufacturers with aircraft currently in development are looking into the concept.[115] Looking back on chevron development and their quick pace through the TRLs from the identification of the basic concept to flight-proven hardware, NASA manager Fay Collier remarked, "We had a bunch of smart NASA people pushing hard, and that gave us the momentum necessary to carry the technology all the way."[116]

There was an additional step beyond the use of physical chevrons to reduce noise. Studies conducted by Boeing and Russian researchers used a series of small jets emerging from the fan nozzle cowl to simulate metal chevrons. They used NASA-developed high-bypass-ratio nozzles as their baseline. Their tests revealed the promise of increased noise reduction and flexibility for a variety of flight conditions and for a wide range of nozzle bypass ratios.[117] NASA's involvement in the development of the chevron opened up new possibilities for innovation in commercial propulsion technology.

Propulsion for Supersonic and Hypersonic Flight

NASA's High-Speed Research Program, 1990–1999

In 1990, NASA initiated the High-Speed Research (HSR) program as a collaborative effort between NASA and industry to overcome the technical barriers that had been plaguing the successful development of a supersonic airliner in the United States since the 1970s. The idea of a 300-passenger High-Speed Civil Transport (HSCT) able to cruise at 1,500 miles per hour across entire oceans in half the time and cost of conventional subsonic airliners was a persistent and provocative concept for American aeronautics. Boeing, GE, and Pratt & Whitney represented the airframe and aircraft engine industries. Langley managed the HSR overall, with Ames, Dryden, and Glenn providing their aeronautical research expertise. Goddard Space Flight Center and the Jet Propulsion Laboratory also contributed to the project overall.[118]

In December 1995, NASA contracted Boeing to undertake a study of the concept, which benefited from Agency-sponsored research from the 1980s. In this iteration, the end product, a Technology Concept Airplane (TCA), would lead to the introduction of an economically viable and environmentally friendly HSCT capable of cruising at Mach 2.4 in the early 21st century. HSR focused on two major development programs centered on airframe structures and propulsion technology. The former included high-temperature composite

The Power for Flight

Figure 6-5. Pictured is a NASA study concept for an HSCT aircraft. (NASA)

materials and structures and the windowless eXternal Visibility System for the cockpit.[119]

The HSR team created the Critical Propulsion Components (CPC) element in 1994 to develop a commercially viable propulsion system for the HSCT. Glenn provided research assistance and managed a joint contract awarded to GE and Pratt & Whitney. They faced two overarching challenges. The first was to reduce NO_x emissions at cruising altitude to an index of less than 5, or by a factor of 10 compared to other engines. Second, the HSCT had to meet the FAA's increasingly stringent FAR 36 Stage 3 airport noise restrictions, which required a reduction of 4–6 EPNdB at the sideline, 8–10 EPNdB at cutback, and 5–6 EPNdB during landing approach.[120] Jet noise research alone accounted for $75 million of the HSR budget.[121]

In 1994, NASA and industry partners chose two concepts to pursue for the HSR program that focused on the development of an economically and environmentally practical American supersonic airliner. The mixed-flow turbofan and fan-on blade (FLADE) concepts promised to be the cheapest, quietest, cleanest, and easiest to develop. Creating an engine that was both quiet at takeoff and clean and efficient during supersonic cruise was a challenge. Both concepts reduced noise by mixing low-energy air with high-energy exhaust flows during takeoff. The mixed-flow turbofan directed a secondary, slower-moving bypass airstream that rejoined the engine airflow before the exhaust nozzle. The FLADE employed an auxiliary fan that added a compression stage in its own flow stream at the fan tip that could be closed to reduce drag during supersonic cruise. The two concepts underwent comparative evaluation with the intention that the winner be selected in 1996. From there, focused development would have led to full-scale testing with enough data to contribute to an operational engine by 2001.[122]

Glenn's capabilities and expertise proved central to the development of the HSCT's propulsion system. Researchers used the promising new computer design tool, NPSS software, to create and run a model of the engine and then compare the results to Pratt & Whitney's own proprietary program. The use of Glenn's Abe Silverstein Supersonic 10- by 10-Foot Wind Tunnel permitted the establishment of fan face pressure profiles in a model inlet. Other propulsion innovations included new turbine blade and disc materials, chevron mixer nozzles to reduce noise, and a powder-metal process used to fabricate the nozzle components. The anticipated conclusion of the CPC was 2002, but the cancellation of the HSR in 1999, due to the age-old debate over the Government's appropriate role in the development of large-scale technologies, prematurely ended the program. That did not stop the GE and Pratt & Whitney engineers from announcing in 2005 that their NASA-funded program had proved that a Mach 2.4 commercial airliner was a practical possibility.[123]

In the wake of the cancellation of the HSCT program, advocates for supersonic air transportation modified their goals. They envisioned that the successful follow-on to the Anglo-French Concorde SST that had been in limited service since 1976 between Europe and North America would be a much smaller business jet. The challenge was the same: the elimination of the effects of the sonic boom. NASA funded the Advanced Supersonic Propulsion and Integration Research (ASPIRE) project in 2000 as a part of the Revolutionary Concepts in Aeronautics (RevCon) Flight Research Project. Under the direction of Principal Investigator David Arend at Glenn, team members aimed to install a mixed-compression supersonic inlet along with a low-sonic-boom nacelle/diverter/wing simulator on the back of NASA's Lockheed SR-71 Blackbird research vehicle for evaluation.

High-Supersonic and Hypersonic Research

In his February 4, 1986, State of the Union Address, President Ronald Reagan announced that the United States was "going forward with a new Orient Express that could, by the end of the next decade, take off from Dulles Airport, accelerate up to 25 times the speed of sound, attaining low earth orbit or flying to Tokyo within two hours."[124] The President was referring to an ambitious study effort by DARPA, in conjunction with the United States Air Force, for a radical single-stage-to-orbit hypersonic vehicle. This led to the formation of a Joint Service and NASA National Aero-Space Plane (NASP) Project Office, the NASP Joint Program Office (JPO), at the Air Force System Command's Aeronautical Systems Division located at Wright-Patterson Air Force Base, OH, headed by Major General Kenneth Staten. Eventually, not quite a decade later, the program came to an end, having failed to overcome the challenge of generating enough power to ensure orbital entry.[125] Nevertheless, while the program ran, it significantly advanced the state of hypersonic knowledge, served as a focal point for facilities development, and encouraged advanced materials and fuels research that had tremendous benefits for subsequent efforts. Different mission profiles included its use as a single-stage-to-orbit vehicle, a transpacific hypersonic airliner, a new military aircraft, experimental research vehicle, or the flying test bed for research and development program.[126]

The Revolutionary Turbine Accelerator Program

Even though the NASP did not succeed, the lure of hypersonic flight was pervasive through the 1990s.[127] Advocates increasingly looked to combined propulsion systems with both gas turbine and turboramjet/scramjet technical approaches. GE began the development of a revolutionary high-speed turbine technology for a new Mach 4 jet engine in conjunction with NASA Glenn in July 2002. The Center selected GE for the development of a Revolutionary Turbine Accelerator (RTA) technology demonstrator for use in a third-generation reusable launch vehicle. Glenn administered the RTA project for the Advanced Space Transportation Program managed by NASA's Marshall Space Flight Center in Huntsville, AL. The goal of the 5-year, $55 million program was to produce a working demonstrator by 2006. Paul Bartolotta, RTA project engineer at Glenn, believed that very-low-cost space access could be realized by an affordable air-breathing propulsion system that provided aircraft-like operations. That capability facilitated expanded versatility beyond just space access to broaden economic revenues. For that reason, NASA selected the GE RTA concept, an engine capable of sustained high supersonic speeds that exhibited a quick-turnaround capability similar to that of a commercial airliner.[128]

NASA intended the RTA to be the first stage of a two-stage vehicle capable of hypersonic flight. At Mach 4, the second stage would take over and propel the vehicle into orbit. The hybrid propulsion system was a way to achieve safe, cost-effective access to space. The RTA featured an augmentor/ramburner, or hyperburner, a key component of the Turbine Based Combined Cycle (TBCC) engine. During takeoff and transition to supersonic flight, the device would serve as a conventional augmentor boosting the turbine engine thrust by approximately 50 percent. The augmentor would transition to a ramburner between Mach 2 and 3 to accelerate the vehicle to speeds above Mach 4. GE worked to construct a fan to demonstrate the performance and efficiency of the new augmentor/ramburner.[129] The advanced propulsion technologies introduced during NASA's UEET and DOD's IHPTET programs contributed to the knowledge base of the RTA project.[130] Glenn engineers completed tests in their W8 facility and documented the results in various NASA and ASME reports.

X-43 and X-51 Prove the Supersonic Combustion Ramjet

NASA's Hyper-X research program investigated hypersonic flight with the X-43A research vehicle, an airframe integrated with a new type of aircraft engine, a supersonic combustion ramjet (scramjet). Previous hypersonic craft—the X-15, the lifting bodies, various reentry vehicles, and the Space Shuttle—had relied upon rocket power for propulsion. A conventional air-breathing jet engine, which relied upon rotating machinery and the mixture of air and atomized fuel for combustion, could propel aircraft only to speeds between Mach 3 and 4. A conventional subsonic combustion ramjet could exceed Mach 4. But a scramjet could operate well past Mach 5 and, in theory, all the way into orbit. Its converging inlet compressed and accelerated the incoming air to supersonic speeds. From there, combustors ignited the fuel and air mixture to produce heat, and a diverging nozzle accelerated the heated air to produce thrust. The disadvantage of a scramjet was twofold. First, like the older subsonic ramjet, it was unable to propel a vehicle at very low subsonic speeds. Second, igniting a scramjet was no easy matter; it has been compared to lighting a match in a hurricane. Programs like the X-43 and the later X-51 addressed two very fundamental questions: (1) Could a scramjet ignite? And if so, (2) could it produce positive thrust (i.e., more thrust than drag)? As the new century opened, the answer was by no means clear.

To get basic answers, Langley tested a spare flight engine on an X-43 wind tunnel model that accurately represented the size and shape of the full-scale vehicle. The model was tested in the Langley 8-Foot High Temperature Tunnel to verify the propulsion system performance at Mach 7 flight conditions. X-43A research vehicles subsequently made three flights from Dryden. For each, NASA used a modified first stage of a Pegasus winged rocket booster to get the X-43A up to speed after being dropped from a B-52 mother ship at 40,000 feet. The first flight on June 2, 2001, was a disaster due to the failure of the booster control surfaces shortly after launch. Following a lengthy flight safety and flight review process, the X-43A flew successfully for the first time on March 27, 2004, on its second flight attempt. During the 10-second flight, the little engine demonstrated the first successful operation of a scramjet in history as it reached Mach 6.83. Later, on November 16, 2004, a second,

The Power for Flight

Figure 6-6. The Pratt & Whitney Rocketdyne SJX61-2 scramjet undergoes ground testing simulating Mach 5 flight conditions in the Langley 8-Foot High Temperature Tunnel. (NASA)

11-second flight achieved Mach 9.68, then the fastest speed ever attained by an air-breathing engine, flying over 6,600 mph. Those flights produced more data on scramjet engines at high Mach numbers than all flights during the previous four decades, including the first free-flight data and the validation of predictive design tools.[131] NASA researchers had demonstrated that hypersonic flight was possible and pointed the way to the future of high-speed flight.

A follow-on program was the Boeing X-51A WaveRider hypersonic flight demonstrator, a cooperative effort between the Air Force, NASA, Boeing, Pratt & Whitney Rocketdyne, and DARPA, with overall management responsibility by the Propulsion Directorate of the Air Force Research Laboratory at Wright-Patterson Air Force Base, OH. The X-43 was simply an experiment to see if a scramjet engine could function. The X-51A aimed at developing a thermally balanced production-quality scramjet engine that was capable of operating for minutes, not just seconds, while using a conventional hydrocarbon fuel rather than an exotic propellant like liquid hydrogen. Indeed, the X-51A used JP-7, the same fuel as the legendary Lockheed Blackbird. Glenn provided the CFD expertise that allowed the accurate prediction of airflow. Langley validated the WaveRider's SJX61-2 scramjet at Mach 5 in its uniquely important 8-Foot High Temperature Tunnel during the period 2006 to 2008.[132]

For its flight on May 26, 2010, the X-51A accelerated over the coast of southern California to Mach 4.87 for 143 seconds, almost two and a half minutes, which became the longest hypersonic flight in history.[133] But subsequent tests were not successful. On June 13, 2011, the X-51A experienced "unstart" following engine ignition, resulting in changes to its fuel-injection system. On August 14, 2012, a fin failed during boost—as had earlier happened on the first X-43 flight—dooming the craft from the outset. Thus, the long-term success of scramjet engines remained in doubt.[134] But then, on May 1, 2013, the X-51A's fourth and final flight, it dropped away from an Air Force Boeing B-52H Stratofortress at an altitude of 50,000 feet over the Pacific Ocean, accelerating under its booster to over Mach 4.5. At that point, the booster separated and the scramjet ignited, accelerating the X-51A from Mach 4.8 to Mach 5.1 in 26 seconds.[135] The powered portion of the flight lasted 240 seconds and constituted a milestone in flight propulsion—what might be termed the "Lindbergh moment" of scramjet propulsion, the point where the scramjet proved itself capable of operating over hundreds of miles in predictable and reliable fashion.[136] Much like the engines created by the Wrights, as well as Frank Whittle and Hans von Ohain's turbojets, scramjets offered the promise of a new revolution in aviation—in this particular case, high-speed global-ranging travel at Mach 5 and above.

Endnotes

1. John C. Freche, "Progress in Superalloys," NASA TN D-2495, October 1964, pp. 1–2, 6.
2. St. Peter, *History of Aircraft Gas Turbine Engine Development*, pp. 416–417; Glenn Research Center, "Turbine Disk Alloys," June 20, 2012, at *http://www.grc.nasa.gov/WWW/StructuresMaterials/AdvMet/research/turbine_disks.html* (accessed September 5, 2012).
3. T.T. Serafini, "PMR Polymide Composites for Aerospace Applications," in *Polymides: Synthesis, Characterization, and Applications*, vol. 2 (New York: Plenum Press, 1984), pp. 957–975.
4. St. Peter, *History of Aircraft Gas Turbine Engine Development*, p. 416.
5. Curt H. Liebert and Francis S. Stepka, "Potential Use of Ceramic Coating as a Thermal Insulation on Cooled Turbine Hardware," NASA TM X-3352, 1976, p. 1.
6. Robert A. Miller, "History of Thermal Barrier Coatings for Gas Turbine Engines: Emphasizing NASA's Role from 1942 to 1990," NASA TM 2003-215459, March 2009, pp. 6, 22.
7. Conway, *High-Speed Dreams*, p. 271; St. Peter, *History of Aircraft Gas Turbine Engine Development*, pp. 417–418.
8. Joseph R. Stevens, "NASA's HITEMP Program for UHBR Engines" (AIAA Paper 90-2395, presented at the 26th AIAA/SAE/ASME/ASEE [American Society for Engineering Education] Joint Propulsion Conference, Orlando, FL, July 16–18, 1990), p. 1; NASA, "Advanced High-Temperature Engine Materials Technology Progresses," n.d. [1995], available at *http://ntrs.nasa.gov/archive/nasa/casi.ntrs.nasa.gov/20050169149.pdf* (accessed October 4, 2014). Superalloys were also part of HITEMP's materials focus.
9. NASA, "Advanced High-Temperature Engine Materials Technology Progresses," n.d. [1995].
10. Ibid.; NASA, "Advanced High-Temperature Engine Materials Technology Progresses," n.d. [1996], available at *http://ntrs.nasa.gov/archive/nasa/casi.ntrs.nasa.gov/20050177123.pdf* (accessed October 5, 2014); "About the R&D 100 Awards," *R&D Magazine* (2014), available at *http://www.rd100awards.com/about-rd-100-awards* (accessed October 5, 2014).
11. Bradley A. Lerch, Susan L. Draper, J. Michael Pereira, Michael V. Nathal, and Curt Austin, "Resistance of Titanium Aluminide to Domestic Object Damage Assessed," in *Research & Technology, 1998*, NASA TM-1999-208815, April 1, 1999, p. 124.

12. Guy Norris, "Power House," *Flight International* (June 13, 2006), available at *http://www.flightglobal.com/news/articles/power-house-207148/* (accessed October 4, 2014).
13. Michael Nathal, "Glenn Takes a Bow for Impact on GEnx Engine," July 11, 2008, *http://www.nasa.gov/centers/glenn/news/AF/2008/July08_GEnx.html* (accessed September 22, 2014).
14. Malcolm Gibson, "Composite Fan Casings: Increasing Safety and Fuel Efficiency for Commercial Aircraft," in *NASA Innovation in Aeronautics: Select Technologies That Have Shaped Modern Aviation*, Clayton J. Bargsten and Malcolm T. Gibson, NASA TM-2011-216987, 2011, pp. 27–28.
15. Gibson, "Composite Fan Casings," p. 24; Banke, "Advancing Propulsive Technology," p. 766.
16. Gibson, "Composite Fan Casings," p. 24.
17. Ibid., pp. 24–25.
18. Gibson, "Composite Fan Casings," p. 25; Nathal, "Glenn Takes a Bow for Impact on GEnx Engine."
19. Membership in the Jet Engine Containment Concepts and Blade-Out Simulation Team included Glenn, the FAA, Rolls-Royce, Boeing, GE, Honeywell, A&P Technology, Williams International, North Coast Composites, North Coast Tool and Mold, Cincinnati Testing Laboratories, MSC Software, the Ohio Aerospace Institute, Ohio State University, and the University of Akron. NASA Aeronautics Research Mission Directorate, "Technical Excellence 2004: New Material Improves Rotor Safety," September 7, 2007, *http://www.aeronautics.nasa.gov/te04_rotor_safety.htm* (accessed August 22, 2013).
20. Gibson, "Composite Fan Casings," p. 26.
21. NASA Office of the Chief Technologist, "NASA Spinoff: Damage-Tolerant Fan Casings for Jet Engines," 2006, *http://spinoff.nasa.gov/Spinoff2006/T_1.html* (accessed August 23, 2013); Gibson, "Composite Fan Casings," p. 26; Gary D. Roberts, J. Michael Pereira, Michael S. Braley, and William A. Arnold, "Design and Testing of Braided Composite Fan Case Materials and Components," ISABE-2009-1201, available at *http://www.braider.com/pdf/Papers-Articles/Design-and-Testing-of-Braided-Composite-Fan-Case-Materials-and-Components.pdf* (accessed March 17, 2013); Bob Griffiths, "Composite Fan Blade Containment Case," *High Performance Composites*, May 2005, available at *http://www.braider.com/pdf/Papers-Articles/Composite-Fan-Blade-Containment-Case.pdf* (accessed March 17, 2013).
22. Gibson, "Composite Fan Casings," p. 28.

23. G.D. Roberts, R.K. Goldberg, W.K. Binienda, W.A. Arnold, J.D. Littell, and L.W. Kohlman, "Characterization of Triaxial Braided Composite Material Properties for Impact Simulation," NASA TM-2009-215660, September 2009; Gibson, "Composite Fan Casings," p. 28.
24. Gibson, "Composite Fan Casings," p. 28.
25. Ibid., p. 28.
26. Ibid., p. 29.
27. Ibid., p. 29; A&P Technology, "GEnx Engine," 2013, *http://www.braider.com/Case-Studies/GEnx-Engine.aspx* (accessed August 23, 2013); GE Aviation, "New GEnx Engine Advancing Unprecedented Use of Composites in Jet Engines," December 14, 2004, *http://www.geaviation.com/press/genx/genx_20041214.html* (accessed August 23, 2013).
28. Gibson, "Composite Fan Casings," p. 30.
29. Ibid., p. 30.
30. *Spinoff 2008: Fifty Years of NASA-Derived Technologies (1958–2008)* (Washington, DC: NASA Center for Aerospace Information, 2008), pp. 178–179.
31. Ibid., pp. 178–179.
32. U.S. Department of Defense, "Integrated High Performance Turbine Engine Technology (IHPTET)," n.d. [2000], available at *http://web.archive.org/web/20060721223255/http://www.pr.afrl.af.mil/divisions/prt/ihptet/ihptet.html* (accessed January 28, 2016); St. Peter, *History of Aircraft Gas Turbine Engine Development*, pp. 383, 385.
33. St. Peter, *History of Aircraft Gas Turbine Engine Development*, pp. 384–385.
34. Ibid., pp. 425–426.
35. Ibid., pp. 396, 398, 402, 410, 425–426.
36. Richard A. Brokopp and Robert S. Gronski, "Small Engine Components Test Facility Compressor Testing Cell at NASA Lewis Research Center" (AIAA Paper 92-3980, presented at the 17th AIAA Aerospace Ground Testing Conference, Nashville, TN, July 6–8, 1992); St. Peter, *History of Aircraft Gas Turbine Engine Development*, pp. 383, 396, 416, 422.
37. IHPTET and VAATE, "Turbine Engine Technology: A Century of Power for Flight," 2002, at *http://web.archive.org/web/20060715190755/http://www.pr.afrl.af.mil/divisions/prt/ihptet/ihptet_brochure.pdf* (accessed January 28, 2016); St. Peter, *History of Aircraft Gas Turbine Engine Development*, pp. 416, 419.
38. Turbomachinery and Heat Transfer Branch, Glenn Research Center, "APNASA: Overview," August 14, 2007, *http://www.grc.nasa.gov/WWW/RTT/Codes/APNASA.html* (accessed September 21, 2014).
39. Banke, "Advancing Propulsive Technology," pp. 761–762.

40. John K. Lytle, "The Numerical Propulsion System Simulation: An Overview," NASA TM 2000-209915, June 2000, p. 1; P.T. Homer and R.D. Schlichting, "Using Schooner To Support Distribution and Heterogeneity in the Numerical Propulsion System Simulation Project," *Concurrency and Computation: Practice and Experience* 6 (June 1994): 271.
41. "Numerical Propulsion System Simulation (NPSS): An Award Winning Propulsion System Simulation Tool," in *Research and Technology, 2001*, NASA TM-2002-211333, 2002, pp. 228–229; Banke, "Advancing Propulsive Technology," pp. 761–762.
42. U.S. Department of Defense, "Integrated High Performance Turbine Engine Technology (IHPTET)," n.d. [2000]; St. Peter, *History of Aircraft Gas Turbine Engine Development*, pp. 383–385.
43. National Research Council of the National Academies, *Recapturing NASA's Aeronautics Flight Research Capabilities* (Washington, DC: National Academies Press, 2012), pp. 58–59.
44. Robert J. Shaw, "NASA's Ultra-Efficient Engine Technology (UEET) Program/Aeropropulsion Technology Leadership for the 21st Century" (paper presented at the 22nd International Congress of Aeronautical Sciences [ICAS], Harrogate, U.K., August 27–September 1, 2000), available at *http://www.dtic.mil/dtic/tr/fulltext/u2/a547568.pdf* (accessed September 5, 2012); Banke, "Advancing Propulsive Technology," pp. 763–765.
45. Timothy P. Gabb, Anita Garg, David L. Ellis, and Kenneth M. O'Connor, "Detailed Microstructural Characterization of the Disk Alloy ME3," NASA TM-2004-213066, May 2004, p. 1.
46. Nathal, "Glenn Takes a Bow for Impact on GEnx Engine."
47. Robert J. Shaw, "Ultra-Efficient Engine Technology Project Integrated into NASA's Vehicle Systems Program," in *Research and Technology 2003*, NASA TM-2004-212729, May 2004, n.p., available at *http://ntrs.nasa.gov/archive/nasa/casi.ntrs.nasa.gov/20050192140.pdf* (accessed August 31, 2013).
48. Sabrina Eaton, "NASA Programs on Jet Efficiency, Noise Survive Attempted Fund Cuts," [Cleveland] *Plain Dealer* (May 20, 1999): 15A.
49. Aircraft Owner's and Pilot's Association, "General Aviation Statistics," March 30, 2011, *http://www.aopa.org/About-AOPA/Statistical-Reference-Guide/General-Aviation-Statistics* (accessed August 31, 2013).
50. The AGATE program also directed considerable effort toward air traffic control, advanced avionics, and the creation of a new and modern infrastructure for general aviation. Langley Research Center, "Affordable Alternative Transportation: AGATE—Revitalizing General Aviation,"

Release FS-1996-07-02-LaRC, July 1996, *http://www.nasa.gov/centers/langley/news/factsheets/AGATE.html* (accessed August 31, 2013); "NASA Langley Research Center: Contributing to the Next 100 Years of Flight," NASA Release FS-2004-02-84-LaRC, February 2004, *http://www.nasa.gov/centers/langley/news/factsheets/FS-2004-02-84-LaRC.html* (accessed July 7, 2013).

51. "Signing Ceremony To Initiate Development of Revolutionary Aircraft Engines," NASA News Release N96-80, December 10, 1996, NASA HRC, file 011151.
52. Glenn Research Center, "Small Aircraft Propulsion: The Future Is Here," Release FS-2000-04-001-GRC, November 22, 2004, *http://www.nasa.gov/centers/glenn/about/fs01grc.html* (accessed July 7, 2013).
53. Dawson, *Engines and Innovation*, p. 37.
54. Glenn Research Center, "Small Aircraft Propulsion: The Future Is Here."
55. Ibid.
56. Glenn Research Center, "General Aviation's New Thrust: Single Lever Technology Promises Efficiency, Simplicity," NASA Release FS-1999-07-002-GRC, July 1999, available at *http://www.nasa.gov/centers/glenn/pdf/84789main_fs02grc.pdf* (accessed July 7, 2013).
57. Glenn Research Center, "Small Aircraft Propulsion: The Future Is Here."
58. Mike Busch, "GAP Engine Update," *AVweb*, July 27, 2000, *http://www.avweb.com/news/reviews/182838-1.html* (accessed August 31, 2013).
59. Alan Dron, "GA Engine Manufacturers Line Up Diesel Options," *Flight International* (April 12, 2011), available at *http://www.flightglobal.com/news/articles/ga-engine-manufacturers-line-up-diesel-options-355474/* (accessed August 31, 2013).
60. Williams International's partners included California Drop Forge, Cessna Aircraft, Chichester-Miles Consultants, Cirrus Design, Forged Metals, New Piper Aircraft, VisionAire, Producto Machine, Scaled Composites, and Unison Industries. "NASA Cooperative Engine Testbed Aircraft on Schedule," *Aerospace Propulsion* (July 7, 1997): 2, NASA HRC, file 011151; Glenn Research Center, "Small Aircraft Propulsion: The Future Is Here."
61. Williams International, "The General Aviation Propulsion (GAP) Program," NASA CR-2008-215266, July 2008, p. 5; Glenn Research Center, "Small Aircraft Propulsion: The Future Is Here."
62. Williams International, "The General Aviation Propulsion (GAP) Program," p. 18.
63. Ibid., pp. 26, 34–38, 44–45.
64. John Adamczyk, quoted in David Noland, "The Little Engine That Couldn't," *Air & Space/Smithsonian* (November 2005), available at

http://www.airspacemag.com/flight-today/the-little-engine-that-couldnt-6865253/?no-ist (accessed January 29, 2016).

65. Williams International, "The General Aviation Propulsion (GAP) Program," pp. 40–43.
66. Noland, "The Little Engine That Couldn't."
67. A reformed Eclipse Aerospace continued to service, modify, and manufacture VLJs in the wake of the original company's bankruptcy beginning in September 2009.
68. Drucella Andersen and Michael Mewhinney, "NASA Testing New, Powerful 'Ducted Fan' Engine for Civil Tests," NASA Release 93-103, June 3, 1993, at *http://www.nasa.gov/home/hqnews/1993/93-103.txt* (accessed September 23, 2014).
69. "Ames Tests New Jet Engine," *NASA News* (June 1993): 4, NASA HRC, file 011151.
70. Guy Norris, "P&W Prepares for Geared Fan Launch," *Flight International* (February 18, 1998), available at *http://www.flightglobal.com/news/articles/pw-prepares-for-geared-fan-launch-32908/* (accessed August 19, 2013).
71. Pratt & Whitney, "PurePower PW1000G Engine," 2010, available at *http://www.purepowerengine.com* (accessed January 29, 2016).
72. "Airlines Continue To Reduce the Use of Noisy Aircraft," FAA News Release 21-93, June 10, 1993, NASA HRC, file 012336; Banke, "Advancing Propulsive Technology," p. 739; U.S. Congress, Office of Technology Assessment, *Federal Research and Technology for Aviation*, OTA-ETI-610 (Washington, DC: U.S. Government Printing Office, 1994), pp. 78–79.
73. Glenn Research Center, "Making Future Commercial Aircraft Quieter: Glenn Effort Will Reduce Engine Noise," NASA Release FS-1999-07-003-GRC, July 1999, available at *http://www.nasa.gov/centers/glenn/pdf/84790main_fs03grc.pdf* (accessed July 7, 2013).
74. John K.C. Low, Paul S. Schweiger, John W. Premo, and Thomas J. Barber, "Advanced Subsonic Technology (AST) Separate-Flow High-Bypass Ratio Nozzle Noise Reduction Program Test Report," NASA CR-2000-210040, December 2000, p. 1.
75. Robert A. Golub, John W. Rawls, Jr., and James W. Russell, "Evaluation of the Advanced Subsonic Technology Program Noise Reduction Benefits," NASA TM-2005-212144, May 2005, p. 1; Glenn Research Center, "Making Future Commercial Aircraft Quieter."
76. Golub et al., "Evaluation of the Advanced Subsonic Technology Program Noise Reduction Benefits," p. 4.
77. Ibid., pp. 4–5, 7.

78. Glenn Research Center, "Making Future Commercial Aircraft Quieter."
79. Jim Banke, "Technology Readiness Levels Demystified," November 19, 2013, *http://www.nasa.gov/topics/aeronautics/features/trl_demystified.html#.VDAA7vldUeg* (accessed October 4, 2014); John C. Mankins, "Technology Readiness Levels: A White Paper," April 6, 1995, available at *http://www.hq.nasa.gov/office/codeq/trl/trl.pdf* (accessed October 4, 2014); Golub et al., "Evaluation of the Advanced Subsonic Technology Program Noise Reduction Benefits," pp. 10, 12.
80. Golub et al., "Evaluation of the Advanced Subsonic Technology Program Noise Reduction Benefits," pp. 4–5, 23.
81. Ibid., p. 23.
82. G. Bielak, J. Gallman, R. Kunze, P. Murray, J. Premo, M. Kosanchick, A. Hersh, J. Celano, B. Walker, J. Yu, H.W. Kwan, S. Chiou, J. Kelly, J. Betts, J. Follet, and R. Thomas, "Advanced Nacelle Acoustic Lining Concepts Development," NASA CR-2002-211672, August 2002, pp. i, 1.
83. Golub et al., "Evaluation of the Advanced Subsonic Technology Program Noise Reduction Benefits," pp. 2, 24, 26. "Cutback" is a sound-reducing procedure performed at takeoff. The flightcrew takes off with full power, climbs rapidly, and then cuts the thrust to a predetermined value at a specified altitude. The airliner continues to climb at a slower, quieter rate until it reaches an altitude where sound is no longer an issue.
84. Ibid., pp. 24–25.
85. Ibid., pp. 25–27.
86. Jia Yu, Hwa-Wan Kwan, and Eugene Chien, "PW 4098 Forward Fan Case Acoustic Liner Design Under NASA EVNRC Program" (AIAA Paper 2010-3827, presented at the 16th AIAA/CEAS Aeroacoustics Conference, Stockholm, Sweden, 2010), p. 2; Douglas C. Matthews, Larry A. Bock, Gerald W. Bielak, R.P. Dougherty, John W. Premo, Dan F. Scharpf, and Jia Yu, "Pratt & Whitney/Boeing Engine Validation of Noise Reduction Concepts: Final Report for NASA Contract NAS3-97144, Phase 1," NASA CR-2014-218088, February 2014, p. iii.
87. Golub et al., "Evaluation of the Advanced Subsonic Technology Program Noise Reduction Benefits," p. 2.
88. Ibid., pp. 3, 13. The interior noise-reduction program was the only element that did not survive the cancellation of the AST program because it only reached a TRL of 4.
89. Ibid., pp. 1, 13.
90. Golub et al., "Evaluation of the Advanced Subsonic Technology Program Noise Reduction Benefits," p. 25; Glenn Research Center, "Making Future Commercial Aircraft Quieter."

91. Glenn Research Center, "Making Future Commercial Aircraft Quieter."
92. Ibid.
93. Golub et al., "Evaluation of the Advanced Subsonic Technology Program Noise Reduction Benefits," p. 25.
94. Kristin K. Wilson, "Quieting the Skies: Cessna Benefits from Lewis Noise Reduction Know How," *Lewis News* (February 1998): 5.
95. Glenn Research Center, "Making Future Commercial Aircraft Quieter."
96. K.B.M.Q. Zaman, J.E. Bridges, and D.L. Huff, "Evolution from 'Tabs' to 'Chevron Technology': A Review," in *Proceedings of the 13th Asian Congress of Fluid Mechanics* (Dhaka, Bangladesh: Engineers Institute of Bangladesh, 2010), pp. 47–63; Springer, "NASA Aeronautics: A Half-Century of Accomplishments," pp. 200–201.
97. Low et al., "Advanced Subsonic Technology (AST) Separate-Flow High-Bypass Ratio Nozzle Noise Reduction Program Test Report," pp. 1, 3.
98. Malcolm T. Gibson, "The Chevron Nozzle: A Novel Approach to Reducing Jet Noise," in *NASA Innovation in Aeronautics*, p. 4.
99. Golub et al., "Evaluation of the Advanced Subsonic Technology Program Noise Reduction Benefits," p. 13.
100. Low et al., "Advanced Subsonic Technology (AST) Separate-Flow High-Bypass Ratio Nozzle Noise Reduction Program Test Report," pp. 1–2; Gibson, "The Chevron Nozzle," p. 5.
101. Clifford Brown and James Bridges, "An Analysis of Model Scale Data Transformation to Full Scale Flight Using Chevron Nozzles," NASA TM-2003-212732, December 2003, pp. 1–7; Zaman et al., "Evolution from 'Tabs' to 'Chevron Technology'," pp. 47–63; Gibson, "The Chevron Nozzle," pp. 6–7.
102. Thomas Edwards, "The Future of Green Aviation" (NASA paper ARC-E-DAA-TN5136, presented at the Future of Flight Foundation Celebrate the Future Conference, Mukilteo, WA, April 22, 2012), p. 67.
103. Peter Bartlett, Nick Humphreys, Pam Phillipson, Justin Lan, Eric Nesbitt, and John Premo, "The Joint Rolls-Royce/Boeing Quiet Technology Demonstrator Programme" (AIAA Paper 2004-2869, presented at the 10th AIAA/CEAS Aeroacoustics Conference, Manchester, U.K., May 10–12, 2004).
104. Guy Norris, "Chevron Tests 'Better Than Expected'," *Flight International* 160 (November 20–26, 2001): 10; Gibson, "The Chevron Nozzle," p. 7.
105. "Alleviating Aircraft Noise: The Quiet Aircraft Technology Program," *NASA Destination Tomorrow*, program 9, 2003, available at *http://www.youtube.com/watch?v=3vFSpW-zC2U* (accessed July 7, 2013); Charlotte E. Whitfield, "NASA's Quiet Aircraft Technology Project," NASA TM-2004-213190, 2004; Gibson, "The Chevron Nozzle," p. 7.

106. Vinod G. Mengle, "Jet Noise Characteristics of Chevrons in Internally Mixed Nozzles" (AIAA Paper 2005-2934, presented at the 11th AIAA/CEAS Aeroacoustics Conference, Monterey, CA, May 23–25, 2005); Gibson, "The Chevron Nozzle," p. 7.
107. Vinod G. Mengle, Ronen Elkoby, Leon Brusniak, and Russ H. Thomas, "Reducing Propulsion Airframe Aeroacoustic Interactions with Uniquely Tailored Chevrons: 1. Isolated Nozzles" (AIAA Paper 2006-2467, presented at the 12th AIAA/CEAS Aeroacoustics Conference, Cambridge, MA, May 8–10, 2006); Gibson, "The Chevron Nozzle," p. 7.
108. Frederick T. Calkins, George W. Butler, and James H. Mabe, "Variable Geometry Chevrons for Jet Noise Reduction" (AIAA Paper 2006-2546, presented at the 12th AIAA/CEAS Aeroacoustics Conference, Cambridge, MA, May 8–10, 2006); Gibson, "The Chevron Nozzle," pp. 7–8.
109. Gibson, "The Chevron Nozzle," p. 9.
110. Zaman et al., "Evolution from 'Tabs' to 'Chevron Technology'," p. 47; General Electric Aviation, "CF34-10E Turbofan Propulsion System," May 2010, available at *http://www.geaviation.com/engines/docs/commercial/datasheet-CF34-10E.pdf* (accessed July 1, 2014); General Electric Aviation, "CF34-10E Engines Outperforming Expectations," May 13, 2014, available at *http://www.geaviation.com/press/cf34/cf34_20140513b.html* (accessed July 1, 2014).
111. Springer, "NASA Aeronautics: A Half-Century of Accomplishments," pp. 200–201.
112. Bob Burnett, "Ssshhh, We're Flying a Plane Around Here: A Boeing-Led Team Is Working To Make Quiet Jetliners Even Quieter," *Boeing Frontiers Online* 4 (December 2005/January 2006), *http://www.boeing.com/news/frontiers/archive/2005/december/ts_sf07.html* (accessed July 13, 2013); Eric J. Bultemeier, Ulrich Ganz, John Premo, and Eric Nesbitt, "Effect of Uniform Chevrons on Cruise Shockcell Noise" (AIAA Paper 2006-2440, presented at the 12th AIAA/CEAS Aeroacoustics Conference, Cambridge, MA, May 8–10, 2006).
113. *Spinoff 2008*, pp. 178–179.
114. Travis L. Turner, Randolph H. Cabell, Roberto J. Cano, and Richard J. Silcox, "Testing of SMA-Enabled Active Chevron Prototypes Under Representative Flow Conditions," in *Proceedings of SPIE*, vol. 6928, *Active and Passive Smart Structures and Integrated Systems 2008*, ed. Mehdi Ahmadian (May 2008), available at *http://ntrs.nasa.gov/archive/nasa/casi.ntrs.nasa.gov/20080014174.pdf* (accessed September 15, 2014); Gibson, "The Chevron Nozzle," p. 8.

115. Gibson, "The Chevron Nozzle," pp. 8, 10; Springer, "NASA Aeronautics: A Half-Century of Accomplishments," pp. 200–201.
116. Banke, "Technology Readiness Levels Demystified."
117. Stanley F. Birch, P.A. Bukshtab, K.M. Khritov, D.A. Lyubimov, V.P. Maslov, A.N. Secundov, and K.Y. Yakubovsky, "The Use of Small Air Jets to Simulate Metal Chevrons" (AIAA Paper 2009-3372, presented at the 15th AIAA/CEAS Aeroacoustics Conference, Miami, FL, May 11–13, 2009); Gibson, "The Chevron Nozzle," p. 9.
118. Langley Research Center, "NASA's High-Speed Research (HSR) Program—Developing Tomorrow's Supersonic Passenger Jet," April 22, 2008, *http://www.nasa.gov/centers/langley/news/factsheets/HSR-Overview2.html* (accessed September 25, 2014).
119. Boeing Commercial Airplanes, "High-Speed Civil Transport Study: Summary," NASA CR-4234, September 1989, pp. 1–10.
120. Pratt & Whitney and General Electric Aircraft Engines, "Critical Propulsion Components, Volume 1: Summary, Introduction, and Propulsion Systems Studies," NASA CR-2005-213584, May 2005, pp. iii, 1–2; Reddy, "Seventy Years of Aeropropulsion Research at NASA Glenn Research Center," p. 212; FAA, "Noise Abatement Departure Profiles," Advisory Circular 91-53A, July 22, 1993, available at *http://rgl.faa.gov/Regulatory_and_Guidance_Library/rgAdvisoryCircular.nsf/list/AC%2091-53A/$FILE/ac91-53.pdf* (accessed September 1, 2012).
121. Gibson, "The Chevron Nozzle," p. 4.
122. "High Speed Engine Cycles Tapped for Further Research," NASA News Release 94-41, March 14, 1994, NASA HRC, file 011151.
123. Pratt & Whitney and General Electric Aircraft Engines, "Critical Propulsion Components, Volume 1," pp. 150, 169; Conway, *High-Speed Dreams*, pp. 297–300.
124. "President Reagan's Speech Before Joint Session of Congress," *New York Times* (February 5, 1986): A20.
125. Larry Schweikart, *The Quest for an Orbital Jet*, vol. 3 of *The Hypersonic Revolution*, ed. R.P. Hallion (Washington, DC: Air Force History and Museums Program, 1996), pp. 48, 363.
126. Curtis Peebles, "Learning from Experience: Case Studies of the Hyper-X Project" (paper presented at the 47th AIAA Aerospace Sciences Meeting, Orlando, FL, January 5–8, 2009), pp. 4–5.
127. Richard P. Hallion, "The History of Hypersonics; or, 'Back to the Future—Again and Again'" (AIAA Paper 2005-0329, presented at the AIAA 43rd Aerospace Sciences Meeting and Exhibit, Reno, NV, January 10–13, 2005).

128. GE Aviation, "GE Aircraft Engines Pursuing Mach 4 Jet Engine at NASA Research Center," July 22, 2002, *http://www.geaviation.com/press/other/other_20020722aa.html* (accessed June 6, 2013).

129. Jinho Lee, Robert J. Buehrle, and Ralph Winslow, "The GE-NASA RTA Hyperburner Design and Development," NASA TM-2005-213803, June 2005, pp. 1–2.

130. GE Aviation, "GE Aircraft Engines Pursuing Mach 4 Jet Engine at NASA Research Center"; Guy Norris, "GE Wins High-Mach Turbine Work," *Flight International* (July 30, 2002), available at *http://www.flightglobal.com/news/articles/ge-wins-high-mach-turbine-work-152266/* (accessed May 1, 2012).

131. Warren E. Leary, "NASA Jet Sets Record for Speed," *New York Times* (November 17, 2004): A24; Curtis Peebles, "Learning from Experience: Case Studies of the Hyper-X Project" (paper presented at the 47th AIAA Aerospace Sciences Meeting, Orlando, FL, January 5–8, 2009); Curtis Peebles, "The X-43A Flight Research Program: Lessons Learned on the Road to Mach 10" (unpublished paper, Dryden Flight Research Center, 2007); Springer, "NASA Aeronautics: A Half-Century of Accomplishments," pp. 198–199.

132. James L. Pittman, John M. Koudelka, Michael J. Wright, and Kenneth E. Rock, "Hypersonics Project Overview" (paper presented at the 2011 Technical Conference, Cleveland, OH, March 15–17, 2011), available at *http://www.aeronautics.nasa.gov/pdf/hypersonics_project.pdf* (accessed December 26, 2012).

133. Boeing Phantom Works, "Boeing X-51A WaveRider Breaks Record in First Flight," May 26, 2010, *http://boeing.mediaroom.com/index.php?s=43&item=1227* (accessed December 26, 2012); Boeing Defense, Space and Security, "Backgrounder: X-51A WaveRider," September 2012, *http://www.boeing.com/assets/pdf/defense-space/military/waverider/docs/X-51A_overview.pdf* (accessed January 29, 2016).

134. Sharon Weinberger, "X-51 WaveRider: Hypersonic Jet Ambitions Fall Short," August 15, 2012, *http://www.bbc.com/future/story/20120815-hypersonic-ambitions-fall-short* (accessed January 29, 2016).

135. Daryl Mayer, "X-51A WaveRider Achieves History in Final Flight," Wright-Patterson Air Force Base News Release, May 3, 2013, *http://www.afmc.af.mil/News/Article-Display/Article/153475/x-51a-waverider-achieves-history-in-final-flight/* (accessed May 31, 2013); Guy Norris, "X-51A WaveRider Achieves Hypersonic Goal on Final Flight," *Aviation Week & Space Technology* (May 2, 2013), available at *http://aviationweek.*

com/defense/x-51a-waverider-achieves-goal-final-flight (accessed May 31, 2013); details also from Dr. Richard P. Hallion.
136. The term "Lindbergh moment" was coined by Air Force Chief Scientist Dr. Mark J. Lewis in a conversation with Dr. Richard P. Hallion.

Seen from the rear in this photo, the DC-8 Airborne Science Laboratory generates exhaust contrails for the ACCESS project in 2013. (NASA)

CHAPTER 7
Toward the Future

On August 14, 2013, NASA Administrator (and former Shuttle astronaut) Charles Bolden addressed the Nation's leading aeronautical engineers, managers, and other professionals at the American Institute of Aeronautics and Astronautics (AIAA) Aviation 2013 conference held in Los Angeles. He announced a "new strategic vision" for NASA's aeronautics work, buttressing his remark with a statistical review of the Nation's aeronautical health. In 2011, civil and general aviation had accounted for $1.3 trillion of American economic activity in general, with 10.2 million jobs generated both directly and indirectly. Passenger revenue and airfreight brought in $636 billion and $1.5 trillion respectively. To meet the Nation's—indeed, the globe's—seemingly insatiable demand for air transport, NASA was tackling six emerging challenges at the "bold, anticipatory edge"

Figure 7-1. Pictured is former NASA Administrator Charles Bolden. (NASA)

that could alter both the perception and use of aviation during the next two to four decades.

Three of those challenges addressed aircraft propulsion directly or indirectly: commercial supersonic aircraft that emitted few or no sonic booms, ultra-efficient commercial transports that incorporated effective and environmentally pioneering technology, and a transition to low-carbon propulsion and alternative fuels that would stimulate the economy and protect Earth. The other three research thrusts were safe, efficient growth in global operations; real-time, systemwide safety assurance; and assured autonomy for aviation transformation. Bolden believed that NASA specifically, and aviation in general, was undergoing a Renaissance in the sky and on the ground and that the critical factors of safety, energy efficiency, the environment, innovation, and responsible management all complemented each other.[1]

Bolden's statements coincided with the Agency's issuing a white paper titled "Transforming Global Mobility" and were a clarion call for the American aviation industry.[2] As he stressed that the changes to advance the future of flight were a communal endeavor, many NASA programs from the early 2000s that reflected those issues were coming to fruition. They also reflected the organizational changes implemented by the former Aeronautics head, Lisa Porter, in early 2006, which had refocused NASA Aeronautics back toward in-house fundamental research, with the addition of strengthening relationships between industry and academia.[3] Some of the Agency's new and renewed initiatives are discussed below.

2007: The Open-Rotor Revival

A revival of prop-fan and unducted fan technology, now called the "open rotor," by NASA, GE, and Rolls-Royce, reflected ongoing concerns over fuel efficiency and new priorities based on reducing the environmental impact of engine emissions in the early 21st century.[4] Since the late 1980s, aircraft configurations, engine technology, noise requirements, and economic requirements have changed. With higher fuel prices and a restricted economic environment, NASA reestablished its interest in open-rotor propulsion. The target application was short- to medium-range, twin-engine, narrow-body jet airliners. The type constituted a considerable section of the global airline fleet in the form of the Boeing 737, the Airbus A320, and what was projected to be 69 percent of new aircraft produced between 2010 and 2030. Work toward the reduction of fuel consumption, noise, and emissions in these aircraft would minimize the future environmental impact of aviation.[5]

Toward the Future

Figure 7-2. The open-rotor concept undergoes testing in the 8- by 6-Foot Supersonic Wind Tunnel at Glenn. (NASA)

The renewed investigations explored methodologies for aircraft-level sizing, performance analysis, and system-level noise analysis in 2012. A Glenn-GE team applied those methods to an advanced single-aisle aircraft using open-rotor engines, where, in the spirit of the UDF from the 1980s, the power plants featured external rotating blade forms from the ATP program. Their results indicated that open-rotor engines had the potential to provide reduced fuel consumption and emissions at high levels. The initial noise analysis indicated that then-current noise regulations could be met with old blade designs, whereas modern, noise-optimized blade designs were expected to result in significantly lower noise levels. There still remained a lot of work to do to bring the correct method of analysis for proper evaluation up to the same level as turbofan capabilities before an actual engine would take to the air.[6]

Other investigations of open rotors stressed the fuel efficiency and emissions merits of open-rotor engines. There were still problems of noise, which was becoming increasingly important as new noise regulations shaped the design of aircraft.[7] One observer simply stated, "Want to save the planet?" Then "put up with noisier airports."[8]

2009: The Environmentally Responsible Aviation Project

While efficiency and environmental concerns had been important since the 1960s in terms of aircraft propulsion systems, the latter became even more important in a new era of green aviation. NASA created the Environmentally Responsible Aviation (ERA) Project in 2009 to explore aircraft concepts and technologies that reduced the impact of aviation on the environment in terms of fuel efficiencies, lower noise levels, and reduced harmful emissions for the next 30 years. ERA was part of the Integrated Systems Research Program sponsored by the Aeronautics Research Mission Directorate (ARMD). The overall goal of ERA was to develop the technologies that made aircraft safer, faster, and more efficient, which would help transform the national air transportation system. Each of the major NASA research centers contributed to the effort. Agency leaders were quick to stress that the key to the success of ERA was industry partnerships. The joint collaboration and funding moved the project forward and gave each member of the team a voice in shaping the technology.[9]

Phase one of ERA evaluated and nurtured new manufacturing techniques, structural materials, and advanced engines in NASA's laboratories. Phase two, which began in late 2012, placed more emphasis on ground and flight tests. The challenge to creating very quiet aircraft with low carbon footprints, according to project manager Fay Collier, was the integration of the various ideas as a practical system. NASA chose eight large-scale, integrated technology demonstrations to generate ERA research. Four of them were propulsion projects:

- The *highly loaded front block compressor demonstration* was to show advanced turbofan efficiency improvements in a transonic high-pressure compressor using two- and three-stage model tests.[10]
- The *second-generation ultra-high-bypass (UHB) propulsor integration* reflected the continued development of a geared turbofan engine to help reduce fuel consumption and noise.
- The *low–nitrogen oxide fuel flexible engine combustor integration* demonstrated a full ring-shaped engine combustor that produced very low emissions.
- Finally, there was work toward *UHB engine integration for a hybrid wing body* that would lead to the verification of power plant and airframe integration concepts that would allow fuel-consumption reductions in excess of 50 percent while reducing noise on the ground.

Within those major projects, NASA and industry researchers delved into five areas of research, three of which centered on propulsion: advanced fuel-efficient and quiet engines; engine combustors with improved emissions; and innovative airframe and engine integration designs that reduced fuel consumption and community noise.[11]

Toward the Future

Figure 7-3. Pratt & Whitney's Geared TurboFan technology became the basis for the PurePower engine series. (Pratt & Whitney)

Funding from ERA reaped benefits quickly and complemented other Government environmental programs. Pratt & Whitney announced that its Geared TurboFan ultra-high-bypass system, marketed as the PurePower engine, which had its origins in the ADP project of the 1990s, had successfully completed 275 hours of fan rig testing in Glenn's 9- by 15-Foot Low Speed Wind Tunnel in June 2013. The manufacturer credited ERA with paving the way for the development of its advanced ultra-high-bypass turbofan technology that reduced fuel consumption, emissions, and noise. The next step was to complete ground and flight testing under the auspices of the FAA's Continuous Lower Energy, Emissions, and Noise (CLEEN) program. Alan Epstein, vice president for technology and environment at Pratt & Whitney, remarked, "Our partnerships with NASA and the FAA are the key to completing the necessary testing to advance the technology for the second generation of the Geared TurboFan system."[12]

Built in 1968, the 9- by 15-Foot Low Speed Wind Tunnel became a premier facility for the aerodynamic and acoustic evaluation of fans, nozzles, inlets, propellers, and STOVL propulsion systems. To this day, researchers can use the tunnel to investigate engine system noise reduction, fan noise prediction codes and measurement methods, low-speed light applications for aircraft, advanced propulsion system components, high-speed and counter-rotating fans, and airport noise at speeds of up to 175 mph. Recent (as of this writing) programs and projects supported in the facility include the Ultra-Efficient Engine Technology (UEET), the Quiet Aircraft Technology (QAT), and the Versatile Affordable Advanced Turbine Engine (VAATE). For the United States, it is the only propulsion research facility capable of simulating takeoff, approach, and landing in a continuous, subsonic flow, wind tunnel environment.[13]

The Subsonic Fixed Wing (SFW) Project of NASA's Fundamental Aeronautics Program and the ERA Project of NASA's Integrated System Research Program established a series of design goals for future subsonic transport technology. They centered on the successive reduction of noise; landing, takeoff, and cruise emissions; and fuel and energy consumption with tiered completion dates of 2015, 2020, and 2025. NASA held the opinion that although manufacturers and airlines always wanted quieter, more efficient, and cleaner engines, the aviation industry's inherently conservative nature prevented it from making the technological and economic investments required to adopt the new and radical innovation necessary to meet those goals.[14]

NASA Generational Goals for Subsonic Aircraft[15]

Benefits	N+1 (2015)	N+2 (2020)	N+3 (2025)
Noise	−32 decibels	−42 decibels	−52 decibels
Landing and Takeoff Emissions	−60 percent	−75 percent	−80 percent
Cruise Emissions	−55 percent	−70 percent	−80 percent
Fuel/Energy Consumption	−33 percent	−50 percent	−60 percent

Toward the Future

2010: Electric Propulsion

The problems confronting environmentally compatible aviation led NASA engineers to go beyond traditional propulsion systems. Echoing changes in the automobile on the road, the electric airplane represented a viable possibility. The challenges involved specific energy and power requirements regarding the development of fuel cells and batteries. Flight-weight electric motors and methods for distributing large amounts of power were needed. Hybrid gas turbine and electric propulsion systems were also under consideration. On the airplane, engineers envisioned a turboelectric distributed propulsion system. Large engines at each wingtip would drive superconducting generators to power small, motor-driven propulsors.[16]

An important precedent was the development of a solar propulsion system for High-Altitude Long-Endurance (HALE) unmanned aerial vehicles (UAVs). Between 1994 and 2003, the joint NASA-industry Environmental Research and Aircraft Sensor Technology (ERAST) program worked to make such craft practical by evaluating their payload capacity and use as a sensor platform in atmospheric research and their overall value to the scientific, Government, and civilian communities. Four generations of flying wing–shaped HALE UAVs (Pathfinder, Pathfinder Plus, Centurion, and Helios), built in collaboration with cutting-edge aeronautical firm AeroVironment, relied upon solar cells, electric motors, and composite construction to achieve flights at record-breaking altitudes of up to 96,000 feet. An important byproduct of the alliance with industry was the availability of more efficient and mass-produced solar cells from SunPower Corporation. Nevertheless, the design of a backup power system that allowed the operation of HALE UAVs in periods of darkness remained a persistent challenge. The Glenn-sponsored Low Emissions Alternative Power program worked to overcome that limitation through the successful demonstration of a lightweight regenerative fuel cell system in September 2003 and July 2005.[17]

Electric propulsion also provides an avenue for a longstanding desire in aviation, the personal air vehicle. The proposed Puffin is a 300-pound, 12-foot-long, 14.5-foot-wingspan personal air vehicle powered by two 30-horsepower electric motors; it appeared in 2010. It resulted from a cooperative program between Langley, the Massachusetts Institute of Technology, the Georgia Institute of Technology, and the National Institute of Aerospace (NIA). The Puffin would have a cruising speed of 150 mph and a battery life of 50 miles.[18]

During the summer of 2011, NASA and Internet corporation Google sponsored the "Green Flight Challenge." Teams of aeronautical engineers competed for a grand prize of $1.35 million. Team Pipistrel-USA.com, the winning group, designed and built an electric-powered aircraft capable of flying four

The Power for Flight

Figure 7-4. The Puffin is designed to be a personal aircraft. (NASA/Mark Moore)

people for approximately 200 miles nonstop. The team's Taurus G4 electric aircraft accomplished the flight on the equivalent of a half-gallon of fuel.[19]

2011: The Future of Green Aviation

NASA also worked with green-energy advocacy groups to promote its new message. Thomas Edwards, director of aeronautics at Ames Research Center, was the keynote speaker at the Future of Flight Foundation's celebration of Earth Day on Sunday, April 22, 2012, north of Seattle.[20] He called for the development of "environmentally progressive" aircraft. He acknowledged that designers could "cherry pick" advances in other modes of transportation, such as electric-car battery technology, for new aircraft. Doing so would accelerate the development of higher-power batteries that would serve as the platform for innovating electric-powered aircraft. Edwards also identified the reduction of airport noise and carbon and other chemical emissions from engines as target areas.[21]

To Edwards and other advocates, the prospect of transforming aviation was intoxicating; as he remarked, "It is a really exciting time for aerospace engineers…. Just when people are saying that aircraft [technology] has been the

Toward the Future

Figure 7-5. The Team Pipistrel-USA Taurus G4 aircraft is shown here. (NASA)

same for the past 50 years…there is a fusion of information technology and traditional aerospace engineering that is making the field evolve very rapidly."[22]

Edwards argued that the aviation industry seriously impacted the environment and energy usage in the United States. Worldwide aviation fuel use made up fully 8 percent of the total 1.3 trillion gallons of refined fossil fuels used in 2011. Fuel accounted for 20 percent of the operating costs for the 18,000 commercial airplanes operated by American-based airlines. Aviation released 600 million tons of CO_2 per year. While aviation contributed only 3 percent of greenhouse gases, it accounted for 13 percent of overall climate impact. The impacts of aviation-produced water vapor and oxides of nitrogen were still unknown. On the ground, communities still complained of aircraft noise despite the FAA's $5 billion investment in abatement programs since 1980.[23]

There were two possible future directions. Echoing the trends that began in the 1970s, the reduction of noise pollution and the improvement of energy efficiency persisted as driving forces in aircraft development. A new priority was alleviating the impact of aviation on global climate change. The nature of the work in the future of green aviation would be an extensive collaborative effort between Government, industry, and academia centered on environmental compatibility and renewable sources of energy.[24]

Edwards also discussed the component development for ever-greener aircraft engines. There would be new and better high-temperature materials used for combustors and liners, electronic controls, and high-output fuel-delivery systems. Advanced adaptive fan blades changed their shape to adapt to required airflow characteristics for an engine embedded within a fuselage.[25]

NASA continued looking into synthetic fuels as it worked toward an understanding of alternative aviation fuels overall. Conducted in partnership with the U.S. Air Force, those investigations led to the first test of synthetic fuel derived by the Fischer-Tropsch process at Dryden in February 2009. Evaluation included burning the new fuel, a combination of carbon monoxide and hydrogen to produce liquid hydrocarbons, in a DC-8 and comparing the data to preexisting information from similar tests using conventional fuel.

Another avenue was biofuels, which offered the promise of even cleaner-burning fuels. Staff at Glenn established the Greenlab Research Facility in 2009 to investigate better ways of growing seawater algae and arid land halophytes, both promising platforms for biofuels. The combination of an indoor laboratory with an outdoor greenhouse permitted the basic study of the biology of renewable energy sources that could lead to customizable solutions to future fuel needs.[26] But the future of the Greenlab was soon left in doubt due to NASA's intention to stay committed to fundamental testing rather than developing outright new biofuels.

2013: Reducing Contrails and Cruise Emissions

Alternate biofuels offered the hope of a safe and effective way to reduce aviation's impact on the environment. NASA initiated the Alternative Fuel Effects on Contrails and Cruise Emissions (ACCESS) study in late 2012. It was a joint project involving researchers at Dryden, Langley, and Glenn. ACCESS was a follow-on study to Alternative Aviation Fuel Experiment studies conducted in 2009 and 2011. For those tests, researchers measured the exhaust emissions of NASA's DC-8 Airborne Science Laboratory as it burned alternative fuels while parked on the ramp at the Palmdale, CA, facility. The Fixed Wing Project within the Fundamental Aeronautics Program of NASA's ARMD managed ACCESS.[27]

The ACCESS study took NASA's investigations into burning biofuels and their effects on engine performance, emissions, and aircraft-generated contrails at altitude in late February 2013. Flying from Dryden's Aircraft Operations Facility in Palmdale, the research team used two aircraft. They filled the tanks of Dryden's DC-8 Airborne Science Laboratory with either conventional JP-8 jet fuel or a 50-50 blend of JP-8 and an alternative fuel of hydroprocessed esters and fatty acids derived from camelina plants. Bruce Anderson, a senior research scientist at Langley, described the new alternative fuel as "flower power." As the DC-8 flew high over the restricted airspace of Edwards Air Force Base,

Toward the Future

Figure 7-6. The researchers of the ACCESS project used Langley's heavily instrumented HU-25 Falcon to measure the chemical composition of the DC-8 Airborne Science Laboratory's exhaust contrail generated by a 50-50 mix of conventional JP-8 jet fuel and a plant-derived biofuel. (NASA/Lori Losey)

researchers in Langley's HU-25 Falcon jet followed in its wake to monitor over 20 scientific and navigation-related instruments designed to detect and record 20 different parameters of the DC-8's exhaust at various distances, altitudes, and engine power settings. Those first flights generated the best methodology to conduct the emissions sampling using the combination of the DC-8 and Falcon jets.

The second phase, called ACCESS II, took place in May 2014. It focused on compiling and adding to the research data obtained during the initial ACCESS experiment. The program included international involvement from the National Research Council (NRC) of Canada and the German Aerospace Center (DLR). The research fleet included NASA's Falcon and DC-8, joined by similar aircraft from both the NRC and the DLR. With the preliminary results, Anderson and his colleagues estimated that alternative fuel blends reduced black carbon emissions by more than 30 percent on the ground. Identifying such dramatic results in the air was more difficult, and the effect of alternative

The Power for Flight

fuels on contrail formation was still unclear. Nevertheless, work toward gaining a broader understanding of fossil fuel substitutes and how they could become more readily available and competitive in cost with conventional jet fuels was worthwhile.[28]

A Challenging Future…

NASA's propulsion specialists have worked for over 50 years to improve the overall operating efficiency of piston engines, propellers, gas turbine engines, and (recently) electric and other hybrid systems. As world events shaped the use of the airplane on the global stage, the specialists learned to recognize the importance of the environmental compatibility of aircraft propulsion systems and their effect on the quality of life on Earth. In many ways, work in the latter areas gave NASA an unprecedented opportunity to contribute to the development of aeronautics.

In the process, the style of NASA's engineering changed over time. During World War II and the early years of the Cold War, NACA researchers got away from wartime-derivative development work on the first generations of subsonic and supersonic gas turbines and ramjets and settled down to innovating and experimenting with the latest turbojet, turboprop, and advanced turbofan and afterburning/ram technology. They worked to make American military aircraft fly higher, faster, and farther as international tensions fueled development.

After the brief distraction of initial participation in the space program, the new NASA led the way in improving the efficiency and reducing the noise and emissions of the airliners of the Jet Age while it tried to do the same for the general aviation community. Federal funding of NASA's propulsion projects provided vital subsidiary funding support to aircraft manufacturers who would not have made the investment otherwise. This was an added benefit on top of the data that they received from NASA testing and analysis. Additionally, large-scale projects like the ACEE marshaled the resources of the Agency as it tried to redefine aircraft propulsion technology through technology transfer.

Throughout its history, NASA has seen its role as promoting and brokering unconventional ideas that could meet shared goals through investment in research and development. NASA's propulsion community still endeavors to change the airplane in ways that will make an impact on modern, everyday life. As aviation progresses onward into the era of scramjets, electric engines, and alternative fuels, it will take the continued vision and energy of NASA to bring new developments to maturity.

Endnotes

1. Charles Bolden, "Embracing a World of Change: NASA's Aeronautics Research Strategy" (address at AIAA Aviation 2013, Los Angeles, CA, August 14, 2013), available at *http://www.nasa.gov/sites/default/files/bolden_aiaa_aviation.pdf* (accessed September 1, 2013). The video of the address can be viewed at *http://www.livestream.com/aiaa/video?clipId=pla_08bf0b46-2019-4db0-997b-aacf7029b61b*.
2. The official press release echoing Bolden's remarks is titled "Aeronautics Research Strategic Vision: A Blueprint for Transforming Global Air Mobility," NASA white paper NF-2013-04-563-HQ, n.d. [2013], available at *http://www.aeronautics.nasa.gov/pdf/armd_strategic_vision_2013.pdf* (accessed September 1, 2013).
3. Lisa Porter, "Reshaping NASA's Aeronautics Program" (presentation given at AIAA Aerospace Sciences Conference, Reno, NV, January 12, 2006).
4. John Croft, "Open Rotor Noise Not a Barrier to Entry," *Flightglobal* (July 5, 2012), available at *http://www.flightglobal.com/news/articles/open-rotor-noise-not-a-barrier-to-entry-ge-373817/* (accessed July 20, 2012).
5. Mark D. Guynn, Jeffrey J. Berton, Eric S. Hendricks, Michael T. Tong, William J. Haller, and Douglas R. Thurman, "Initial Assessment of Open Rotor Propulsion Applied to an Advanced Single-Aisle Aircraft" (paper presented at the 11th AIAA Aviation Technology, Integration, and Operations [ATIO] Conference, Virginia Beach, VA, September 20–22, 2011), pp. 2–3.
6. Ibid., pp. 2–3.
7. Bruce Dorminey, "Prop Planes: The Future of Eco-Friendly Aviation?" *Pacific-Standard Magazine* (February 9, 2012), available at *http://www.psmag.com/business-economics/prop-planes-the-future-of-eco-friendly-aviation-39649/* (accessed June 7, 2012).
8. Lewis Page, "NASA Working on 'Open Rotor' Green (But Loud) Jets," *Register* (June 12, 2009), available at *http://www.theregister.co.uk/2009/06/12/nasa_open_rotor_trials/* (accessed June 21, 2012).
9. Kathy Barnstorff, "NASA Researchers Work To Turn Blue Skies Green," February 27, 2013, at *http://www.nasa.gov/topics/aeronautics/features/blue_skies_green.html#.Udi-Z_mThDA* (accessed August 11, 2013).
10. The airflow through the engine casing is 10 times greater than the flow of air going through the compressor and combustion chamber on an ultra-high-bypass turbofan. Banke, "Advancing Propulsive Technology," p. 751.

11. The other four technology demonstrations were the Active Flow Control Enhanced Vertical Tail Flight Experiment, the Damage Arresting Composite Demonstration, the Adaptive Compliant Trailing Edge Flight Experiment, and the Flap and Landing Gear Noise Reduction Flight Experiment. The other two research areas were aircraft drag reduction through innovative flow control concepts and weight reduction from advanced composite materials. Barnstorff, "NASA Researchers Work To Turn Blue Skies Green."
12. Pratt & Whitney, "Pratt & Whitney and NASA Demonstrate Benefits of Geared TurboFan System in Environmentally Responsible Aviation Project," June 19, 2013, http://www.pw.utc.com/Press/Story/20130619-0800/2013/All%20Categories (accessed August 19, 2013).
13. Glenn Research Center, "9'×15' Low-Speed Wind Tunnel," November 30, 2011, http://facilities.grc.nasa.gov/9x15/ (accessed August 20, 2013); NASA Aeronautics Test Program, "9- by 15-Foot Low-Speed Wind Tunnel," n.d. [2013], http://facilities.grc.nasa.gov/documents/TOPS/Top9x15.pdf (accessed August 20, 2013).
14. Guynn et al., "Initial Assessment of Open Rotor Propulsion Applied to an Advanced Single-Aisle Aircraft," pp. 1–2; Banke, "Advancing Propulsive Technology," p. 770.
15. Ed Envia, "Progress Toward N+1 Noise Goal" (paper presented at the Fundamental Aeronautics Program 12-Month Program Preview, Washington, DC, November 5–6, 2008); National Academy of Engineering of the National Academies, *Technology for a Quieter America* (Washington, DC: National Academies Press, 2010), p. 58.
16. Thomas Edwards, "The Future of Green Aviation" (NASA paper ARC-E-DAA-TN5136, presented at the Future of Flight Foundation Celebrate the Future Conference, Mukilteo, WA, April 22, 2012), pp. 20, 36, 42.
17. Bruce I. Larrimer, "Good Stewards: NASA's Role in Alternative Energy," in *NASA's Contributions to Flight*, vol. 1, *Aerodynamics*, ed. Richard P. Hallion (Washington, DC: Government Printing Office, 2010), pp. 849–872.
18. Edwards, "The Future of Green Aviation," p. 51; Kathy Barnstorff, "The Puffin: A Passion for Personal Flight," February 8, 2010, http://www.nasa.gov/topics/technology/features/puffin.html (accessed May 5, 2012); Charles Q. Choi, "Electric Icarus: NASA Designs a One-Man Stealth Plane," *Scientific American* (January 19, 2010), available at http://www.scientificamerican.com/article.cfm?id=nasa-one-man-stealth-plane (accessed May 5, 2012).

19. Clay Dillow, "NASA Awards the Largest Prize in Aviation History to an All-Electric, Super-Efficient Aircraft," *Popular Science* (October 4, 2011), available at *http://www.popsci.com/technology/article/2011-10/nasa-awards-largest-prize-aviation-history-all-electric-super-efficient-aircraft* (accessed August 31, 2013).
20. Established in 2003, the Future of Flight Foundation provides exhibits, education programs, and student exchange programs using commercial aviation to inspire innovative thinking and to explore solutions to critical global issues. The foundation changed its name to the Institute of Flight in December 2015. Institute of Flight, "About," 2016, *https://www.futureofflight.org/about* (accessed December 2, 2016).
21. Elizabeth Griffin, "Earth Day Celebration Highlights Environmentally Progressive Aircraft," *Journal: Your Community Magazine* 37 (April 2012): 18.
22. Ibid.
23. Edwards, "The Future of Green Aviation," pp. 7, 60.
24. Ibid., pp. 9, 52.
25. Ibid., pp. 19, 28.
26. Kathleen Zona, "Experience NASA's Green Lab Research Facility," January 27, 2012, *http://www.nasa.gov/centers/glenn/events/tour_green_lab.html* (accessed October 6, 2014); Edwards, "The Future of Green Aviation," p. 29; Harrington, "Leaner and Greener: Fuel Efficiency Takes Flight," pp. 815–817.
27. Jim Banke, "NASA Researchers Sniff Out Alternate Fuel Future," May 13, 2013, *http://www.nasa.gov/topics/aeronautics/features/access_fuel.html#.Udi9WvmThDA* (accessed August 11, 2013).
28. Karen Northon, "NASA Signs Agreement with German, Canadian Partners To Test Alternative Fuels," NASA News Release 14-089, April 10, 2014, at *http://www.nasa.gov/press/2014/april/nasa-signs-agreement-with-german-canadian-partners-to-test-alternative-fuels/#.VB9MJPldUeh* (accessed September 21, 2014); Banke, "NASA Researchers Sniff Out Alternate Fuel Future."

Abbreviations

AATE	Advanced Affordable Turbine Engine
ACCESS	Alternative Fuel Effects on Contrails and Cruise Emissions
ACEE	Aircraft Energy Efficiency
ACTIVE	Advanced Control Technology for Integrated Vehicles
ADECS	Adaptive Engine Control System
ADP	Advanced Ducted Propulsor
ADVENT	Adaptive Versatile Engine Technology
AEC	Atomic Energy Commission
AERL	Aircraft Engine Research Laboratory
AGARD	Advisory Group for Aerospace Research and Development
AGATE	Advanced General Aviation Transportation Experiments
AGT	Advanced Gas Turbine
AIAA	American Institute of Aeronautics and Astronautics
ALCM	air-launched cruise missile
ARMD	Aeronautics Research Mission Directorate
ASEE	American Society for Engineering Education
ASME	American Society for Mechanical Engineers
ASPIRE	Advanced Supersonic Propulsion and Integration Research
AST	Advanced Subsonic Technology
ATEGG	Advanced Turbine Engine Gas Generator
ATF	Advanced Tactical Fighter
ATIO	Aviation Technology, Integration, and Operations
ATP	Advanced Turboprop Project
ATT	Advanced Transport Technology
ATTAP	Advanced Turbine Technology Applications
avgas	aviation gasoline
AWT	Altitude Wind Tunnel
BAC	British Aircraft Corporation
C-MAPSS	Commercial Modular Aero-Propulsion System Simulation
CARD	Civil Aviation Research and Development
CEMCAN	ceramic matrix composite analyzer
CFD	computational fluid dynamics
CFRP	carbon fiber–reinforced polymer

CIA	Central Intelligence Agency
CLEEN	Continuous Lower Energy, Emissions, and Noise
CO	carbon monoxide
CO_2	carbon dioxide
CPC	Critical Propulsion Components
DARPA	Defense Advanced Research Projects Agency
dB	decibels
DEEC	Digital Electronic Engine Control
DEFCS	Digital Electronic Flight Control System
DLR	German Aerospace Center
DOD	Department of Defense
DOE	Department of Energy
DOT	Department of Transportation
E^3, or EEE	Energy Efficient Engine
ECCP	Experimental Clean Combustor Program
ECI	Engine Component Improvement
EPA	Environmental Protection Agency
EPM	Enabling Propulsion Materials
EPNdB	effective perceived noise in decibels
ERA	Environmentally Responsible Aviation
ERAST	Environmental Research and Aircraft Sensor Technology
ERB	Engine Research Building
ETAF	Energy Trends and Alternate Fuels
ETOPS	Extended Range Twin Operations
EVNRC	Engine Validation of Noise Reduction Concepts
FAA	Federal Aviation Administration
FADEC	full-authority digital electronic control
FAR	Federal Aviation Regulations
FCC	flight control computer
FLADE	fan on blade
FPS	Flight Propulsion System
FSER	Full-Scale Engine Research
GAP	General Aviation Propulsion
GE	General Electric
HALE	High-Altitude Long-Endurance
HARV	High Alpha Research Vehicle
HATP	High-Angle-of-Attack Technology Program
HIDEC	Highly Integrated Digital Electronic Control
HISTEC	High Stability Engine Control
HITEMP	Advanced High Temperature Engine Materials Technology Program

Abbreviations

HPT	High Pressure Turbine
HQ	Herschel-Quincke tubes
HRC	Historical Reference Collection
HSCT	High-Speed Civil Transport
HSR	High-Speed Research
ICAO	International Civil Aviation Organization
ICAS	International Congress of Aeronautical Sciences
IDEEC	Improved Digital Electronic Engine Controller
IHPTET	Integrated High Performance Turbine Engine Technology
ILS	instrument landing system
ILTV	Inner Loop Thrust Vectoring
IPCS	integrated propulsion control system
IRAC	Integrated Resilient Aircraft Control
JPO	Joint Program Office
JTDE	Joint Technology Demonstrator Engine
LAP	Large-Scale Advanced Propfan
LDV	Laser Doppler Velocimetry
MATV	Multi-Axis Thrust Vectoring
MIT	Massachusetts Institute of Technology
mph	miles per hour
NAA	National Aeronautic Association
NACA	National Advisory Committee for Aeronautics
NAFEC	National Aviation Facilities Experimental Center
NARA	National Archives and Records Administration
NASA	National Aeronautics and Space Administration
NASM	National Air and Space Museum
NASP	National Aero-Space Plane
NATO	North Atlantic Treaty Organization
NATR	Nozzle Acoustic Test Rig
NEPA	Nuclear Energy for the Propulsion of Aircraft
NFAC	National Full-Scale Aerodynamics Complex
NIA	National Institute of Aerospace
NiAl	nickel aluminide
NO_x	oxides of nitrogen
NPSS	Numerical Propulsion System Simulation
NRC	National Research Council
NTSB	National Transportation Safety Board
NVFEL	National Vehicle and Fuel Emissions Laboratory
OART	Office of Advanced Research and Technology
OAST	Office of Aeronautics and Space Technology
ODS	Oxide Dispersion Strengthened

ONERA	Office National d'Etudes et Recherches Aérospatiales
OPEC	Organization of Petroleum Exporting Countries
Pan Am	Pan American World Airways
PCA	propulsion-controlled aircraft
PI	Performance Improvement
PMR	Polymerization of Monomer Reactants
PNdB	perceived noise in decibels
POI	power-off-immersed
POR	power-off-retracted
PRT	Propeller Research Tunnel
PSC	Performance Seeking Control
PSL	Propulsion Systems Laboratory
PTA	Propfan Test Assessment
Q-Fan	Quiet-Fan
Q/STOL	Quiet-STOL
QAT	Quiet Aircraft Technology
QCGAT	Quiet, Clean, General Aviation Turbofan
QCSEE	Quiet, Clean, Short-Haul Experimental Engine
QEP	Quiet Engine Program
QSRA	Quiet Short-Haul Research Aircraft
QTD	Quiet Technology Demonstrator
QTD2	Quiet Technology Demonstrator 2
QUESTOL	Quiet Experimental Short Takeoff and Landing
RBSN	reaction-bonded silicon nitride
RevCon	Revolutionary Concepts in Aeronautics
RG	record group
rpm	revolutions per minute
RTA	Revolutionary Turbine Accelerator
SAE	Society of Automotive Engineers
SAM	Sound Absorption Material
SBIR	Small Business Innovation Research
SCAR	Supersonic Cruise Aircraft Research
scramjet	supersonic combustion ramjet
SECTF	Small Engine Component Test Facility
SFC	Specific Fuel Consumption
SFN	Separate-Flow Nozzle (Jet Noise Reduction Test Program)
SFW	Subsonic Fixed Wing
SLPC	single-lever power control
SMA	shape memory alloy
SPO	System Program Office
SR	single-rotation (propfan)

SST	supersonic transport
STOL	Short Takeoff and Landing
STOVL	Short Takeoff and Vertical Landing
TBCC	Turbine Based Combined Cycle
TCA	Technology Concept Airplane
TFX	Tactical Fighter Experimental
THC	unburned hydrocarbons
TiAl	titanium aluminide
TN	Technical Note
TOC	Throttles Only Control
TR	Technical Report
TRL	Technology Readiness Level
UAL	United Airlines
UART	universal asynchronous receiver/transmitter
UAV	unmanned aerial vehicle
UDF	Unducted Fan
UEET	Ultra-Efficient Engine Technology
UHB	ultra-high-bypass
UTRC	United Technologies Research Center
V-CAP	Vehicle Charging And Potential
V/STOL	Vertical and Short Takeoff and Landing
VAATE	Versatile Affordable Advanced Turbine Engine
VCE	variable cycle engine
VDT	Variable-Density Tunnel
Virginia Tech	Virginia Polytechnic Institute and State University
VLJ	very light jet
Vorbix	vortex burning and mixing

Bibliography

Books and Monographs

Aerospace Industries Association. *Aerospace Facts and Figures 2008*. Arlington, VA: Aerospace Industries Association of America, 2008.

Anderson, Jr., John D. *Introduction to Flight*. 7th edition. New York: McGraw-Hill, 2012.

Arrighi, Robert S. *Pursuit of Power: NASA's Propulsion Systems Laboratory No. 1 and 2*. Washington, DC: NASA, 2012.

Becker, John V. *The High Speed Frontier: Case Histories of Four NACA Programs, 1920–1950*, NASA SP-445. Washington, DC: NASA, 1980.

Bilstein, Roger E. *Orders of Magnitude: A History of the NACA and NASA, 1915–1990*, NASA SP-4406. Washington, DC: NASA Scientific and Technical Information Division, 1989.

Bowles, Mark D. *The "Apollo" of Aeronautics: NASA's Aircraft Energy Efficiency Program, 1973–1987*, NASA SP-2009-574. Washington, DC: NASA, 2010.

———. *Science in Flux: NASA's Nuclear Program at Plum Brook Station, 1955–2005*, NASA SP-2006-4317. Washington, DC: NASA, 2006.

Boyne, Walter J., and Donald S. Lopez, eds. *The Jet Age: Forty Years of Jet Aviation*. Washington, DC: Smithsonian Institution, 1979.

Conner, Margaret. *Hans von Ohain: Elegance in Flight*. Reston, VA: AIAA, 2001.

Connors, Jack. *The Engines of Pratt & Whitney: A Technical History*. Reston, VA: AIAA, 2010.

Constant, Edward W. *The Origins of the Turbojet Revolution*. Baltimore: Johns Hopkins University Press, 1980.

Conway, Erik M. *High-Speed Dreams: NASA and the Technopolitics of Supersonic Transportation, 1945–1999*. Baltimore: Johns Hopkins University Press, 2005.

Coulam, Robert F. *Illusions of Choice: The F-111 and the Problem of Weapons Acquisition Reform*. Princeton: Princeton University Press, 1977.

Crouch, Tom D. *A Dream of Wings: Americans and the Airplane, 1875–1905*. New York: Norton, 1981; reprint, Washington, DC: Smithsonian Institution Press, 1989.

Dawson, Virginia P. *Engines and Innovation: Lewis Laboratory and American Propulsion Technology*, NASA SP-4306. Washington, DC: NASA, 1991.

Dick, Steven J., ed. *NASA's First 50 Years: Historical Perspectives*, NASA SP-2010-4704. Washington, DC: Government Printing Office, 2010.

Ethell, Jeffrey L. *Fuel Economy in Aviation*, NASA SP-462. Washington, DC: Government Printing Office, 1983.

Ferguson, Robert G. *NASA's First A: Aeronautics from 1958 to 2008*, SP-2012-4412. Washington, DC: NASA, 2013.

Garvin, Robert R. *Starting Something Big: The Commercial Emergence of GE Aircraft Engines*. Reston, VA: AIAA, 1998.

Gray, George. *Frontiers of Flight: The Story of NACA Research*. New York: Knopf, 1948.

Gunston, Bill. *The Development of Jet and Turbine Aero Engines*. 2nd edition. Somerset, England: Patrick Stephens, Ltd., 1997.

Hager, Roy D., and Deborah Vrabel. *Advanced Turboprop Project*, NASA SP-495. Washington, DC: NASA, 1988.

Hallion, Richard P., ed. *NASA's Contributions to Flight*, vol. 1. Washington, DC: Government Printing Office, 2010.

———. *On the Frontier: Flight Research at Dryden, 1946–1981*, NASA SP-4303. Washington, DC: NASA, 1984.

―――. *Supersonic Flight: Breaking the Sound Barrier and Beyond—The Story of the Bell XS-1 and the Douglas D-558*. New York: The Macmillan Co., in association with the National Air and Space Museum, Smithsonian Institution, 1972.

Hansen, James R. *Engineer in Charge: A History of the Langley Aeronautical Laboratory, 1917–1958*, NASA SP-4305. Washington, DC: NASA, 1987.

Jakab, Peter L. *Visions of a Flying Machine: The Wright Brothers and the Process of Invention*. Shrewsbury, England: Airlife, 1990.

Jenkins, Dennis R. *Hypersonics Before the Shuttle: A Concise History of the X-15 Research Airplane*, NASA SP-2000-4518. Washington, DC: NASA, 2000.

Kinney, Jeremy R. *Airplanes: The Life Story of a Technology*. New York: Greenwood, 2006; reprint, Baltimore: Johns Hopkins, 2008.

Launius, Roger D., ed. *Innovation and the Development of Flight*. College Station, TX: Texas A&M University Press, 1999.

Leyes, Richard A., II, and William A. Fleming. *The History of North American Small Gas Turbine Aircraft Engines*. Reston, VA: AIAA, 1999.

Mead, Cary Hoge. *Wings over the World: The Life of George Jackson Mead*. Wauwatosa, WI: Swannet Press, 1971.

Merlin, Peter W. *Mach 3+: NASA/USAF YF-12 Flight Research, 1969–1979*. Washington, DC: NASA History Division, 2002.

NASA Acoustically Treated Nacelle Program: A Conference Held at Langley Research Center, Hampton, Virginia, October 15, 1969, NASA SP-220. Washington, DC: NASA, 1969.

National Academy of Engineering of the National Academies. *Technology for a Quieter America*. Washington, DC: National Academies Press, 2010.

National Research Council of the National Academies. *Recapturing NASA's Aeronautics Flight Research Capabilities*. Washington, DC: National Academies Press, 2012.

Nayler, J.L., and E. Ower. *Aviation: Its Technical Development*. London: Peter Owen/Vision Press, 1965.

Roland, Alex. *Model Research: The National Advisory Committee for Aeronautics, 1915–1958*, NASA SP-4103. Washington, DC: NASA Scientific and Technical Information Division, 1985.

Rolt, R.T.C. *The Aeronauts: A History of Ballooning, 1783–1903*. New York: Walker and Co., 1966.

Rosen, George. *Thrusting Forward: A History of the Propeller*. Windsor Locks, CT: United Technologies Corporation, 1984.

Schlaifer, Robert, and Samuel D. Heron. *Development of Aircraft Engines and Aviation Fuels*. Boston: Harvard University Graduate School of Business Administration, 1950.

Spinoff 2008: Fifty Years of NASA-Derived Technologies (1958–2008). Washington, DC: NASA Center for Aerospace Information, 2008.

St. Peter, James. *The History of Aircraft Gas Turbine Engine Development in the United States: A Tradition of Excellence*. Atlanta: International Gas Turbine Institute of the American Society of Mechanical Engineers, 1999.

Taylor, C. Fayette. *Aircraft Propulsion: A Review of the Evolution of Aircraft Piston Engines*. Washington, DC: Smithsonian Institution Press, 1971.

Tucker, Tom. *Touchdown: The Development of Propulsion Controlled Aircraft at NASA Dryden*. Washington, DC: NASA History Office, 1999.

Weick, Fred E. *Aircraft Propeller Design*. New York: McGraw-Hill Book Co., 1930.

Williams, Louis J. *Small Transport Aircraft Technology*. Washington, DC: NASA, 1983; reprint, Honolulu: University Press of the Pacific, 2001.

Reports and Papers

Abdalla, K.L., and J.A. Yuska. "NASA REFAN Program Status," NASA TM-X-71705. 1975.

"Aeropropulsion '87: Proceedings of a Conference Held at NASA Lewis Research Center, Cleveland, Ohio, November 17–19, 1987," NASA CP-3049. 1990.

Albers, James A. "Status of the NASA YF-12 Propulsion Research Program," NASA TM-X-56039. 1976.

Albright, A.J., D.J. Lennard, and J.A. Ziemanski. "NASA/General Electric Engine Component Improvement Program," paper presented at the 14th AIAA/SAE Joint Propulsion Conference, Las Vegas, NV, July 25–27, 1978.

Bargsten, Clayton J., and Malcolm T. Gibson. "NASA Innovation in Aeronautics: Select Technologies That Have Shaped Modern Aviation," NASA TM-2011-216987. 2011.

Bartlett, Peter, Nick Humphreys, Pam Phillipson, Justin Lan, Eric Nesbitt, and John Premo. "The Joint Rolls-Royce/Boeing Quiet Technology Demonstrator Programme," AIAA Paper 2004-2869, presented at the 10th AIAA/CEAS Aeroacoustics Conference, Manchester, U.K., May 10–12, 2004.

Bielak, G., J. Gallman, R. Kunze, P. Murray, J. Premo, M. Kosanchick, A. Hersh, J. Celano, B. Walker, J. Yu, H.W. Kwan, S. Chiou, J. Kelly, J. Betts, J. Follet, and R. Thomas. "Advanced Nacelle Acoustic Lining Concepts Development," NASA CR-2002-211672. August 2002.

Birch, Stanley F., P.A. Bukshtab, K.M. Khritov, D.A. Lyubimov, V.P. Maslov, A.N. Secundov, and K.Y. Yakubovsky. "The Use of Small Air Jets To Simulate Metal Chevrons," AIAA Paper 2009-3372, a presented at the 15th AIAA/CEAS Aeroacoustics Conference, Miami, FL, May 11–13, 2009.

Bloomer, Harry E., and Carl E. Campbell. "Experimental Investigation of Several Afterburner Configurations on a J79 Turbojet Engine," NACA RM E57I18. 1957.

Boeing Commercial Airplanes. "High-Speed Civil Transport Study: Summary," NASA CR-4234. 1989.

Bolden, Charles. "Embracing a World of Change: NASA's Aeronautics Research Strategy," address at AIAA Aviation 2013, Los Angeles, CA, August 14, 2013. *http://www.nasa.gov/sites/default/files/bolden_aiaa_aviation.pdf*. Accessed September 1, 2013.

Bowers, Albion H., and Joseph W. Pahle. "Thrust Vectoring on the NASA F-18 High Alpha Research Vehicle," NASA TM-4771. 1996.

Bowers, Albion H., Joseph W. Pahle, R. Joseph Wilson, Bradley C. Flick, and Richard L. Rood. "An Overview of the NASA F-18 High Alpha Research Vehicle," NASA TM-4772. 1996.

Braithwaite, Willis M., John H. Dicus, and John E. Moss. "Evaluation with a Turbofan Engine of Air Jets as a Steady-State Inlet Flow Distortion Device," NASA TM-X-1955. 1970.

Brokopp, Richard A., and Robert S. Gronski. "Small Engine Components Test Facility Compressor Testing Cell at NASA Lewis Research Center," AIAA Paper 92-3980, presented at the 17th AIAA Aerospace Ground Testing Conference, Nashville, TN, July 6–8, 1992.

Brown, Clifford, and James Bridges. "An Analysis of Model Scale Data Transformation to Full Scale Flight Using Chevron Nozzles," NASA TM-2003-212732. December 2003.

Bull, John, Robert Mah, Gloria Davis, Joe Conley, Gordon Hardy, Jim Gibson, Matthew Blake, Don Bryant, and Diane Williams. "Piloted Simulation Tests of Propulsion Control as a Backup to Loss of Primary Flight Controls for a Mid-Size Jet Aircraft," NASA TM-110374. 1995.

Bultemeier, Eric J., Ulrich Ganz, John Premo, and Eric Nesbit. "Effect of Uniform Chevrons on Cruise Shockcell Noise," AIAA Paper 2006-2440, presented at the 12th AIAA/CEAS Aeroacoustics Conference, Cambridge, MA, May 8–10, 2006.

Burcham, Frank W., Jr., and C. Gordon Fullerton. "Controlling Crippled Aircraft—With Throttles," NASA TM-104238. 1991.

Bibliography

Burcham, Frank W., Jr., C. Gordon Fullerton, and Trindel A. Maine. "Manual Manipulation of Engine Throttles for Emergency Flight Control," NASA TM-2004-212045. 2004.

Burcham, Frank W., Jr., John J. Burken, Trindel A. Maine, and C. Gordon Fullerton. "Development and Flight Test of an Emergency Flight Control System Using Only Engine Thrust on an MD-11 Transport Airplane," NASA TP-97-206217. 1997.

Burcham, Frank W., Jr., and Peter G. Batterton. "Flight Experience with a Digital Integrated Propulsion Control System on an F-111E Airplane," AIAA Paper 76-653, presented at the 12th Joint AIAA/SAE Propulsion Conference, Palo Alto, CA, July 26–29, 1976.

Burcham, Frank W., Jr., Ronald J. Ray, Timothy R. Conners, and Kevin R. Walsh. "Propulsion Flight Research at NASA Dryden from 1967 to 1997," NASA TP-1998-206554. 1998.

Burcham, Frank W., Jr., Trindel A. Maine, John J. Burken, and John Bull. "Using Engine Thrust for Emergency Flight Control: MD-11 and B-747 Results," NASA TM-1998-206552. 1998.

Burcham, Frank W., Jr., Trindel Maine, and Thomas Wolf. "Flight Testing and Simulation of an F-15 Airplane Using Throttles for Flight Control," NASA TM-104255. 1992.

Burrus, D.L., C.A. Chahrour, H.L. Foltz, P.E. Sabla, S.P. Seto, and J.R. Taylor. "Combustion System Component Technology Performance Report," NASA CR-168274. 1984.

Bushman, Mark, and Steven G. Nobbs. "F-15 Propulsion System: PW1128 Engine and DEEC," NASA Dryden Flight Research Center Document N95-33012. 1995.

Calkins, Frederick T., George W. Butler, and James H. Mabe. "Variable Geometry Chevrons for Jet Noise Reduction," AIAA Paper 2006-2546, presented at the 12th AIAA/CEAS Aeroacoustics Conference, Cambridge, MA, May 8–10, 2006.

Ciepluch, C.C. "Preliminary QCSEE Program Test Results," SAE Paper 771008. 1977.

———. "A Review of the QCSEE Program," NASA TM-X-71818. 1975.

Crowl, R., W.R. Dunbar, and C. Wentworth. "Experimental Investigation of Marquardt Shock-Positioning Control Unit on a 28-Inch Ramjet Engine," NACA RM E56E09. 1956.

Davis, D.Y., and E.M. Stearns. "Energy Efficient Engine: Flight Propulsion System Final Design and Analysis [GE Design]," NASA CR-168219. 1985.

Draley, Eugene C., Blake W. Corson, Jr., and John L. Crigler. "Trends in the Design and Performance of High-Speed Propellers," in *NACA Conference on Aerodynamic Problems of Transonic Airplane Design: A Compilation of Papers Presented, September 27–29, 1949*, NASA TM-X-56649. 1950.

Dryden, Hugh L. "Jet Engines for War," January 23, 1951. NASA Historical Reference Collection, file 40958.

———. "Research and Development in Aeronautics," Publication No. L49-25. October 4, 1948.

Durand, William F. "Experimental Research on Air-Propellers," NACA TR 14. 1917.

———. "Interaction Between Air-Propellers and Airplane Structures," NACA TR 235. 1926.

Durand, William F., and E.P. Lesley. "Experimental Research on Air-Propellers, II," NACA TR 30. 1918.

———. "Experimental Research on Air-Propellers, V," NACA TR 141. 1922.

———. "Comparison of Tests on Airplane Propellers in Flight with Wind Tunnel Model Tests on Similar Forms," NACA TR 220. 1925.

Edwards, Thomas. "The Future of Green Aviation," NASA paper ARC-E-DAA-TN5136, presented at the Future of Flight Foundation Celebrate the Future Conference, Mukilteo, WA, April 22, 2012.

Ellis, Jr., Macon C., and Clinton E. Brown. "NACA Investigation of a Jet-Propulsion System Applicable to Flight," NACA TR-802. 1943.

Envia, Ed. "Progress Toward N+1 Noise Goal," paper presented at the Fundamental Aeronautics Program 12-Month Program Review, Washington, DC, November 5–6, 2008.

Federal Aviation Administration. "Noise Abatement Departure Profiles," Advisory Circular 91-53A, July 22, 1993. *http://rgl.faa.gov/Regulatory_and_Guidance_Library/rgAdvisoryCircular.nsf/list/AC%2091-53A/$FILE/ac91-53.pdf.* Accessed September 1, 2012.

———. "Noise Standards: Aircraft Type and Airworthiness Certification," Advisory Circular 36-4C, July 15, 2003.

Federal Aviation Agency, Department of Defense, and NASA. "A National Program for a Commercial Supersonic Transport," NASA TMX-50927. 1960.

Fiock, Ernest F., and H. Kendall King. "The Effect of Water Vapor on Flame Velocity in Equivalent $CO-O_2$ Mixtures," NACA TR 531. 1935.

Fishbach, L.H., and Michael J. Caddy. "NNEP: The Navy-NASA Engine Program," NASA TM-X-71857. 1975.

Freche, John C. "Progress in Superalloys," NASA TN D-2495. 1964.

Gabb, Timothy P., Anita Garg, David L. Ellis, and Kenneth M. O'Connor. "Detailed Microstructural Characterization of the Disk Alloy ME3," NASA TM-2004-213066. May 2004.

Gaffin, W.O. "NASA ECI Programs: Benefits to Pratt & Whitney Engines," paper presented at the 27th ASME International Gas Turbine Conference and Exhibit, London, England, April 18–22, 1982. *http://ntrs.nasa.gov/search.jsp?R=19820051913.* Accessed August 18, 2013.

Garg, Sanjay. "Aircraft Turbine Engine Control Research at NASA Glenn Research Center," NASA TM-2013-217821. 2013.

German, J., P. Fogel, and C. Wilson. "Design and Evaluation of an Integrated Quiet Clean General Aviation Turbofan (QCGAT) Engine and Aircraft Propulsion System," NASA CR-165185. 1980.

Golub, Robert A., John W. Rawls, Jr., and James W. Russell. "Evaluation of the Advanced Subsonic Technology Program Noise Reduction Benefits," NASA TM-2005-212144. May 2005.

Guyn, Mark D., Jeffrey J. Berton, Eric S. Hendricks, Michael T. Tong, William J. Haller, and Douglas R. Thurman. "Initial Assessment of Open Rotor Propulsion Applied to an Advanced Single-Aisle Aircraft," paper presented at the 11th AIAA Aviation Technology, Integration, and Operations (ATIO) Conference, Virginia Beach, VA, September 20–22, 2011.

Hallion, Richard P. "The History of Hypersonics: Or, 'Back to the Future—Again and Again,'" AIAA Paper 2005-0329, presented at the AIAA 43rd Aerospace Sciences Meeting and Exhibit, Reno, NV, January 10–13, 2005.

Hammack, Jerome B., Max C. Kurbjun, and Thomas C. O'Bryan. "Flight Investigation of a Supersonic Propeller on a Propeller Research Vehicle at Mach Numbers to 1.01," NACA RM L57E20. 1957.

Hammack, Jerome B., and Thomas C. O'Bryan. "Effect of Advance Ratio on Flight Performance of a Modified Supersonic Propeller," NACA TN 4389. 1958.

Harper, Charles W. "Introductory Remarks," in *Progress of NASA Research Relating to Noise Alleviation of Large Subsonic Jet Aircraft: A Conference Held at Langley Research Center, Hampton, Virginia, October 8–10, 1968*, NASA SP-189. 1968.

Hassell, Patrick. "A History of the Development of the Variable Pitch Propeller," paper presented before the Royal Aeronautical Society Hamburg Branch, Hamburg, Germany, April 26, 2012.

Heldenbrand, R.W., and W.M. Norgren. "AiResearch QCGAT Program," NASA CR-159758. 1979.

Hughes, Donald L., Jon K. Holzman, and Harold Johnson. "Flight-Determined Characteristics of an Air Intake System on an F-111A Airplane," NASA TN D-6679. 1972.

Jaw, Link C., and Sanjay Garg. "Propulsion Control Technology Development in the United States: A Historical Perspective," NASA TM-2005-213978. 2005.

Johnsen, Irving A., and Robert O. Bullock, eds. *Aerodynamic Design of Axial-Flow Compressors*, NASA SP-36. Washington, DC: NASA, 1965.

Ketchum, James R., and R.T. Craig. "Simulation of Linearized Dynamics of Gas-Turbine Engines," NACA TN 2826. 1952.

Kinney, J.R. "Starting from Scratch?: The American Aero Engine Industry, the Air Force, and the Jet, 1940–1960," AIAA Report 2003-2671, July 2003.

Klineberg, J.M. "Technology for Aircraft Energy Efficiency," American Society of Civil Engineers Paper A79-14126 03-03, in *Proceedings of the International Air Transportation Conference, Washington, D.C., April 4–6, 1977*. New York: American Society of Civil Engineers, 1977.

Koenig, R.W., and G.K. Sievers. "Preliminary QCGAT Program Test Results," NASA TM-79013. 1979.

Kramer, J.J., and F.J. Montegani. "The NASA Quiet Engine Program," NASA TM-X-67988. 1972.

Krebs, J.N. "Advanced Supersonic Technology Study Engine Program Summary: Supersonic Propulsion, 1971 to 1976," in *Proceedings of the SCAR Conference Held at Langley Research Center, Hampton, Virginia, November 9–12, 1976*, NASA CP-001. 1977.

Lee, Jinho, Robert J. Buehrle, and Ralph Winslow. "The GE-NASA RTA Hyperburner Design and Development," NASA TM-2005-213803. 2005.

Lewis Research Center. "The Propulsion Systems Laboratory," Brochure B-0363, March 1991. *http://pslhistory.grc.nasa.gov/PSL_Assets/History/E%20PSL%20No%203%20and%204/Propulsion%20Systems%20Lab%20No.%203-4%20brochure%20(1991).pdf*. Accessed August 15, 2013.

Lesley, E.P., and B.M. Woods. "The Effect of Slipstream Obstructions on Air-Propellers," NACA TR 177. 1923.

Liebert, Curt H., and Francis S. Stepka. "Potential Use of Ceramic Coating as a Thermal Insulation on Cooled Turbine Hardware," NASA TM-X-3352. 1976.

Litt, Jonathan S., Dean K. Frederick, and Ten-Huei Guo. "The Case for Intelligent Propulsion Control for Fast Engine Response," NASA TM-2009-215668/AIAA 2009-1876, 2009.

Low, John K.C., Paul S. Schweiger, John W. Premo, and Thomas J. Barber. "Advanced Subsonic Technology (AST) Separate-Flow High-Bypass Ratio Nozzle Noise Reduction Program Test Report," NASA CR-2000-210040. December 2000.

Lytle, John K. "The Numerical Propulsion System Simulation: An Overview," NASA TM-2000-209915. June 2000.

Mankins, John C. "Technology Readiness Levels: A White Paper," April 6, 1995. *http://www.hq.nasa.gov/office/codeq/trl/trl.pdf*. Accessed October 4, 2014.

Martin, Richard A. "Dynamic Analysis of XB-70-1 Inlet Pressure Fluctuations During Takeoff and Prior to a Compressor Stall at Mach 2.5," NASA TN D-5826. 1970.

Matthews, Douglas C., Larry A. Bock, Gerald W. Bielak, R.P. Dougherty, John W. Premo, Dan F. Scharpf, and Jia Yu. "Pratt & Whitney/Boeing Engine Validation of Noise Reduction Concepts: Final Report for NASA Contract NAS3-97144, Phase 1," NASA CR-2014-218088. February 2014.

McAulay, John. E. "Engine Component Improvement Program: Performance Improvement," AIAA Paper 80-0223, presented at the 12th AIAA Aerospace Sciences Meeting, January 14–16, 1980.

Mehalic, Charles M., and Roy A. Lottig. "Inlet Temperature Distortion on the Stall Limits of J85-GE-13," NASA TM-X-2990. 1974.

Mengle, Vinod G. "Jet Noise Characteristics of Chevrons in Internally Mixed Nozzles," AIAA Paper 2005-2934, presented at the 11th AIAA/CEAS Aeroacoustics Conference, Monterey, CA, May 23–25, 2005.

Mengle, Vinod G., Ronen Elkoby, Leon Brusniak, and Russ H. Thomas. "Reducing Propulsion Airframe Aeroacoustic Interactions with Uniquely Tailored Chevrons: 1. Isolated Nozzles," AIAA Paper 2006-2467, presented at the 12th AIAA/CEAS Aeroacoustics Conference, Cambridge, MA, May 8–10, 2006.

Merlin, Peter W. "Design and Development of the Blackbird: Challenges and Lessons Learned," AIAA Paper 2009-1522, presented at the 47th AIAA Aerospace Sciences Meeting, Orlando, FL, January 5–8, 2009.

Miller, Cearcy D. "A Study by High Speed Photography of Combustion and Knock in a Spark-Ignition Engine," NACA TR 727. 1942.

Miller, George P. "NASA," in Congressional Record, 92nd Cong., 1st sess., vol. 117s. December 14, 1971, p. H12545.

Miller, Robert A. "History of Thermal Barrier Coatings for Gas Turbine Engines: Emphasizing NASA's Role from 1942 to 1990," NASA TM-2003-215459. 2009.

Munk, Max M. "Analysis of W.F. Durand's and E.P. Lesley's Propeller Tests," NACA TR 175. 1923.

National Aeronautics and Space Administration. "Aeronautics Research Strategic Vision: A Blueprint for Transforming Global Air Mobility," NASA White Paper NF-2013-04-563-HQ, n.d. [2013]. *http://www.aeronautics. nasa.gov/pdf/armd_strategic_vision_2013.pdf*. Accessed September 1, 2013.

National Aeronautics and Space Administration. *NASA Conference on V/STOL Aircraft: A Compilation of the Papers Presented at Langley Research Center, Langley Field, Virginia, November 17–18, 1960*. http://ntrs.nasa.gov/archive/nasa/casi.ntrs.nasa.gov/19630004807.pdf. Accessed August 21, 2015.

Niedzwiecki, Richard W. "The Experimental Clean Combustor Program: Description and Status to November 1975," NASA TM-X-71849. 1975.

Niedzwiecki, Richard W., and C.C. Gleason. "Results of the NASA/General Electric Experimental Clean Combustor Program," AIAA Paper 76-76, presented at the 12th AIAA/SAE Joint Propulsion Conference, Palo Alto, CA, July 26–29, 1976.

Niedzwiecki, Richard W., and R. Jones. "The Experimental Clean Combustor Program: Description and Status," SAE Technical Paper 740485. February 1, 1974.

"Numerical Propulsion System Simulation (NPSS): An Award Winning Propulsion System Simulation Tool," in *Research & Technology, 2001*, NASA TM-2002-211333, 2002.

Otto, Edward W., and Burt L. Taylor III. "Dynamics of a Turbojet Engine Considered as a Quasi-Static System," NACA TR 1011. 1950.

Peebles, Curtis. "Learning from Experience: Case Studies of the Hyper-X Project," paper presented at the 47th AIAA Aerospace Sciences Meeting, January 5–8, 2009.

———. "The X-43A Flight Research Program: Lessons Learned on the Road to Mach 10," unpublished paper, Dryden Flight Research Center, CA, 2007.

Pittman, James L., John M. Koudelka, Michael J. Wright, and Kenneth E. Rock. "Hypersonics Project Overview," paper presented at the 2011 Technical Conference, Cleveland, OH, March 15–17, 2011. *http://www.aeronautics.nasa.gov/pdf/hypersonics_project.pdf*. Accessed December 26, 2012.

Porter, Lisa. "Reshaping NASA's Aeronautics Program," presentation given at AIAA Aerospace Sciences Conference, Reno, NV, January 12, 2006.

Pratt & Whitney and General Electric Aircraft Engines. "Critical Propulsion Components, Volume 1: Summary, Introduction, and Propulsion Systems Studies," NASA CR-2005-213584. 2005.

Putnam, Terrill W., and Ronald H. Smith. "XB-70 Compressor-Noise Reduction and Propulsion Performance for Choked Inlet Flow," NASA TN D-5692. 1970.

Rhines, Thomas B. "Summary of United Aircraft Wind Tunnel Tests of Supersonic Propellers," in *NACA Conference on Aerodynamic Problems of Transonic Airplane Design: A Compilation of Papers Presented, September 27–29, 1949*, NASA TM-X-56649. 1949.

Richey, G. Keith. "F-111 Systems Engineering Case Study," Center for Systems Engineering at the Air Force Institute of Technology, March 10, 2005. *http://www.afit.edu/docs/0930AFIT14ENV125%202-2.pdf*. Accessed February 2, 2016.

———. Interview by Squire Brown, August 31, 2006. Interview no. 2, transcript. Cold War Aerospace Technology Project, Wright State University Libraries Special Collections and Archives.

Roberts, G.D., R.K. Goldberg, W.K. Binienda, W.A. Arnold, J.D. Littell, and L.W. Kohlman. "Characterization of Triaxial Braided Composite Material Properties for Impact Simulation," NASA TM-2009-215660. 2009.

Roberts, R., A. Peduzzi, and R.W. Niedzwiecki. "Low Pollution Combustor Designs for CTOL Engines: Results of the Experimental Clean Combustor Program," AIAA Paper 76-762, presented at the 12th AIAA/SAE Joint Propulsion Conference, Palo Alto, CA, July 26–29, 1976.

Rothrock, Addison M., Alois Krsek, and Anthony W. Jones. "The Induction of Water to the Inlet Air as a Means of Internal Cooling in Aircraft Engine Cylinders," NACA TR 756. 1943.

Rothrock, A.M., and Arnold E. Biermann. "The Knocking Characteristics of Fuels in Relation to Maximum Permissible Performance of Aircraft Engines," NACA TR 655. 1940.

Ryan, William F. "Aircraft Noise Abatement at JFK International Airport," in Congressional Record, 92nd Cong., 1st sess., vol. 117s. June 14, 1971, pp. E5783–E5784.

Schey, Oscar, Benjamin Pinkel, and Herman H. Ellerbrock, Jr. "Correction of Temperatures of Air-Cooled Engine Cylinders for Variation in Engine and Cooling Conditions," NACA TR 645. 1939.

Schmittman, Craig. "Ultra High Bypass (UHB)." *Aerospace-Defense News*, 1989. *https://www.youtube.com/watch?v=zxVAaIsfPIY*. Accessed May 3, 2016.

Schwartz, Ira, ed. "Third Conference on Sonic Boom Research Held at NASA, Washington, D.C., October 29–30, 1970," NASA SP-255. 1971.

Shaw, Robert J. "NASA's Ultra-Efficient Engine Technology (UEET) Program/ Aeropropulsion Technology Leadership for the 21st Century," paper presented at the 22nd International Congress of Aeronautical Sciences (ICAS), August 27–September 1, 2000, Harrogate, U.K. *http://www.dtic.mil/dtic/tr/fulltext/u2/a547568.pdf*. Accessed September 5, 2012.

———. "Ultra-Efficient Engine Technology Project Integrated into NASA's Vehicle Systems Program," in *Research and Technology 2003*, NASA TM-2004-212729. May 2004, n.p. *http://ntrs.nasa.gov/archive/nasa/casi.ntrs.nasa.gov/20050192140.pdf.* Accessed February 2, 2016.

Shovlin, M.D., and J.A. Cochrane. "An Overview of the Quiet Short-Haul Research Aircraft Program," NASA TM-78545. November 1978.

Silverstein, Abe, and Newell D. Sanders. "Concepts on Turbojet Engines for Transport Application," paper presented at the SAE Aeronautic Meeting, New York, April 10, 1956. NASA HRC, file 41567.

Sinnette, John T., Oscar W. Schey, and J. Austin King. "Performance of NACA Eight-Stage Axial-Flow Compressor Designed on the Basis of Airfoil Theory," NACA TR 758. 1943.

Smolka, James W., Laurence A. Walker, Gregory H. Johnson, Gerard S. Schkolnik, Curtis W. Berger, Timothy R. Conners, John S. Orme, Karla S. Shy, and C. Bruce Wood. "F-15 ACTIVE Flight Research Program," in *Report to the Aerospace Profession: Fortieth Symposium Proceedings of the Society of Experimental Test Pilots*, September 1996. *https://www.nasa.gov/centers/dryden/pdf/89247main_setp_d6.pdf.* Accessed February 2, 2016.

Sokolowski, Daniel E., and John E. Rohde. "The E^3 Combustors: Status and Challenges," NASA TM-82684. 1981.

Stack, John. "Tests of Airfoils Designed to Delay the Compressibility Burble," NACA TN 976. December 1944.

Staff of the NASA Research Centers. "Summary of NASA Support of the F-111 Development Program, Part I-December 1962-December 1965," NASA Langley Working Paper 246. October 10, 1966.

"Statement of Dr. Mike J. Benzakein," in Subcommittee on Space and Aeronautics, Committee on Science, House of Representatives, *The Future of Aeronautics at NASA*, 109th Cong., 1st sess., Serial No. 109-8. March 16, 2005. Available at *http://commdocs.house.gov/committees/science/hsy20007.000/hsy20007_0.HTM.* Accessed February 2, 2016.

Stevens, Joseph R. "NASA's HITEMP Program for UHBR Engines," AIAA Paper 90-2395, presented at the 26th AIAA/SAE/ASME/ASEE Joint Propulsion Conference, Orlando, FL, July 16–18, 1990.

Stewart, James F. "Integrated Flight Propulsion Control Research Results Using the NASA F-15 HIDEC Flight Research Facility," NASA TM-4394. June 1992.

Stewart, Warner L. "Introduction: Session III-Propulsion," in *Proceedings of the SCAR Conference held at Langley Research Center, Hampton, Virginia, November 9–12, 1976*, NASA CP-001. 1977.

Szuch, John R., James F. Soeder, Kurt Seldner, and David S. Cwynar. "F100 Multivariable Control Synthesis Program—Evaluation of a Multivariable Control Using a Real-Time Engine Simulation," NASA TP 1056. October 1977.

Taylor, D.W. "Some Aspects of the Comparison of Model and Full-Scale Tests," NACA AR 1925. 1926.

Turner, Travis L., Randolph H. Cabell, Roberto J. Cano, and Richard J. Silcox. "Testing of SMA-Enabled Active Chevron Prototypes Under Representative Flow Conditions," in *Proceedings of SPIE, Volume 6928: Active and Passive Smart Structures and Integrated Systems 2008*, ed. Mehdi Ahmadian, May 2008. *http://ntrs.nasa.gov/archive/nasa/casi.ntrs.nasa.gov/20080014174.pdf.* Accessed September 15, 2014.

U.S. Congress. House Committee on Science and Technology. "Aircraft Noise Abatement Technology." 94th Cong., 2nd sess., 1976. S. Rept. W.

——. Office of Technology Assessment. *Federal Research and Technology for Aviation*, OTA-ETI-610. Washington, DC: Government Printing Office, 1994.

——. Senate. *Committee on Aeronautical and Space Sciences, United States Senate, 1958–1976*. Washington: Government Printing Office, 1977.

——. Senate Committee on Government Operations. *TFX Contract Investigation (Second Series)*, Part 2, 91st Cong., 2nd sess., March 25–26, April 7, 9, 14, 1970.

———. Office of Aeronautics and Space Technology. "Aircraft Fuel Conservation Technology Task Force Report," NASA TM-X-74295. 1975.

U.S. National Transportation Safety Board. *Aircraft Accident Report: United Airlines Flight 232, July 19, 1989*, NTSB/AAR-SO/06. Washington, DC: NTSB, 1990.

Vasu, George, Clint E. Hart, and William R. Dunbar. "Preliminary Report on Experimental Investigation of Engine Dynamics and Controls for a 48-Inch Ramjet Engine," NACA RM E55J12. 1956.

Ware, Marsden. "Description and Laboratory Tests of a Roots Type Aircraft Engine Supercharger," NACA TR 230. 1920.

———. "Description of the NACA Universal Test Engine and Some Test Results," NACA TR 250. 1920.

Waters, M.H., T.L. Galloway, C. Rohrbach, and M.G. Mayo. "Shrouded Fan Propulsors for Light Aircraft," SAE Paper 730323, presented at the Society of Automotive Engineers Business Aircraft Meeting, Wichita, KS, April 3–6, 1973.

Weick, Fred E., and Donald H. Wood. "The Twenty-Foot Propeller Research Tunnel of the National Advisory Committee for Aeronautics," NACA TR 300. 1929.

Werner, Roger A., Mahmood Abdelwahab, and Willis M. Braithwaite. "Performance and Stall Limits of an Afterburner-Equipped Turbofan Engine With and Without Inlet Flow Distortion," NASA TM X-1947. 1970.

Whitfield, Charlotte E. "NASA's Quiet Aircraft Technology Project," NASA TM-2004-213190. 2004.

Williams International. "The General Aviation Propulsion (GAP) Program," NASA CR-2008-215266. 2008.

Witcofski, Robert D. "Comparison of Alternate Fuels for Aircraft," NASA TM-80155. 1979.

Wydler, John W. "Noise," in Congressional Record, 91st Cong., 2nd sess., vol. 116. June 1, 1970, pp. E4989–E4991.

Yu, Jia, Hwa-Wan Kwan, and Eugene Chien. "PW 4098 Forward Fan Case Acoustic Liner Design Under NASA EVNRC Program," AIAA Paper 2010-3827, presented at the 16th AIAA/CEAS Aeroacoustics Conference, Stockholm, Sweden, 2010.

Articles and Chapters

"1962–1972: The First Decade of Centaur," *Lewis News* (October 20, 1972): 4–5. *http://pslhistory.grc.nasa.gov/PSL_Assets/History/C%20Rockets/First%20Decade%20of%20Centaur%20article%20(1972).pdf*. Accessed August 15, 2013.

A&P Technology. "GEnx Engine," 2013. *http://www.braider.com/Case-Studies/GEnx-Engine.aspx*. Accessed August 23, 2013.

"About the R&D 100 Awards," *R&D Magazine* 2014. *http://www.rd100awards.com/about-rd-100-awards*. Accessed October 5, 2014.

"Active F-15," *The X-Press* (October 17, 1997): 4.

Aircraft Owner's and Pilot's Association. "General Aviation Statistics," March 30, 2011. *http://www.aopa.org/About-AOPA/Statistical-Reference-Guide/General-Aviation-Statistics*. Accessed August 31, 2013.

"Airlines Continue To Reduce the Use of Noisy Aircraft," FAA News Release 21-93, June 10, 1993. NASA HRC, file 012336.

"Alleviating Aircraft Noise: The Quiet Aircraft Technology Program," *NASA Destination Tomorrow*, program 9, 2003. *http://www.youtube.com/watch?v=3vFSpW-zC2U*. Accessed July 7, 2013.

"Ames Tests New Jet Engine," *NASA News* (June 1993): 4.

Andersen, Drucella, and Michael Mewhinney. "NASA Testing New, Powerful 'Ducted Fan' Engine for Civil Tests," NASA Release 93-103, June 3, 1993. *http://www.nasa.gov/home/hqnews/1993/93-103.txt*. Accessed September 23, 2014.

Arrighi, Robert S. "Propulsion Systems Laboratory: PSL No. 1 & 2, 1952–1979," December 3, 2012. *http://pslhistory.grc.nasa.gov/Default.aspx*. Accessed September 25, 2013.

Banke, James. "Advancing Propulsive Technology," in *NASA's Contributions to Aeronautics*, vol. 1, *Aerodynamics*, ed. Richard Hallion, pp. 734–783. Washington, DC: NASA, 2010.

———. "NASA Researchers Sniff Out Alternate Fuel Future," May 13, 2013. *http://www.nasa.gov/topics/aeronautics/features/access_fuel.html#. Udi9WvmThDA*. Accessed August 11, 2013.

———. "Technology Readiness Levels Demystified," November 19, 2013. *http://www.nasa.gov/topics/aeronautics/features/trl_demystified.html#. VDAA7vldUeg*. Accessed October 4, 2014.

Barnstorff, Kathy. "NASA Researchers Work To Turn Blue Skies Green," February 27, 2013. *http://www.nasa.gov/topics/aeronautics/features/blue_skies_green.html#.Udi-Z_mThDA*. Accessed August 11, 2013.

———. "The Puffin: A Passion for Personal Flight," February 8, 2010. *http://www.nasa.gov/topics/technology/features/puffin.html*. Accessed May 5, 2012.

Benzakein, M.J., S.B. Kazin, and F. Montegani. "NASA/GE Quiet Engine 'A'," *Journal of Aircraft* 10 (February 1973): 67–73.

Boeing. "Boeing 777-300ER Performs 330-Minute ETOPS Flight," October 15, 2003. *http://www.prnewswire.com/news-releases/boeing-777-300er-performs-330-minute-etops-flight-72530852.html*. Accessed June 30, 2013.

Boeing Defense, Space, and Security. "Backgrounder: X-51A WaveRider," September 2012. *http://www.boeing.com/assets/pdf/defense-space/military/waverider/docs/X-51A_overview.pdf*. Accessed December 26, 2012.

Boeing Phantom Works. "Boeing X-51A WaveRider Breaks Record in First Flight," May 26, 2010. *http://boeing.mediaroom.com/index.php?s=43&item=1227*. Accessed December 26, 2012.

"Boeing Studies Refanned 727 Benefits," *Aviation Week & Space Technology* 102 (January 13, 1975): 24–25.

Bibliography

Bowles, Mark D., and Virginia P. Dawson. "The Advanced Turboprop Project: Radical Innovation in a Conservative Environment," in *From Engineering Science to Big Science: The NACA and NASA Collier Trophy Research Project Winners*, ed. Pamela E. Mack. Washington, DC: NASA, 1998, pp. 320–343.

Burnett, Bob. "Ssshhh, We're Flying a Plane Around Here: A Boeing-Led Team Is Working To Make Quiet Jetliners Even Quieter," *Boeing Frontiers Online* 4 (December 2005/January 2006). *http://www.boeing.com/news/frontiers/archive/2005/december/ts_sf07.html*. Accessed July 13, 2013.

Busch, Mike. "GAP Engine Update," *AVweb*, July 27, 2000. *http://www.avweb.com/news/reviews/182838-1.html*. Accessed August 31, 2013.

Choi, Charles Q. "Electric Icarus: NASA Designs a One-Man Stealth Plane," *Scientific American* (January 19, 2010). *http://www.scientificamerican.com/article.cfm?id=nasa-one-man-stealth-plane*. Accessed May 5, 2012.

Ciepluch, Carl C., Donald Y. Davis, and David E. Gray. "Results of NASA's Energy Efficient Engine Program," *Journal of Propulsion and Power* 3 (November–December 1987): 560–568.

Croft, John. "Open Rotor Noise Not a Barrier to Entry," *Flightglobal* (July 5, 2012). *http://www.flightglobal.com/news/articles/open-rotor-noise-not-a-barrier-to-entry-ge-373817/*. Accessed July 20, 2012.

Dillow, Clay. "NASA Awards the Largest Prize in Aviation History to an All-Electric, Super-Efficient Aircraft," *Popular Science* (October 4, 2011). *http://www.popsci.com/technology/article/2011-10/nasa-awards-largest-prize-aviation-history-all-electric-super-efficient-aircraft*. Accessed August 31, 2013.

Dorminey, Bruce. "Prop Planes: The Future of Eco-Friendly Aviation?" *Pacific-Standard Magazine* (February 9, 2012). *http://www.psmag.com/business-economics/prop-planes-the-future-of-eco-friendly-aviation-39649/*. Accessed June 7, 2012.

"DOT and NASA and Noise," *Washington Post* (August 29, 1971): B6.

"DOT/NASA Noise Abatement Office Established," NASA News Release 71-213, October 21, 1971, NASA HRC, file 012336.

Douglas, Donald W. "The Development and Reliability of the Modern Multi-Engine Air Liner," *Journal of the Royal Aeronautical Society* 40 (November 1935): 1042–1056.

Dron, Alan. "GA Engine Manufacturers Line Up Diesel Options," *Flight International* (April 12, 2011). http://www.flightglobal.com/news/articles/ga-engine-manufacturers-line-up-diesel-options-355474/. Accessed August 31, 2013.

Dryden Flight Research Center. "F-15 ACTIVE Nozzles," November 13, 1995. NASA HRC, file 011664.

Eaton, Sabrina. "NASA Programs on Jet Efficiency, Noise Survive Attempted Fund Cuts," [Cleveland] *Plain Dealer* (May 20, 1999): 15A.

"F-15 ACTIVE Achieves First-Ever Mach 2 Thrust-Vectoring," NASA Dryden News Release 96-62, November 8, 1996. http://www.nasa.gov/centers/dryden/news/NewsReleases/1996/96-62_pf.html. Accessed September 4, 2012.

"F-15 Flight Research Facility," Dryden Flight Research Center Fact Sheet, December 3, 2009. http://www.nasa.gov/centers/dryden/news/FactSheets/FS-022-DFRC.html. Accessed February 19, 2012.

"F-18 High Angle-of-Attack (Alpha) Research Vehicle," Dryden Flight Research Center Fact Sheet, December 3, 2009. http://www.nasa.gov/centers/dryden/news/FactSheets/FS-002-DFRC.html#.UiNUsjaTiHM. Accessed September 1, 2012.

"FAA Issues Noise Regulations for Older Jets," Department of Transportation News Release 76-121, December 28, 1976. NASA HRC, file 012336.

"The Fan-Compressor Flutter Team," *Lewis News* (February 3, 1978). http://pslhistory.grc.nasa.gov/PSL_Assets/History/F%20Turbofan%20Engines/Full%20Scale%20Engine%20Program%20article%20(1978).pdf. Accessed August 16, 2013.

Feaver, Douglas B. "Ford Pledges To Abate Airport Noise," *Washington Post* (October 22, 1976): A19.

Field, David. "Engine Makers in Three-Sided 'War' To Provide Power for Boeing 777," *Washington Times* (May 17, 1994): B7, B12.

"First Quiet Engine Noise Tests," NASA News Release 71-156, August 27, 1971. NASA HRC, file 012336.

Fitzpatrick, Mary. "Highlights of 1979 Activities: Year of the Planets," NASA News Release 79-179, December 27, 1979.

"Flight Research Reveals Outstanding Flying Qualities Result from Integrating Thrust Vectoring with Flight Controls," NASA Dryden Flight Research Center News Release, January 4, 1999. NASA HRC, file 011664.

Friedlander, Paul J.C. "Jet Age Prospect," *New York Times* (October 26, 1958): X25.

"General Aviation Technology Program," NASA News Release 75-65, March 1975.

General Electric Aviation. "Aircraft Engine History and Technology," 2009. *http://www.youtube.com/watch?v=4lip8lPWFLo*. Accessed July 2, 2013.

———. "CF34-10E Engines Outperforming Expectations," May 13, 2014. *http://www.geaviation.com/press/cf34/cf34_20140513b.html*. Accessed July 1, 2014.

———. "CF34-10E Turbofan Propulsion System," May 2010. *http://www.geaviation.com/engines/docs/commercial/datasheet-CF34-10E.pdf*. Accessed July 1, 2014.

———. "The CF6 Engine Family," 2012. *http://www.geaviation.com/engines/commercial/cf6/*. Accessed August 18, 2013.

———. "GE Aircraft Engines Pursuing Mach 4 Jet Engine at NASA Research Center," July 22, 2002. *http://www.geaviation.com/press/other/other_20020722aa.html*. Accessed June 6, 2013.

———. "The GE90 Engine Family," 2012. *http://www.geaviation.com/engines/commercial/ge90/*. Accessed June 30, 2013.

———. "New GEnx Engine Advancing Unprecedented Use of Composites in Jet Engines," December 14, 2004. *http://www.geaviation.com/press/genx/genx_20041214.html*. Accessed August 23, 2013.

Glenn Research Center. "9' × 15' Low Speed Wind Tunnel," November 30, 2011. *http://facilities.grc.nasa.gov/9x15/*. Accessed August 20, 2013.

——. "Flying on the Ground: The Wind Tunnels of Glenn Research Center," NASA Release FS-2002-06-005-GRC, 2002. *http://www.nasa.gov/centers/glenn/about/fs05grc.html*. Accessed July 7, 2013.

——. "General Aviation's New Thrust: Single Lever Technology Promises Efficiency, Simplicity," NASA Release FS-1999-07-002-GRC, July 1999. *http://www.nasa.gov/centers/glenn/pdf/84789main_fs02grc.pdf*. Accessed July 7, 2013.

——. "Making Future Commercial Aircraft Quieter: Glenn Effort Will Reduce Engine Noise," NASA Release FS-1999-07-003-GRC, July 1999. *http://www.nasa.gov/centers/glenn/pdf/84790main_fs03grc.pdf*. Accessed July 7, 2013.

——. "Small Aircraft Propulsion: The Future Is Here," Release FS-2000-04-001-GRC, November 22, 2004. *http://www.nasa.gov/centers/glenn/about/fs01grc.html*. Accessed July 7, 2013.

——. "Turbine Disk Alloys," June 20, 2012. *http://www.grc.nasa.gov/WWW/StructuresMaterials/AdvMet/research/turbine_disks.html*. Accessed September 5, 2012.

Griffin, Elizabeth. "Earth Day Celebration Highlights Environmentally Progressive Aircraft," *Journal: Your Community Magazine* 37 (April 2012): 18.

Griffiths, Bob. "Composite Fan Blade Containment Case," *High Performance Composites* (May 2005). *http://www.braider.com/pdf/Papers-Articles/Composite-Fan-Blade-Containment-Case.pdf*. Accessed March 17, 2013.

"Halving of Fuel Needs for Jetliners Is Seen," *Washington Post* (February 7, 1975): A9.

Hamilton, Martha M. "Firms Give Propellers a New Spin," *Washington Post* (February 8, 1987): H1.

Henderson, Breck W. "Dryden Completes First Flights of F/A-18 HARV with Thrust Vectoring," *Aviation Week & Space Technology* (July 29, 1991): 25.

"High Speed Engine Cycles Tapped for Further Research," NASA News Release 94-41, March 14, 1994. NASA HRC, file 011151.

Hines, William. "NASA's Quiet Engine Still a Long Way in Future," *Chicago Sun-Times* (June 7, 1970). Reprinted in *NASA Current News* (June 10, 1970). NASA HRC, file 012336.

Homer, P.T., and R.D. Schlichting. "Using Schooner To Support Distribution and Heterogeneity in the Numerical Propulsion System Simulation Project," *Concurrency and Computation: Practice and Experience* 6 (June 1994): 271–287.

Huff, Dennis L. "NASA Glenn's Contributions to Aircraft Noise Research," *Journal of Aerospace Engineering* 26 (April 2013): 218–250.

IHPTET and VAATE. "Turbine Engine Technology: A Century of Power for Flight," 2002. *http://web.archive.org/web/20060715190755/http://www.pr.afrl.af.mil/divisions/prt/ihptet/ihptet_brochure.pdf.* Accessed August 25, 2013.

Johnsen, Katherine. "Congress Leaves Noise Control to FAA," *Aviation Week & Space Technology* (October 23, 1972): 19–20.

Johnson, Greg. "Something New for Airliners: Propellers," *Los Angeles Times* (June 16, 1986): SD-C1.

Kocivar, Ben. "QUESTOL: A New Kind of Jet for Short Hops," *Popular Science* 200 (May 1972): 78, 80, 84.

"Kramer Heads Jet Noise Research Program," NASA News Release 71-170, September 7, 1971. NASA HRC, file 012336.

Langley Research Center. "Affordable Alternative Transportation: AGATE—Revitalizing General Aviation," Release FS-1996-07-02-LaRC, July 1996. *http://www.nasa.gov/centers/langley/news/factsheets/AGATE.html.* Accessed August 31, 2013.

———. "NASA's High-Speed Research (HSR) Program—Developing Tomorrow's Supersonic Passenger Jet," April 22, 2008. *http://www.nasa.gov/centers/langley/news/factsheets/HSR-Overview2.html*. Accessed September 25, 2014.

———. "Wind Tunnels at NASA Langley Research Center," FS-2001-04-64-LaRC, April 2008. *http://www.nasa.gov/centers/langley/news/factsheets/windtunnels.html*. Accessed September 17, 2002.

———. "World War II and the National Advisory Committee for Aeronautics," FS-LaRC-95-07-01, July 1995. *http://www.nasa.gov/centers/langley/news/factsheets/WWII.html*. Accessed May 1, 2012.

Leary, Warren E. "NASA Jet Sets Record for Speed," *New York Times* (November 17, 2004).

Lerch, Bradley A., Susan L. Draper, J. Michael Pereira, Michael V. Nathal, and Curt Austin. "Resistance of Titanium Aluminide to Domestic Object Damage Assessed," in *Research & Technology, 1998*, NASA TM-1999-208815, April 1, 1999.

Levintan, R.M. "Q-Fan Demonstrator Engine," *Journal of Aircraft* 12 (1975): 658–663.

"Lowenstein Says FAA Won't Curb Jets' Noise," *New York Times* (February 21, 1970).

Matthews, Jim. "Thrust Vectoring," *Air & Space Smithsonian* (July 2008). *http://www.airspacemag.com/flight-today/Thrust_Vectoring.html*. Accessed September 4, 2012.

Mayer, Daryl. "X-51A WaveRider Achieves History in Final Flight," Wright-Patterson Air Force Base News Release, May 3, 2013. *http://www.afmc.af.mil/News/Article-Display/Article/153475/x-51a-waverider-achieves-history-in-final-flight/*. Accessed May 31, 2013.

"NACA: The Force Behind Our Air Supremacy," *Aviation* 43 (January 1944): 175, 364.

NASA. "Advanced High-Temperature Engine Materials Technology Progresses," n.d. [1995]. *http://ntrs.nasa.gov/archive/nasa/casi.ntrs.nasa.gov/20050169149.pdf*. Accessed October 4, 2014.

———. "Advanced High-Temperature Engine Materials Technology Progresses," n.d. [1996]. *http://ntrs.nasa.gov/archive/nasa/casi.ntrs.nasa.gov/20050177123.pdf*. Accessed October 5, 2014.

NASA Aeronautics Research Mission Directorate. "Technical Excellence 2004: New Material Improves Rotor Safety," September 7, 2007. *http://www.aeronautics.nasa.gov/te04_rotor_safety.htm*. Accessed August 22, 2013.

NASA Aeronautics Test Program. "9- by 15-Foot Low Speed Wind Tunnel," n.d. [2013]. *http://facilities.grc.nasa.gov/documents/TOPS/Top9x15.pdf*. Accessed August 20, 2013.

"NASA Aircraft Test Engines Promise Noise, Pollution Reduction," NASA News Release 78-138, September 11, 1978. NASA HRC, file 012336.

"NASA Chief Welcomes Jet Noise Challenge," NASA News Release 71-222, November 3, 1971. NASA HRC, file 012336.

"NASA Cooperative Engine Testbed Aircraft on Schedule," *Aerospace Propulsion* (July 7, 1997): 2. NASA HRC, file 011151.

"NASA F-15 Being Readied for Advanced Maneuvering Flight," NASA News Release 93-115, June 16, 1993. NASA HRC, file 011664.

"NASA F-15 Makes First Engine-Controlled Touchdown," NASA News Release 93-75, April 22, 1993. NASA HRC, file 011664.

"NASA Langley Research Center: Contributing to the Next 100 Years of Flight," NASA Release FS-2004-02-84-LaRC, February 2004. *http://www.nasa.gov/centers/langley/news/factsheets/FS-2004-02-84-LaRC.html*. Accessed July 7, 2013.

"NASA Noise Conference Planned," NASA News Release No. 77-221, October 14, 1977. NASA HRC, file 012336.

NASA Office of the Chief Technologist. "NASA Spinoff: Damage-Tolerant Fan Casings for Jet Engines," 2006. *http://spinoff.nasa.gov/Spinoff2006/T_1.html*. Accessed August 23, 2013.

"NASA Researching Engine Airflow Controls To Improve Performance, Fuel Efficiency," NASA News Release 97-183, August 27, 1997. NASA HRC, file 011664.

"NASA Tests New Nozzle To Improve Performance," NASA News Release 96-59, March 27, 1996. NASA HRC, file 011664.

Nathal, Michael. "Glenn Takes a Bow for Impact on GEnx Engine," July 11, 2008. *http://www.nasa.gov/centers/glenn/news/AF/2008/July08_GEnx.html*. Accessed September 22, 2014.

"Noise Suit," *Washington Post* (September 10, 1970): A24.

Noland, David. "The Little Engine That Couldn't," *Air & Space Smithsonian* (November 2005). *http://www.airspacemag.com/flight-today/the-little-engine-that-couldnt-6865253/?no-ist*. Accessed April 1, 2016.

Nored, D.L. "Propulsion," *Astronautics and Aeronautics* 16 (July–August 1978): 47–54, 119.

Norris, Guy. "Chevron Tests 'Better Than Expected'," *Flight International* 160 (November 20–26, 2001): 10.

———. "GE Wins High-Mach Turbine Work," *Flight International* (July 30, 2002). *http://www.flightglobal.com/news/articles/ge-wins-high-mach-turbine-work-152266/*. Accessed May 1, 2012.

———. "P&W Prepares for Geared Fan Launch," *Flight International* (February 18, 1998). *http://www.flightglobal.com/news/articles/pw-prepares-for-geared-fan-launch-32908/*. Accessed August 19, 2013.

———. "Power House," *Flight International* (June 13, 2006). *http://www.flightglobal.com/news/articles/power-house-207148/*. Accessed October 4, 2014.

———. "X-51A WaveRider Achieves Hypersonic Goal on Final Flight," *Aviation Week & Space Technology* (May 2, 2013). *http://aviationweek.com/defense/x-51a-waverider-achieves-goal-final-flight*. Accessed May 31, 2013.

Northon, Karen. "NASA Signs Agreement with German, Canadian Partners To Test Alternative Fuels," NASA Release 14-089, April 10, 2014. *http:// www.nasa.gov/press/2014/april/nasa-signs-agreement-with-german-canadian-partners-to-test-alternative-fuels/#.VB9MJPldUeh*. Accessed September 21, 2014.

"OMB Impounds NASA Noise Research Funds," *Aviation Week & Space Technology* 97 (October 23, 1972): 19.

Page, Lewis. "NASA Working on 'Open Rotor' Green (But Loud) Jets," *Register* (June 12, 2009). *http://www.theregister.co.uk/2009/06/12/nasa_open_rotor_trials/*. Accessed June 21, 2012.

Povinelli, L.A., and B.H. Anderson. "Investigation of Mixing in a Turbofan Exhaust Duct, Computer Code Application and Verification," *AIAA Journal* 22 (April 1984): 518–525.

Pratt & Whitney. "F100 Engine," n.d. [2013]. *http://www.pw.utc.com/F100_Engine*. Accessed August 22, 2013.

———. "JT8D Engine," n.d. [2013]. *http://www.pw.utc.com/JT8D_Engine*. Accessed August 30, 2013.

———. "Pratt & Whitney and NASA Demonstrate Benefits of Geared TurboFan System in Environmentally Responsible Aviation Project," June 19, 2013. *http://www.pw.utc.com/Press/Story/20130619-0800/2013/All%20Categories*. Accessed August 19, 2013.

———. "PurePower PW1000G Engine," 2010. *http://www.purepowerengine.com*. Accessed April 1, 2016.

Pratt & Whitney Media Relations. "JT8D Engine Family: The Low-Cost Performer," October 2012. *http://www.pw.utc.com/Content/JT8D_Engine/pdf/B-1-7_commercial_jt8d.pdf*. Accessed August 18, 2013.

———. "Pratt & Whitney's JT9D Engine Family," October 2012. *http://www.pw.utc.com/Content/JT9D_Engine/pdf/B-1-8_commercial_jt9d.pdf*. Accessed August 18, 2013.

"President Reagan's Speech Before Joint Session of Congress," *New York Times* (February 5, 1986).

"Press Briefing: NASA Aircraft Noise Abatement Research," *NASA News* (March 31, 1972). NASA HRC, file 012336.

"Problems in Propeller Design," *Aviation and Aeronautical Engineering* 4 (June 1918): 108.

"Q-Fan Undergoing Aerodynamic Testing," *Aviation Week & Space Technology* (October 16, 1972): 59.

"Quiet-STOL Program Started," NASA News Release 71-146, August 4, 1971. NASA HRC, file 012336.

"Quieter Engines for Small Aircraft Possible, Tests Show," *Aviation Week & Space Technology* 16 (March 16, 1979): 1.

Rachul, Lori. "HISTEC To Boost Performance, Fuel Efficiency," *Lewis News* (November 1997): 5.

Reddy, Dhanireddy, R. "Seventy Years of Aeropropulsion Research at NASA Glenn Research Center," *Journal of Aerospace Engineering* 26 (April 2013): 202–217.

Roberts, Gary D., J. Michael Pereira, Michael S. Braley, and William A. Arnold. "Design and Testing of Braided Composite Fan Case Materials and Components," ISABE-2009-1201, 2009. *http://www.braider.com/pdf/Papers-Articles/Design-and-Testing-of-Braided-Composite-Fan-Case-Materials-and-Components.pdf.* Accessed March 17, 2013.

Salpukas, Agis. "Advances: Rebuilding Planes To Cut Noise," *New York Times* (November 18, 1987): D8.

Schefter, Jim. "New Blades Make Prop Liners as Fast as Jets," *Popular Science* 226 (March 1985): 66–69.

Scott, Phil. "Busting the Boom," *New Scientist* (August 18, 2001): 26–29.

Serafini, T.T. "PMR Polymide Composites for Aerospace Applications," in *Polymides: Synthesis, Characterization, and Applications*, vol. 2. New York: Plenum Press, 1984, pp. 957–975.

"Shhh…" *Forbes* (March 15, 1970): 31.

"Signing Ceremony To Initiate Development of Revolutionary Aircraft Engines," NASA News Release N96-80, December 10, 1996. NASA HRC, file 011151.

Smith, George E., and David E. Mindell. "The Emergence of the Turbofan Engine," in *Atmospheric Flight in the Twentieth Century*, ed. Peter Galison and Alex Roland, Boston: Kluwer Academic Publishers, 2000, pp. 107–155.

Smithsonian National Air and Space Museum. "Lockheed SR-71 Blackbird," November 11, 2001. *https://airandspace.si.edu/collection-objects/lockheed-sr-71-blackbird*. Accessed September 4, 2012.

———. "Pratt & Whitney PW 4098 Turbofan Engine," 2007. Registrar's File 20070002000.

Springer, Anthony M. "NASA Aeronautics: A Half-Century of Accomplishments," in *NASA's First 50 Years: Historical Perspectives*, ed. Steven J. Dick, NASA SP-2010-4704. Washington, DC: Government Printing Office, 2010.

"STOL Conference at Ames," NASA News Release 72-201, October 11, 1972. NASA HRC, file 012336.

Stuart, Peter C. "Congress Raises Voice To Soften Noise," *Christian Science Monitor* (March 3, 1972): n.p. Copy available in the NASA HRC, file 012336.

Syvertson, C.A. "Civil Aviation Research and Development (CARD) Policy Study," in *Proceedings of the Northeast Electronics Research and Engineering Meeting, Boston, Massachusetts, November 2–5, 1971*. Newton, MA: Institute of Electrical and Electronics Engineers, Inc., 1971.

Tanners, James C. "Fueling Inflation: Sharp Increases Seen in Prices of Gasoline and Most Other Fuels," *Wall Street Journal* (October 31, 1973): 1.

Terman, Frederick E. "William Frederick Durand (1859–1958)," in *Aeronautics and Astronautics: Proceedings of the Durand Centennial Conference Held at Stanford University, 5–8 August 1959*, ed. Nicholas J. Huff and Walter G. Vincenti. New York: Pergamon Press, 1960.

"Test Engines Promise Less Noise from Small Aircraft," NASA News Release 79-24, March 2, 1979. NASA HRC, file 012336.

"Three Contracts Let for Quiet Engines," NASA News Release 72-166, August 15, 1972. NASA HRC, file 012336.

"Too Many Decibels Are Making Us Ding-a-Lings," *Washington News* (June 8, 1972): 2.

Tracy, Charles. "New Quiet Engine To Cut Jet Noise," *Cleveland Press* (May 15, 1972): 8.

Turbomachinery and Heat Transfer Branch, Glenn Research Center. "APNASA: Overview," August 14, 2007. *http://www.grc.nasa.gov/WWW/RTT/Codes/APNASA.html*. Accessed September 21, 2014.

U.S. Department of Defense. "Integrated High Performance Turbine Engine Technology (IHPTET)," n.d. [2000], *http://web.archive.org/web/20060721223255/http://www.pr.afrl.af.mil/divisions/prt/ihptet/ihptet.html*. Accessed July 12, 2012.

Vincenti, Walter G. "Air-Propeller Tests of W.F. Durand and E.P. Lesley: A Case Study in Technological Methodology," *Technology and Culture* 20 (1979): 712–751.

Wald, Matthew L. "A Cleaner, Leaner Jet Age Has Arrived," *New York Times* (April 9, 2008): H2.

Weinberger, Sharon. "X-51 WaveRider: Hypersonic Jet Ambitions Fall Short," August 15, 2012. *http://www.bbc.com/future/story/20120815-hypersonic-ambitions-fall-short*. Accessed April 1, 2016.

"Whatever Happened to Propfans?" *Flight International* (June 12, 2007). *http://www.flightglobal.com/news/articles/whatever-happened-to-propfans-214520/*. Accessed July 2, 2013.

Wheeler, Keith. "Building the XB-70," *LIFE* 58 (January 15, 1966): 74–80, 84.

———. "The Full Story of the 2.8 Seconds That Killed the XB-70," *LIFE* 61 (November 11, 1966): 126–130, 132, 135–136, 139, 141, 143.

Wilkinson, Stephan. "ZWRRWWWBRZR: That's the Sound of the Prop-Driven XF-84H, and It Brought Grown Men to Their Knees," *Air & Space* (July 2003). *http://www.airspacemag.com/how-things-work/zwrrwwwbrzr-4846149/*. Accessed April 1, 2016.

Wilson, Kristin K. "Quieting the Skies: Cessna Benefits from Lewis Noise Reduction Know How," *Lewis News* (February 1998): 5.

Wilson, Michael. "Hamilton Standard and the Q-Fan," *Flight International* 103 (April 19, 1973): 617–619.

Winter, William. "Has the Propeller a Future?" *Popular Mechanics* 89 (February 1948): 171–173.

Witkin, Richard. "Government Control of Jet Fuel May Lead to 10 Percent Cut in Flights," *New York Times* (October 13, 1973): 36.

Young, David, and George Papajohn. "3 Pilots Struggled To Keep Jet Aloft," *Chicago Tribune* (July 22, 1989): 4.

Zaman, K.B.M.Q., J.E. Bridges, and D.L. Huff. "Evolution from 'Tabs' to 'Chevron Technology': A Review," in *Proceedings of the 13th Asian Congress of Fluid Mechanics*. Dhaka, Bangladesh: Engineers Institute of Bangladesh, 2010, pp. 47–63.

Zona, Kathleen. "Experience NASA's Green Lab Research Facility," January 27, 2012. *http://www.nasa.gov/centers/glenn/events/tour_green_lab.html*. Accessed October 6, 2014.

About the Author

Jeremy R. Kinney curates the air racing, aircraft propulsion, and American military aviation (1919–1945) collections at the Smithsonian National Air and Space Museum. Part of that work includes overseeing and interpreting one of the world's largest and most significant collections of aircraft piston and gas turbine engines and propellers. His research and exhibitions focus on technology in the United States and Europe during the 20th century. He is the author of *Reinventing the Propeller: Aeronautical Specialty and the Triumph of the Modern Airplane* (Cambridge University Press, 2017) and *Airplanes: The Life Story of a Technology* (Johns Hopkins University Press, 2008). Kinney is a contributor to the multivolume work *The Wind and Beyond: A Documentary Journey into the History of Aerodynamics in America*, published by NASA. He was a corecipient of the inaugural Eugene Ferguson Prize (given by the Society for the History of Technology) for volume 1 of *The Wind and Beyond*. Kinney earned a B.A. from Greensboro College and an M.A. and Ph.D. in history from Auburn University.

Index

Page numbers in **bold** indicate pages with illustrations.
A reference to an endnote is indicated with an "n" after an entry's page number.

8-Foot High Speed Tunnel, 10, 12
16-Foot High Speed Tunnel, 10
24-Inch Jet Tunnel, 10
500 VLJ aircraft, 185
720 simulator, 149
727 airliner, 75, 81–84, 107, **108**, 114, 122
727 demonstrator, **114**, 122
737 airliner, 75, 81–84, 107, 112, 188, 218
747 jumbo jet, 77, 80, 86, 107–108, **109**, 187–188
747 jumbo jet (Japan Airlines Flight 123), 146
747 jumbo jet (PCA program), 153
747-8 Dreamliner, 197
747-8 jumbo jet, 197
747-9 Dreamliner, 197
747-100 jumbo jet, 84
747-200 jumbo jet, 75
757 airliner, 75, 108, 143
757 airliner (PCA program), 153
767 airliner, 108, 110, 113, **113**
777 airliner, 112–113, 125, 171, 194, 196
777-200 Extended Range, 113
777-300 Extended Range, 113
787 Dreamliner, 174, 196

A

A&P Technology, Inc., 172–173, 175, 205n19
A3J-1 Vigilante, 36
A-12 aircraft, 56
A-37 Dragonfly, 63
A300/A300-600 airliner, 107, 113
A310 airliner, 108, 113
A319 airliner, 75
A320 airliner, 218
A321 airliner, 197
A330 airliner, 113, 188
A350 airliner, 174
accidents, 145–147, 153
Adamczyk, John, 177, 184
Adamson, A.P., 92
Adaptive Engine Control System (ADECS), 143–144, 160n11
Advanced Control Technology for Integrated Vehicles (ACTIVE) program, **156**, 156–160
Advanced Gas Turbine (AGT), 168
Advanced General Aviation Transportation Experiments (AGATE), 180–181, 208n50
Advanced Space Transportation Program, 200
Advanced Subsonic Technology (AST) program, 169, 187–190, 194, 211n88
Advanced Supersonic Propulsion and Integration Research (ASPIRE) project, 199
Advanced Transport Technology (ATT), 132n8
Advanced Turbine Engine Gas Generator (ATEGG) program, 168
Advanced Turbine Technology Applications (ATTAP) program, 168
Advanced Turboprop Project (ATP), 179, 219
aerial congestion, 91–92, 187
Aeronautical Society of Great Britain, 6–7, 9
Aeronautics Research Mission Directorate (ARMD), 220, 226

275

Aeroproducts Propeller Company, 35, 46n119
Aerosance, 181–182
Aerospace Industry Technology Program, 169
air defense, 32, 61, 103
Air Force, U.S., 30, 35, 117
    Advanced Control Technology for Integrated Vehicles (ACTIVE) program, 156
    Advanced Tactical Fighter (ATF), 159
    Advanced Turbine Engine Gas Generator (ATEGG) program, 168
    Aero Propulsion Laboratory, 65, 94, 176
    Arnold Engineering Development Center, 62
    C-5/C-5A Galaxy transport, 106–107, 146
    collaboration with NASA, 57, 66
    F-15 series, 143
    F-22 Raptor, 158
    F-111A, **59**, 59–61
    F-111/TF-30 Propulsion Program Review Committee, 60–61
    Flight Dynamics Laboratory, 159
    Flight Test Center, 156
    Full-Scale Engine Research (FSER), 65
    High Stability Engine Control (HISTEC) project, 157
    Joint Technology Demonstrator Engine (JTDE) program, 168
    NPSS team, 177
    propulsion-controlled aircraft (PCA) program, 152
    Quiet-STOL (Q/STOL) aircraft program, 89
    single-stage-to-orbit hypersonic flight, 200
    SR-71 Blackbird, 56
    synthetic fuels, 226
    Systems Command, 66
    TFX program, 58
    Triple Plow II inlet, 61
    turbofan preference over turbojet, 64
    Versatile Engine Technology (ADVENT) program, 179
    V/STOL development, 88

X-5 research aircraft, 59
X-51A WaveRider, 203
XB-70 Valkyrie program, **53**
XJ79-GE-1 afterburning turbojet, 36
air pollution, 86–87, 92, 126–127, 224–225
Airbus Industrie, 73
    A300/A300-600 airliner, 107, 113
    A310 airliner, 108, 113
    A319 airliner, 75
    A320 airliner, 218
    A321 airliner, 197
    A330 airliner, 113, 188
    A350 airliner, 174
    CF6 turbofan, 107
    propulsion-controlled aircraft (PCA) program, 152
    PurePower PW1000G engine, 186
Aircraft Energy Efficiency (ACEE) program, 104–126, 125, 131–132, 228. *See also* Energy Efficient Engine (E^3); Engine Component Improvement (ECI) Program
Aircraft Engine Research Laboratory (AERL), **xii, 19, 20, 28**
    creation of, 16
    development teams, 20
    jet propulsion research, 30
    piston engine design, 18–22, 25
    propeller research, 25
    renamed Lewis Flight Propulsion Laboratory, 32
    schematic drawing, **21**
    V-1710 V-12 engine research, 20
airfoils, 125–127
    16-series, 10, 12
    Clark Y, 10, 12
    cuffs, 11
    DC-1 aircraft, 15
    defined, 4
    laminar-flow, 29
    and propellers, 3–4

Index

RAF-6, 10, 12
    supercritical, 132n8
air-launched cruise missile (ALCM), 179
airships, 3, 40n14
air-to-air interceptors, 54
All Nippon Airways, 196
AlliedSignal, Inc., 169, 176
Allis-Chalmers, 27
Allison Engine Company, 20, 27
    570 engine, 118
    571 industrial gas turbine engine, 123
    578-DX demonstrator engine, 123–124, **123**
    Advanced Turboprop Project (ATP), 115
    relationship with NACA, 19
    stator design, 190–191
    Ultra-Efficient Engine Technology Program (UEET), 179
    XT38 turboprop engine, 46n119
alloys, 159, 167–168, 175
Alternative Aviation Fuel Experiment studies, 226
Alternative Fuel Effects on Contrails and Cruise Emissions (ACCESS) study, **216**, 226–227, **227**
Altitude Wind Tunnel (AWT), 18, **21**, **24**, 30, **34**, 35–36
American Airlines, 103, 152, 194
American Institute of Aeronautics and Astronautics (AIAA), 217
American Society for Mechanical Engineers (ASME), 201
Ames Research Center, 66, 224
    Advanced Ducted Propulsor (ADP), 185–186, **186**
    Advanced Subsonic Technology (AST) program, 188
    Advanced Turboprop Project (ATP), 115–116
    collaboration with Lockheed Aircraft Corporation, 57
    Concepts and Missions Division, 91
    F-111A/F-111B design, 60
    F-111/TF-30 Propulsion Program Review Committee, 61
    High-Angle-of-Attack Technology Program (HATP), 154
    High-Speed Research (HSR) program, 197
    National Full-Scale Aerodynamics Complex (NFAC), 185, **186**
    "PCA Lite"/"PCA Ultralite," 153
    propulsion-controlled aircraft (PCA) program, 153
    STOL conference, 90
    Ultra-Efficient Engine Technology Program (UEET), 179
    YF-12A/C program, 58
Anderson, Bruce, 226–227
Arab-Israeli war, 103
Arend, David, 199
Armstrong, Neil A., 37
Armstrong Flight Research Center, 54. *See also* Dryden Flight Research Center; NASA Flight Research Center
Army, U.S., 29
    Advanced Affordable Turbine Engine (AATE) program, 179
    Air Forces, 19, 21, 26, 30
    airfoils (16-series), 10
    Engineering Division, 8
    Integrated High Performance Turbine Energy Technology (IHPTET) program, 176
Arnold, Henry H. ("Hap"), 19, 27–28, 30
AT-5A Hawk fighter (Curtiss), 15
ATK Space Systems, 173
Atlantic Richfield Company, 94
Atomic Energy Commission (AEC), 38
Austin, Curt, 170
Avco-Lycoming, 91, 93, 127, 129
Aviation, 22
Aviation and Aeronautical Engineering, 7
Aviation Week, 152
Aviation Week & Space Technology, 124

B

B-1 Lancer, 59, 61
B-2 stealth aircraft, 103
B-17 Flying Fortress, 26
B-29 Superfortress, 11, 21, 25, 32
B-47 Stratojet, 32, 140
B-52/B-52H Stratofortress, 33, 50, 201, 203
B-58/B-58A aircraft series, 36, 51
B-70 aircraft series, 51
Bartolotta, Paul, 200
Beech Aircraft, 116
Beggs, James. M., 81
Bell Aircraft Corporation, 22–23, 28–30, 44n78, 59, 88
Bellis, Benjamin, 66
Benzakein, Meyer J., 125
Bibliography of Aeronautics, 1, 39n4
"bizjets," 128–130. *See also* business jets
Blanchard, Jean-Pierre, 40n14
Bloomer, Harry E., 85
Boeing Company
    707 airliner, 34, 75–77, 79–81, 84, 86, 93
    720 airliner, 84
    720 simulator, 149
    727 airliner, 75, 81–82, 84, 107, **108**, 114
    727 demonstrator, **114**, 122
    737 airliner, 75, 81–84, 107, 112, 188, 218
    747 jumbo jet, 77, 86, 93, 107–108, **109**, 187–188
    747 jumbo jet (PCA program), 153
    747-8/9 jumbo jet, 197
    747-100 jumbo jet, 84
    747-200 airliner, 75
    757 airliner, 75, 108, 143, 153
    767 airliner, 108, 110, **113**
    777 airliner, 112–113, 125, 194
    787 Dreamliner, 174, 196
    Advanced Subsonic Technology (AST) program, 188
    Advanced Turboprop Project (ATP), 115–117
    B-17 Flying Fortress, 26
    B-29 Superfortress, 11, 21, 25, 32
    B-47 Stratojet, 32, 140
    B-52 Stratofortress, 50
    chevron studies, 195–197
    Energy Efficient Engine (E^3), 111
    Engine Component Improvement (ECI) Program, 106–107
    Engine Validation of Noise Reduction Concepts (EVNRC) program, 191
    F-111A/F-111B design, 60
    High-Speed Research (HSR) program, 197
    IM-99 Bomarc, 37, **37**
    integrated propulsion control system (IPCS) program, 142
    Jet Engine Containment Concepts and Blade-Out Simulation, 205n19
    Model 2707, 52–53, **52**
    Nacelle Aeroacoustic System Technology Assessment, 190
    noise abatement, 76
    noise frequencies, 193
    NPSS team, 177
    propulsion-controlled aircraft (PCA) program, 151–152
    QSRA development, 93
    Quiet Technology Demonstrator 2 (QTD2) program, 196
    Quiet Technology Demonstrator (QTD) program, 194
    Refan program, 82–83
    Ultra-Efficient Engine Technology Program (UEET), 179
    wind testing, 120, 121
    X-51A WaveRider, 203
    YF-22 (ATF program), 159
Boeing Phantom Works, 157
Bolden, Charles, **217**, 217–218
Bombardier, 186
    Regional, CRJ700 jet, 197

Index

Bradley International Airport, 78
Bresket, David, 86
British Aerospace, 130
British Aircraft Corporation (BAC), 84
Brocket, Paul, 1
Bruckmann, Bruno, 54
Bull, John, 153
Burcham, Frank W. ("Bill"), 147–154, **148**
Bureau of Standards, 8, 27
Burkhardt, Leo, 181
Busch, Mike, 182
Bush, Vannevar, 18
business jets, 116, 118–119, 126–130, 188, 194, 199. *See also* "bizjets"

C

C-5A Galaxy transport, 85, 106–107
C-17 Globemaster III transport, 153
California Institute of Technology, 3
Campini, Secondo, 29
Cape Kennedy Regional Airport, 78
Capp, Al, 56
Caproni N.1 aircraft, 29
CARD Policy Study, 80–81, 89, 126
Castner, Ray, **184**
Central Intelligence Agency (CIA), 56
Centurion unmanned aerial vehicles, 223
Cessna, 63, 128, 192, 208n60
Champion, C.C., Jr., 14
Chance Vought, 27
chevrons, 175, 192–197, 199
Chicago Sun-Times, 84–85
Ciepluch, Carl C., 110, 112
Cirrus, 181, 208n60
Civil Aviation Research and Development (CARD) Policy Study, 80–81, 89, 126
Clark Y (airfoil), 10, 12
Clarke Thomson Research Group, 2
Cold War, 31–33, 35–37, 49, 56–57, 228
"coldwall" heat transfer pod, **48**
Collier, Fay, 197, 220

Collier Trophy
    Eclipse 500 VLJ aircraft, 185
    General Electric and Lockheed Aircraft Corporation, 36
    NACA, 15
    NASA, 114–115, 123
    Packard Motor Car Company, 14
    Stack, John, 88
    Williams, Sam, 184
"Compressor Bible," 33
computational fluid dynamics (CFD), 55, 63, 154, 176–177, 183, 193, 203
Concorde supersonic transport, 199
Congress (U.S.)
    aeronautics research, 73
    aircraft noise legislation, 75
    Aviation Noise and Capacity Act, 187
    engine noise, 74–78, 80
    House Committee on Science and Astronautics, 73, 78, 82
    House Committee on Science and Technology, 83
    House Subcommittee on Space and Aeronautics, 125
    Jet Aircraft Noise Control Bill, 78
    NACA appropriations, 1, 18
    Senate Committee on Aeronautical and Space Sciences, 104–105
    sonic boom regulation, 52–53
    Subcommittee on Aviation and Transportation Research and Development, 83
    supersonic program, 52–54
    TFX program, 60
    XB-70 Valkyrie program, 54
Conley, Joe, 149
Continental Motors, Inc., 181
Continuous Lower Energy, Emissions, and Noise (CLEEN) program, 222
Continuum Dynamics, Inc., 196
Convair, 32–33, 36, 51, 58, 62–63, 84
Craig, R.T., 141

279

Crawford, Frederick C., 18
Critical Propulsion Components (CPC), 198–199
CRJ700 jet, 197
Cross, Carl, 54
Curtiss Aeroplane and Motor Company, 15, 20, 25
Curtiss-Wright Corporation, 33
"cutback," 190–193, 198, 210n83

D

Dankhoff, Walter F., 81
Dassault Falcon, 130
David, Edward E., Jr., 81
Davis, Donald Y., 111–112
DC-1 aircraft, 15, **17**
DC-2 aircraft, 16
DC-3 aircraft, 16
DC-8 airliner (noise of), 84–86
DC-8 Airborne Science Laboratory, **216**, 226–227, **227**
DC-8 airliner, 34, 63, 75–77, 79–81
DC-9 airliner, 75, 81–84, 107, 109, 116
DC-10 airliner, 75, 77, 85–86, 93, 106–109, **107**, 151
DC-10 airliner (loss of), 146–147, 170
decibels (db), 73, 76, 128, 196
  defined, 74
  effective perceived noise in decibels (EPNdB), 78–79, 82, 84–86, 91, 93, 188–197
  effective perceived noise in decibels (EPNdB), defined, 74
  perceived noise in decibels (PNdB), 84, 129
  perceived noise in decibels (PNdB), defined, 74
  subsonic (chart), 222
Deets, Dwain, 150–151
Defense Advanced Research Projects Agency (DARPA), 176, 200, 203
Del Balzo, Joseph M., 187
DeLaat, John, 158

Delta Air Lines, 146, 152–153
Department of Defense (DOD), 51–52, 58, 94, 176–179, 201
Department of Energy (DOE), 67, 168, 180
Department of Transportation (DOT), 78, 80–81, 83
Digital Electronic Engine Control (DEEC), **138**, 141–143, 149
Digital Electronic Flight Control System (DEFCS), 142, 147, 160n11
Doolittle, James ("Jimmy"), 25
Dornheim, Michael A., 124
Douglas Aircraft Company, 15–16, 34
drag, 12, 32, 159, 201
  lift-to-drag characteristics, 120
  low-drag wing shapes, 104
  reduction of, 15, 60, 78, 156, 199, 230n11
Draper, Susan, 170
Dreamliner series, 174, 196–197
Drinkwater, Fred, 88
Dryden, Hugh L., 32–33
Dryden Flight Research Center, 54, 57–58. *See also* Armstrong Flight Research Center; NASA Flight Research Center
  Adaptive Engine Control System (ADECS), 144
  Advanced Control Technology for Integrated Vehicles (ACTIVE) program, 156
  Advanced Turboprop Project (ATP), 115–116
  Alternative Fuel Effects on Contrails and Cruise Emissions (ACCESS) study, 226
  Digital Electronic Engine Control (DEEC), **138**
  flight and propulsion control system development, 147
  High-Angle-of-Attack Technology Program (HATP), 154–155
  High-Speed Research (HSR) program, 197
  instrument landing system (ILS), 149
  integrated propulsion control system (IPCS) program, 142
  "PCA Ultralite," 153

Index

propulsion-controlled aircraft (PCA) program, 147–154, **148**, 150–151, 153
synthetic fuels, 226
X-43A research aircraft flights, 201
Drzewiecki, Stefan, 3–4, 7
Dulles Airport, 199
Durand, William F., 9, 42n42
   background, 5
   career of, 40–41n19
   NACA Chair, 5–6
   propeller experiments, 7–10
   Special Committee on Jet Propulsion, 27, 29
Dvorak, Dudley J., 146–147

E

E.28/39 aircraft, 22–23, 28, 44n78
Eastern Airlines, 106
Eckert, Ernst, 33
Eclipse Aerospace, 185, 209n67
Edwards, Thomas, 224–225
Edwards Air Force Base, 22, 142, 152, 226
EF-111A "Sparkvark," 59
Eiffel, Gustave, 4, 6, 41n25
Eisenhower, Dwight D., 96n10
E-Jets, 197
electric propulsion, 49–50, 223–224
Embraer, 186, 195, 197
Enabling Propulsion Materials (EPM), 168–169, 170–175
Energy Crisis, 104–106. *See also* fuel, supply of
Energy Efficient Engine (E³), **102**, 105–108, 110–113, 115, 125, 179
Energy Efficient Transport, 105–106, 132n12. *See also* Aircraft Energy Efficiency (ACEE) program
Energy Trends and Alternative Fuels (ETAF) program, 132n8
Engine Component Improvement (ECI) Program, 105–110, 125, 179. *See also* Aircraft Energy Efficiency (ACEE) program

engine knock, 13–14, 20
Engine Noise Reduction Element, 188–190
engines, 20, 114
   570 engine series, 118
   571 industrial gas turbine, 123
   578-DX demonstrator engine, 123, **123**
   Advanced Ducted Propulsor (ADP), 185–186
   Advanced Turboprop Project (ATP), 105–106
   afterburner, 30
   afterburning turbofan design, 228
   air-cooled opposed, 181
   augmentor, 200–201
   axial-flow, 140
   CF6 turbofan, 85, 87, 93, 105–107, **109**
   CF6 turbofan tri-jet, 146
   CF6-50C engine, 111
   CF34 turbofan, 171, 195
   CJ805-23 aft-fan, 63
   combustion (defined), 13
   compound, 31
   control systems, **4**, 139–145, 154
   cooling, 20, 33, 108, 127, 168, 176, 181
   CRP-X1 tractor propfan, 120–121
   diagnostic systems, 139
   diesel, 13–14, 181–182
   double annular axially parallel engine design, 87
   ducted fan (variable-pitch), 128
   E³ demonstrator, **102**
   E³ turbofan program, 105–108, 110
   efficiencies, 104, 167
   EJ22, 184–185
   electric, 223–224, 228
   emissions, 99n55, 167, 218
   emissions reduction, 86–87, 92–93, 104–105, 126–127
   Energy Efficient Transport, 105, 132n12
   Engine Alliance GP7200, 180
   Engine Component Improvement (ECI) Program, 105–110

281

The Power for Flight

"engine war," 112
F-1, 49
F100, 65–66, **67**, 143, 178
F-100-229 engine, 168
F100-PW-229, 158
F101, 92, 178
F119, 159–160, **178**
F119-PW-100, 145
F404 turbofan, 121, 168
fan-on-blade (FLADE) concept, 198–199
fast engine response, 154
GAP diesel engine, 181–182
GAP FJX-2 turbofan, 183–184, **184**
GAP small gas turbine initiative, 183–185
gas turbine, 27–29, 33, 175, 200, 228
gas turbine (hybrid), 223
GE F404 turbofan, 155
GE4 turbojet, 52–53, 55
GE36 UDF UHB, 114, **114**, 121, 123
GE90 engine series, 112–113, 125, 168, 189
GE90-115B, 113
Geared TurboFan, 185–186, **221**, 222
GEnx turbofan, 169, 174–175, 196
GEnx-2B engine, **174**
helicopter, 89
HeS 3/3B turbojet, 22, 24
high-bypass-ratio turbofan, 187, 189, 194
higher-thrust, 93, 187
high-lift flap-and-shot turbofan, 191
Highly Integrated Digital Electronic Control (HIDEC), 142–145
hydrogen injection, 127
I-16, 30
I-40, 30
I-A (Whittle), 28, **31**, 139–140
inlet incompatibility, 60
J-5 Whirlwind, 25–26
J42, 33
J47 turbojet, 140
J48, 33

J57 turbojet, 33–34, 141
J58 turbo-ramjet, 56–57, 104
J75, 58
J79, **36**, 54
J85 turbojet, 58, 63, 65
J85-21, 65
J93, 50
"Jake's Jeep," 29–30
jet noise test, **72**
jet performance, 143
Joint Technology Demonstrator Engine (JTDE) program, 168
JP-4, 94
JP-5/Jet A, 94
JT3 turbojet, 34, 63
JT3D turbofan, 63, 86
JT3D turbofan (noise from), 75, 81–82
JT8D turbofan, 86, 98n34, 105–106, **108–109**, 122
JT8D turbofan (low-bypass), 107–108
JT8D turbofan (noise from), 81–83
JT8D-115 refan, 83
JT8D-200 refan series, 84
JT8D-217 refan, 83–84
JT8D-219 refan, 83
JT9D turbofan, 87, 105–108, **109**, 112
JT9D-7A engine, 111
JT15D, 190
Jumo 004 axial-flow turbojet, 24, **28**
L-1000 turbojet, 26
laminar flow control, 105
Large-Scale Advanced Propfan (LAP) project, 116
life of, 145
low-bypass-ratio, 176
Merlin, 20
modification program, 86, 127
Nene turbojet, 33
"open rotor," 218–219, **219**
optimized rotor, 92

Index

overwing pod installation, 93
Pegasus, 87
piston, 2, 13–22, 25, 29, 31, 115, 181–182
piston (emissions), 127
propfan, 114–125, **116**, 126
propfan (gearless), 121–122
propfan (single-rotation), 115, 117, 120, 125
Propfan Test Assessment (PTA) project, 118–119
propfans, perceptions of, 124
PT-1, 26
PurePower engine, 186, 221–222
PW100 series, 126
PW1128 turbofan, 142
PW2000 turbofan, 108
PW2037 turbofan, 123
PW4000 engine series, 112–113, 125
PW4000 turbofan, **113**
PW4084, 112–113
PW4096, 192
PW4098, 113
Quiet, Clean, General Aviation Turbofan (QCGAT) program, 128–131
Quiet, Clean, Short-Haul Experimental Engine (QCSEE), 92–93
Quiet Engine Program (QEP), 76, 78, 81, 82, 84–86, 92
Quiet Experimental Short Takeoff and Landing (QUESTOL), 92
Quiet-Fan (Q-Fan), 91
R-3350 turbosupercharged radials, 25
radial air cooled, **xii**, 2, 15–17, 21, 33
ramburner, 200–201
ramjet, 31–33, **37**, 37–38, 201, 228
reciprocating, 2, 31, 92
Refan program, 82–83
reverse-thrust capability, 122
Revolutionary Turbine Accelerator (RTA), 200–201
RJ43 engine, **37**, 38

rocket, 32–33, 33, 35, 37
Roots supercharger, 13–14, 19
rotary combustion, 92
scramjet, 200–201, 203, 228
SJX61-2 scramjet, **202**
SR-7L propfan, 117–118
STOL quiet engine, 89
supercharged, 18, 20, 29
supersonic, 31
T55-L-11 turboshaft, 91
TF30, 60–61, 64–66
TF30 afterburning turbofan, 64
TF39 turbofan, 85, 106
TFE731 turbofan, 130
TG-100 turboprop, 27
TG-180 (J35), 29
Throttles Only Control (TOC), 148–150
Trent 800 turbofan, 194
Trent engine series, 112
TSX-2 turboprop, 184
Turbine Based Combined Cycle (TBCC), 200–201
Turbodyne, 26
turbofan, 60, 66, 87, 89, 94–95, 103, 107, 171
turbofan (advanced), 132, 179, 220, 228, 229n10
turbofan (afterburning), 142, 145
turbofan (Distortion Tolerant Control program), 158
turbofan (general aviation), 129
turbofan (high-bypass), 110–113, 141
turbofan (hydrogen-injected), 127
turbofan (mixed-flow), 198–199
turbofan (modified), 189–190
turbofan (weight of), 174
turbofan characteristics, 55
turbofan market, 73
turbofan noise, 74–76
turbofan reduction gearbox, 92

283

turbojet, 25, 27–33, 36, 60, 89, 115, 194, 228
turbojet (modified), 110–113
turbojet (single-shaft), 141
turbojet (single-spool), 140
turbojet (twin-spool), 141
turbojet (von Ohain), 203
turbojet characteristics, 55
turbojets, commercial airlines, 63
turboprop, 27, 31–35, **34**, 115–116, 124, 126, 228
turboprop (counter-rotating), 120–121
turboprop (tilt-wing), 89
turboramjet, 200
turbosupercharged, 14, 19–21, 28, 31
ultra-high-bypass (UHB), 114, 121, 124, 169–170, 220–222, 229n10
underwing pod installation, 93
Unducted Fan (UDF), 114, 121–123, 122–124, 126
Universal Test Engine, 2
"unstart," 58, 203
V-1400, 25
V-1710 V-12, 20
variable-cycle (VCE), 55
V-Jet II, 184
Vorbix axial series, 87
V/STOL (noise from), 89
W.1X (Whittle), 28
W.2B (Whittle), 28
Wasp radial, 26
Wasp supercharger, 14–15
water injection, 13–15
Whittle, 27–28, 30
wind testing, 18
XJ-37, 26
XJ79-GE-1 afterburning turbojet, 36
XLR99 rocket, 50
XRJ47, 38
XT38, 46n119
YF-102 high-bypass turbofans, 93
YJ93 turbojet, **53**, 54

Enhanced Fighter Maneuverability program, 158
Environmental Protection Agency (EPA), 74, 93, 180
    1979 Standard Parameters, 86
    emissions requirements, 112, 183–184
    NASA compliance, 127
    National Vehicle and Fuel Emissions Laboratory (NVFEL), 127
Environmentally Responsible Aviation (ERA) Project, 220–222
Epstein, Alan, 222
Euler solution, 121
Experimental Aircraft Association (EAA), 180, 182, 184
Experimental Clean Combustor Program (ECCP), 55, 87, 111, 179
"Experimental Research in Air-Propellers," 8–9
Extended Range Twin Operations (ETOPS), 112–113

F

F4H-1 Phantom II, 36
F4U-1 Corsair aircraft, 27
F-5 fighter series, 63
F-5E/F Tiger II, 65
F9F Panther, 33
F-14/F-14A Tomcat, 59, **64**, 64–66
F-15 Eagle, 61, **138**, 149–150
F-15 HIDEC, 142–145, **144**
F-15 series, 65–66, **138**, 143, **150**, 159, 178
F-15 simulator, 148–149
F-15A Eagle, 65–66, **65**
F-15B two-seat aircraft, 156–158
F-16 (MATV program), **159**
F-16 Fighting Falcon, 61, 65–66, 158, 178
F-16A Fighting Falcon, 65
F-22 Raptor, **145**, 145, 158–160, 178
F-86 Sabre, 32, **140**, 140
F-86D, K, and L interceptors, 140
F-94 interceptor, 30
F-100 Super Sabre, 33

Index

F-101A Voodoo, 35
F-102 Delta Dagger, 33
F-104/F-104A Starfighter, 36, 54
F-106 Delta Dart, 58, 62
F-111 supersonic fighter bomber, 58–59, 64, 142
F-111 twin engine, 159
F-111A supersonic fighter bomber, 59–61
F-111B supersonic fighter bomber, 59–61
F-111C bomber, 61
F-111E test aircraft, **143**
F-111F Aardvark, 59
F-111/TF-30 Propulsion Program Review Committee, 60–61
F-117 aircraft, 103
F119-PW-100 turbofan, 145, 158
F/A-18 Hornet High Alpha Research Vehicle (HARV), 155, **159**, 163n51
F/A-18 Hornet (PCA program), 153
F/A-18E/F, 178
Fairey Gannet antisubmarine aircraft, 12
fan-on-blade (FLADE) concept, 198–199
FB-111A bomber, 61
Federal Aviation Administration (FAA), 180
    Advanced General Aviation Transportation Experiments (AGATE), 180–181
    Advanced Subsonic Technology (AST) program, 188
    certification, 185
    collaboration with NASA, 83, 89, 120, 126–128
    Continuous Lower Energy, Emissions, and Noise (CLEEN) program, 222
    FAR 36 noise regulations, 75, 82–84, 110, 123, 129, 198
    Jet Aircraft Noise Control Bill, 78
    Jet Engine Containment Concepts and Blade-Out Simulation, 205n19
    Joint Office of Noise Abatement, 81–82
    National Aviation Facilities Experimental Center (NAFEC), 127
    noise abatement, 186–187, 225

noise abatement retrofitting, 76
noise regulations, 73–75, 77–79, 84, 93, 183–185
propulsion-controlled aircraft (PCA) program, 152–153
Quiet-STOL (Q/STOL) aircraft program, 89
Refan program, 83
supersonic transport (SST), 51–53
two-segment landing approach, 78–80
Ferguson, James, 66
FG-1 Corsair aircraft, 27
FH-1 Phantom aircraft, 27
Fiat Avio, 185
First National Defense Appropriations Act, 18
Fishbach, Larry H., 55
Fitch, Dennis E., 146–147
"flameout," 55, 57, 60, 140
Fletcher, James C., 66, 77, 81, 104
flight control, 141, 143, 145–159
flight control computer (FCC), 149
Flight Propulsion System (FPS), 110–111
Fokker, 75
Forbes, 76
Ford, Gerald R., 84
Ford Motor Company, 169
fuel, 203
    alternative, 218, 224–228
    biofuels, 226–227
    consumption of, 36, 49, 63, 93, 119
    consumption rate, 134n22, 145
    control system, 38
    distribution/preparation, 87
    efficiency, 32, 63–64, 73–74, 83–84, 94–95, **102**, 103–106, 131–132
    efficiency (chevron studies), 195
    efficiency (general aviation), 126–127
    engine knock, 13, 20
    fossil, 225
    gasoline, 23–24
    high octane, 13–14
    hydrogen, 23

injection, 19
metering valve, 140
prices, 103–106, 124, 218
Specific Fuel Consumption (SFC), 108
supply of, 103–106. *See also* Energy Crisis
synthetic, 226
full-authority digital electronic control (FADEC), 112, 123, 144, 152, 182
Fullerton, C. Gordon, 149–152, 154
Full-Scale Engine Research (FSER), 65
Full-Scale Tunnel, 11, 88
Future of Flight Foundation, 224, 231n20

G

Gallup, D.L., 4
Garg, Anita, 170
Garrett AiResearch Manufacturing Company, 129–130
Gates, Thomas S., Jr., 51
Gatlin, Donald H., 155–156
General Aviation Propulsion (GAP) diesel engine, 181–182
General Aviation Propulsion (GAP) gas turbine initiative, 180, 183–184
General Aviation Propulsion (GAP) program, 180–184
General Dynamics
    EF-111A "Sparkvark," 59
    F-16/F-16A Fighting Falcon, 61, 65–66, 143
    F-111 supersonic fighter bomber, 58–60, 64, 142
    F-111A, **59**
    F-111A/F-111B design, 60
    F-111F "Aardvark," 59
    TF30 engine, 60–61
    Triple Plow I inlet, 61
    YF-22 (ATF program), 159
General Electric (GE), 22, 25, 64
    Advanced Subsonic Technology (AST) program, 188–193, 192–193

Advanced Turboprop Project (ATP), 115
afterburning turbofan design, 140–141
aircraft testing, 188
airliner market, 112
alloys, 168, 180
CF6 engine, 85
CF6 turbofan, 87, 93, 105–110, **109**
CF6 turbofan tri-jet, 146
CF6-50C engine, 111
CF34 turbofan, 171, 195
chevron-enhanced engines, 197
CJ805-23 aft-fan, 63
computational fluid dynamics (CFD), 63
Critical Propulsion Components (CPC), 198
double annular axially parallel engine design, 87
Energy Efficient Engine (E^3), 112, 125
Energy Efficient Engine (E^3) demonstrator, **102**
Engine Alliance GP7200 engine, 180
Engine Component Improvement (ECI) Program, 106, 110–113
engine-component noise nomenclature, 190
Experimental Clean Combustor Program (ECCP), 111
F101 engine, 92, 178
F404 turbofan, 121, 168
F414 engine, 178
"fan" (propfan), 124
Flight Propulsion System (FPS), 110–111
fuel efficiency, 132n8
GE F404 turbofan, 155
GE4 turbojet, 52–53
GE36 UDF UHB engine, 114, **114**, 121, 123
GE90 engine series, 112–113, 125, 168
GEnx turbofan, 169, 174–175, 196
GEnx-2B engine, **174**
High Temperature Engine Materials Technology Program (HITEMP), 169
high-bypass-ratio engines, 93, 187, 192

Index

High-Speed Research (HSR) program, 197–199
I-A (Whittle) engine, 139–140
Integrated High Performance Turbine Energy Technology (IHPTET) program, 176
J47 turbojet, 140–141
J75 engine, 58
J79 engine, 54
J85 turbojet, 58, 63, 65
J93 engine, 50
Jet Engine Containment Concepts and Blade-Out Simulation, 205n19
NPSS team, 177
nuclear propulsion research, 38
"open rotor" engine, 218–219
Polymerization of Monomer Reactants (PMR), 168
propfan, 121–122, 124
QCSEE program, 92–93
Quiet Engine Program (QEP), 84, 87
Quiet Technology Demonstrator 2 (QTD2) program, 196
Revolutionary Turbine Accelerator (RTA), 200
stator design, 190–191
TF39 turbofan, 85, 106
turbofan market, 73
turbofan reduction gearbox, 92
turboprop (counter-rotating), 120–121
Ultra-Efficient Engine Technology Program (UEET), 179
underwing pod installation, 92–93
Unducted Fan (UDF), 126
Unducted Fan (UDF) engine, 122–124
variable-cycle engine (VCE), 55–56
YJ93 turbojet, **53**, 54
General Electric (Schenectady), 29
I-16, 30
I-A engine (Whittle), 28, **31**
J79, **36**
Special Committee on Jet Propulsion, 27
TG-100 turboprop, 27
TG-100A, **34**
TG-180 (J35), 29
Whittle engine, 30
XJ79-GE-1 afterburning turbojet, 36
Georgia Institute of Technology, 223
German Aerospace Center (DLR), 227
Germany
  influence on American technology, 33, 37, 62, 117
  jet propulsion, 22, 25, 29, 30
  Kochel high speed wind tunnels, 62
  ramjet design, 37
  rockets, 27
GKN Aerospace, 174
Glenn Research Center, 93, **184**, 189. *See also* Lewis Flight Propulsion Laboratory
  Advanced Ducted Propulsor (ADP), 185
  alloys, 180, 196
  Alternative Fuel Effects on Contrails and Cruise Emissions (ACCESS) study, 226
  chevron studies, 194
  Commercial Modular Aero-Propulsion System Simulation (C-MAPSS), 154
  computational fluid dynamics (CFD), 177, 203
  computer software, 173
  Critical Propulsion Components (CPC), 198–199
  engine-component noise nomenclature, 190
  GAP diesel engine, 181
  GAP FJX-2 turbofan, 183, **184**
  Greenlab Research Facility, 226
  High-Speed Research (HSR) program, 197–199
  Jet Engine Containment Concepts and Blade-Out Simulation, 172, 205n19
  Low Emissions Alternative Power program, 223
  Low Speed Wind Tunnel, 221–222
  Megabraider, 172–173

287

Numerical Propulsion System Simulation
   (NPSS), 177
"open rotor" engine, 219
Quiet Aircraft Technology (QAT), 195
Revolutionary Turbine Accelerator (RTA),
   200–201
stator design, 190–191
Supersonic Wind Tunnel, 199, **219**
Ultra Safe Propulsion Project, 171–172
Ultra-Efficient Engine Technology Program
   (UEET), **166**, 179–180
Glennan, T. Keith, 51
Gloster Aircraft Company, 22–23, 28, 44n78
Goddard Space Flight Center, 179, 197
Goldin, Daniel S., 180–181
Goldwater, Barry M., 104
Goodrich Corporation, 196
Goodyear, 27
Google, 223–224
Gough, Melvin N., 10
Gray, David, 112
Gray, George, 12
Great Britain, 22, 25, 30, 88. *See also* United
   Kingdom
Grubb, H. Dale, 76
Grumman Corporation. *See also* Northrop
   Grumman
   Advanced Turboprop Project (ATP), 116–117
   F9F Panther, 33
   F-14 Tomcat, 61
   F-14A series, 66
   F-14A Tomcat, 64, **64**
   Gulfstream business jet, **116**
   Gulfstream II business jet, 116–119,
      118–119, **119**
   XF10F-1 Jaguar, 59
GTD-21 reconnaissance drone, 56
Guckian, William, **184**
Guinness Book of World Records, 113
Gulf War (first), 59
Gulfstream, 115–119

H

Hamilton Standard, 35
   578-DX demonstrator engine, 123–124, **123**
   Advanced Turboprop Project (ATP), 115
   CRP-X1 tractor propfan engine, 120–121
   Large-Scale Advanced Propfan (LAP) project,
      126
   propellers, **xii**, 115
   propellers (variable-pitch), 125
   QCSEE program, 92
   Quiet-Fan (Q-Fan) program, 91–92
   SR-7L propfan, 117–118
   turboprop (counter-rotating), 120
Hartzell Propeller, 125, 181–182
Hawker Aircraft, 87–88
Hawker Siddeley Harrier reconnaissance/strike-
   fighter, 87–88
Hawks, Frank, 15
Haynes, Alfred C., 146–147
He 178 aircraft, 22
Heinkel, Ernst, 22–23
Helios unmanned aerial vehicles, 223
Henderson, Bill, 159
Herschel-Quincke (HQ) tubes, 190–191
Hibbard, Hall L., 26
High Alpha Research Vehicle (HARV), **155**,
   155–156, 158, **159**
high supersonic flight, 57, 200. S*ee also*
   supersonic flight
High Temperature Engine Materials Technology
   Program (HITEMP), 169–170
High-Altitude Long-Endurance (HALE) vehicles,
   223
High-Angle-of-Attack Technology Program
   (HATP), 154–155
Highly Integrated Digital Electronic Control
   (HIDEC) program, 142–145, 147, 149
High-Speed Civil Transport (HSCT), 156, 197–199
High-Speed Research (HSR) program, 168,
   197–199, **198**
Hines, Williams, 84–85

Index

HITEMP Review, 169
Hobbs, Leonard S., 26
Hobbs, Luke, 14
Holmes, Bruce A., 180
Honeywell Inc.
  Advanced Subsonic Technology (AST)
    program, 188
  business jet engines, 188
  Engine Validation of Noise Reduction
    Concepts (EVNRC) program, 191
  engine-component noise nomenclature, 190
  integrated propulsion control system (IPCS)
    program, 142
  Jet Engine Containment Concepts and
    Blade-Out Simulation, 205n19
  NPSS team, 177
  propulsion-controlled aircraft (PCA) program,
    152–153
  stator design, 191
  Ultra-Efficient Engine Technology Program
    (UEET), 179
Honeywell International Inc., 172
Hooker, Stanley, 87
Horne, Clifton, 185
Housing and Urban Development and
  Transportation, 78
Howes, Rob, 192
HU-25 Falcon jet, **227**
"hush kit," 79
hybrid propulsion, 37–38
hypersonic flight, 50, 154, 199–203

I

IM-99 Bomarc, **37**, 37–38,
inlets, 144, 154, **166**, 192, 222
  engine incompatibility, 60
  F-111 design, 60
  guide vanes, 83
  multidimensional, 158–159
  noise from, 75, 190–192
  scarf, 190

temperature/pressure, 63, 145
Triple Plow I/II, 61
"unstart" (defined), 58
variable-geometry, 142
vortex generators, 60
Inner Loop Thrust Vectoring (ILTV), 156
instrument landing system (ILS), 149, 152
Integrated High Performance Turbine Energy
  Technology (IHPTET) program, 168–170,
  176–180, 201
integrated propulsion control system (IPCS)
  program, 142–143
intercontinental ballistic missiles, 38
International Civil Aviation Organization (ICAO),
  86, 187
International Nickel Company, 167
Irkut, 186
Israeli-Arab war, 103

J

Jacobs, Eastman N., 10, 29–30
"Jake's Jeep," 29–30, 33
Japan Airlines, 146, 152
Jet Propulsion Static Laboratory, 29, 197
jet-propelled aircraft, 22, 22–32, **31**, 34
Jetstar aircraft, 116
John F. Kennedy International Airport, 77–78
Journal of Defense Software Engineering, 178
Journal of the Franklin Institute, 5
Junkers Motoren Werke, 24
  Jumo 004 axial-flow turbojet, **28**

K

Kahler, Jeff, 153
Kalitinsky, Andrew, 26
Keirn, David, 28–29
Kemper, Carlton, 2–3
Kennedy, John F., 52
Kennedy International Airport, 77–78
Ketchum, James R., 141
Klimov VK-1 aircraft, 33

289

knock, 13–14, 20
Korean War, 32, 35, 140
Kramer, James J., 85–86
Kramer Commission, 115
Kucinich, Dennis, 180

L

L-133 aircraft, 26
Lancair International, Inc., 181
Langley Memorial Aeronautical Laboratory, 66, 180
   24-Inch Jet Tunnel, 10
   Advanced Ducted Propulsor (ADP), 185
   Advanced Subsonic Technology (AST) program, 188
   Advanced Turboprop Project (ATP), 115–116
   AERL, 2–3
   aircraft noise conference, 75
   Alternative Fuel Effects on Contrails and Cruise Emissions (ACCESS) study, 226
   axial-flow compressor research, 29
   chevron studies, 195–196, 196
   drag reduction, 159
   8-Foot High Temperature Tunnel, 201, **202**, 203
   F-111A/F-111B design, 60
   fuels research, 94
   Full-Scale Tunnel, 11, 88, 128
   High-Angle-of-Attack Technology Program (HATP), 154
   HU-25 Falcon jet, **227**
   Nacelle Acoustic Treatment program, 76–77, 81
   Power Plants Division, 19
   propeller research airplane, 35
   Propeller Research Tunnel (PRT) creation, 9–10
   propulsion-controlled aircraft (PCA) program, 150–151

Puffin (personal air vehicle), 223, **224**
Quiet Aircraft Technology (QAT), 194–195
radial air-cooled engines, 15
relationship with NACA, 2–3, 7–10
role, 2–3
Supercharger Division, 19
Supersonic Cruise Aircraft Research (SCAR) Program, 55
transonic region testing, 88
Ultra-Efficient Engine Technology Program (UEET), 179
VDT tests, 10
V/STOL development, 89
XV-6A Kestrel testing, 88
Learjet, 120, 122, **128**, 128–129, 188
Lerch, Bradley, 170
Lesley, Everett P., 5, 7–10, 41n25
Levintan, Richard M., 91
Lewis, George W., 2, 16, 30, 32
Lewis Flight Propulsion Laboratory, 30–31, 35, 171, 180. *See also* Glenn Research Center
   Active Noise Control Fan, 192
   Advanced Ducted Propulsor (ADP), 185
   Advanced Subsonic Technology (AST) program, 188–189
   Advanced Turboprop Project (ATP), 115–116, 125
   afterburning turbofan design, 140–141
   Aircraft Energy Efficiency (ACEE) program, 131
   Aircraft Noise Reduction Conference, 85
   alloys, 168
   Altitude Wind Tunnel (AWT), **xii**, 140–141
   Anechoic Low Speed Wind Tunnel, 121
   computer-based aeronautics, 141
   Distortion Tolerant Control, 158
   emissions reduction, 127
   Energy Efficient Engine (E^3), 110, 112, 125
   Engine Component Improvement (ECI) Program, 106, 110, 125–126

Index

Engine Noise Reduction Element, 188–190
Experimental Clean Combustor Program (ECCP), 87
F-15 design, 66
F-111/TF-30 Propulsion Program Review Committee, 61
fuels research, 94, 94–95
GAP FJX-2 turbofan, **184**
General Aviation Propulsion (GAP) program, 181
High Stability Engine Control (HISTEC) project, 157
High Temperature Engine Materials Technology Program (HITEMP), 169–170
independence, loss of, 67
Joint Office of Noise Abatement, 81–82
Low-Speed Centrifugal Compressor Facility, **130**, 130–131
Low-Speed Wind Tunnel, 192
Nozzle Acoustic Test Rig (NATR), 193
nuclear propulsion research, 38
propfan, 115
PSL propulsion program, 66–67
QCSEE funding, 92–93
QUESTOL funding, 91
Quiet and Clean Turbofan Engine (QCGAT) program, 129–131
Quiet Engine Program (QEP), 76, 81, 85
rocket research, 49
Separate-Flow Nozzle (SFN) Jet Noise Reduction Test Program, 193
successes of, 67
Supersonic Cruise Aircraft Research (SCAR) Program, 55
Supersonic Wind Tunnel, 128
Transonic Wind Tunnel, 119, 121
turbojet design, 33
turbojet engines (single-shaft), 141
wind testing, 121
YF-12A/C program, 58

Lewis Research Center, 49, 62–63, **72**, 105
Liebert, Curt H., 168
Li'l Abner, 56
Lincoln Composites, 169
Lindbergh, Charles, 17, 25
"Lindbergh moment," 203, 215n136
Locci, Ivan, 170
Lockheed Air Express (1929), 15
Lockheed Aircraft Corporation, 15
   A-12, 56
   Advanced Cargo Aircraft Study, 116
   Advanced Development Projects division, 56. *See also* "Skunk Works"
   Advanced Turboprop Project (ATP), 115–117
   C-5/C-5A Galaxy transport, 85, 106–107, 146
   collaboration with Ames Research Center, 57
   Energy Efficient Engine (E^3), 111
   F-94 interceptor, 30
   F-104/F-104A Starfighter, 36, 54
   F-117 aircraft, 103
   GTD-21 reconnaissance drone, 56
   Jetstar aircraft, 116
   L-133 aircraft, 26
   L-1000 turbojet, 26
   L-1011 aircraft, 75, 77–78, 146
   M-12 aircraft, 56
   P-38 Lightning, 20
   P-80 Shooting Star, 30, 32
   SR-71 Blackbird, 56–58, 104, 199, 203
   SR-71 series, 58
   T-33 trainer, 30
   U-2, 56
   Ultra-Efficient Engine Technology Program (UEET), 179
   Vega, 15
   XJ-37, 26
   YF-12 Blackbird, **48**, **57**
   YF-12A/C program, 56–58
   YP-80A Shooting Star, 30

Lockheed Georgia
   570 engine series, 118
   Propfan Test Assessment (PTA) project, 118–119, 126
Lockheed Martin
   Advanced Subsonic Technology (AST) program, 188
   F-22A Raptor, **145**
   F-22/F-22A Raptor, 145, 158
   Martin F-35 Lightning II STOVL aircraft, 178–179
   YF-22 (ATF program), 159
Logan International Airport, 77
Lowenstein, Allard K., 77–78
Lundin, Bruce, 67
Lycoming Engines, 181

M

M-12 aircraft, 56
Marquardt Aircraft, RJ43 engine, **37**, 38
Martin F-35 Lightning II STOVL aircraft, 178–179
Martin-Baker Aircraft Company Ltd., 64
Massachusetts Institute of Technology (MIT), 3, 17, 26, 128, 223
Massachusetts Port Authority, 77
materials, 167–175, 225–226
McAulay, John E., 106
McCook Field, 17
McDonnell Aircraft
   F4H-1 Phantom II, 36
   F-101A Voodoo, 35
   FH-1 Phantom, 27
   XF-88B, 35
McDonnell Douglas
   Advanced Control Technology for Integrated Vehicles (ACTIVE) program, 156
   Advanced Turboprop Project (ATP), 115–116
   C-17 Globemaster III transport, 153
   DC-8 airliner, 75–77, 79–81, 84, 86
   DC-9 airliner, 75, 81–84, 107, 116

   DC-10 airliner, 75, 77, 85–86, 93, 106–108, **107**, 151
   DC-10 airliner (loss of), 146–147
   Energy Efficient Engine (E^3), 111
   Engine Component Improvement (ECI) Program, 106–107
   F-15A Eagle, **65**
   F-15/F-15A Eagle, 61, 65–66, 149–150
   F/A-18 Hornet, 153, 155
   MD-11 airliner, 113, **151**, 151–152
   MD-80 airliner, 75, 83–84, 107, 110, 116, 122, **123**, 124
   MD-91 aircraft, 114
   noise abatement, 76
   propulsion-controlled aircraft (PCA) program, 151–152
   "propulsor" (propfan), 124
   ultra-high-bypass (UHB) demonstrator program, 122–123
   Unducted Fan (UDF) engine, 122–124
McKinney, Marion ("Mack"), 88
McNamara, Robert, 58, 60
MD-11 airliner, 113, **151**, 151–152
MD-80 airliner, 75, 83–84, 107, 110, 122, **123**, 124
MD-91 aircraft, 114
Me 262 aircraft series, 22, 29
Me 262-A-1a Schwalbe ("Swallow"), 24
Mead, George, **17**, 25
Menasco Manufacturing Company, 26
Messerschmitt AG, 22, 24, 29
Meusnier, Jean Baptiste Marie, 40n14
MiG-15 series, 33
MiG-23 aircraft, 59
Mikkelson, Daniel, 115
Milford, Dale, 83
Miller, Cearcy D., 13–15
Miller, George P., 49, 73
Miller, Robert, 170
Mitsubishi, 186
Moore, Charles Stanley, 19–20

Morrill, William, 81
Moss, Frank E., 104
MTU Aero Engines, 185
Multi-Axis Thrust Vectoring (MATV) program, 158
Munk, Max M., 8

N

NA-17 aircraft, 20
Nahigyan, Kervork K., 29–30
Nathal, Michael, 170
National Advisory Committee for Aeronautics (NACA), 85
    aerodynamics research, 15
    aeronautics research, 73
    afterburning turbofan design, 140–141
    airfoil theory, 29
    airfoils (16-series), 10
    annual report, 5
    axial-flow compressor research, 29
    Collier Trophy, 15
    committee structure, 1, 8
    cuffs, 11, **11**
    DC-1 aircraft, 15
    diesel engine research, 181
    dissolved, 49
    Durand at, 5
    engineering parameter variation, 15
    fuels research, 94
    fundamental research, 1–2, 16, 18, 31, 35
    jet propulsion, 25–30
    Jet Propulsion Static Laboratory, 30
    laboratory plan for agency research, 31
    Lewis at, 2, 16
    Mead at, 17–18
    NACA cowling, 15, **16**
    National Aeronautical Research Policy, 31
    noise-reduction studies, 96n10
    piston engine design, 13–22
    propeller experiments, 7–10
    propeller research, 1–12
    propulsion research, 1–38, 228
    relationship with Langley, 2–3, 7–10
    relationship with Stanford University, 5–10
    role, 1
    Special Committee on Jet Propulsion, 27
    Subcommittee on Propellers for Aircraft, 35
    successes of, 22
    turbojet research, 30
    Universal Test Engine, 2
    water injection tests, 13–14
    wind testing, 60
    women at, **20**, **22**
    X-5 research aircraft, 59
National Aeronautic Association (NAA), 15, 124
National Aeronautics and Space Administration (NASA)
    Adaptive Engine Control System (ADECS), 143–144
    Advanced Control Technology for Integrated Vehicles (ACTIVE) program, 156–160
    Advanced Ducted Propulsor (ADP), 185
    Advanced Gas Turbine (AGT), 168
    Advanced General Aviation Transportation Experiments (AGATE), 180–181
    Advanced Subsonic Technology (AST) program,169 188–190
    Advanced Supersonic Propulsion and Integration Research (ASPIRE) project, 199
    Advanced Turbine Technology Applications (ATTAP) program, 168
    Advanced Turboprop Project (ATP), 105–106, 114–118, 179
    aeronautics research, 73, 217, 228
    Aeronautics Research Mission Directorate (ARMD), 226
    Aerospace Industry Technology Program, 169
    afterburning turbofan design, 66
    aircraft analytics, 65
    Aircraft Energy Efficiency (ACEE) program, 104–126, 131

293

Aircraft Fuel Conservation Technology Task Force, 104–105
aircraft noise conference, 75
Airworthiness Committee, 119
alloys, 167–168, 179–180
Alternative Fuel Effects on Contrails and Cruise Emissions (ACCESS) study, 226–227
Alternative Fuel Effects on Contrails and Cruise Emissions II (ACCESS II) program, 227
APNASA code, 176, 183
Aviation Safety Program, 154, 172
Bell X-14 aircraft, 88
budget cuts, 55–56, 67, 180
chevron studies, 192–197
Civil Aviation Research and Development (CARD) Policy Study, 80–81, 89
collaboration with FAA, 83, 89, 120, 126–128
collaboration with U.S. Air Force, 57, 66
Collier Trophy, 114–115, 123
composite fan casings, **174**, 174–175
computational fluid dynamics (CFD), 63
Concepts and Missions Division, 91
contact with EPA, 127
creation of, 49
Digital Electronic Engine Control (DEEC), **138**
Distortion Tolerant Control, 158
electric propulsion, 49–50, 223–224
emissions reduction, 73, 104–105, 125, 224–225
Enabling Propulsion Materials (EPM), 168–169, 171–175
Energy Efficient Engine (E³), 105–108, 110–111, 115, 125
Energy Efficient Transport, 105
Engine Component Improvement (ECI) Program, 105–110, 125
engine development, 61

engine modification, 86
engine noise, 105
Engine Validation of Noise Reduction Concepts (EVNRC) program, 191
engine-component noise nomenclature, 190
Environmental Research and Aircraft Sensor Technology (ERAST) program, 223
Environmentally Responsible Aviation (ERA) Project, 220–222, 230n11
EPA emissions standards, 86–87
F-15 aircraft, 142, **150**, 160n11
F-15 HIDEC, 143–145
F-15 simulator, 148–149
F-111A/F-111B design, 60
F-111E test aircraft, **143**
five-stage compressor design, 63
flight and propulsion control system development, 147
free-flight research, 88
fuel efficiency, 104–106, 110, 125
fuel prices, 104
fuels research, 94–95
Full-Scale Engine Research (FSER), 65
Fundamental Aeronautics Program, 222
fundamental research, 169
GAP diesel engine, 182
GAP small gas turbine initiative, 183–185
gas turbine research, 74, 130, 175
general aviation, 126
General Aviation Propulsion (GAP) program, 181–185
Green Flight Challenge, 223–224
Gulfstream business jet, 116
High Temperature Engine Materials Technology Program (HITEMP), 169–170
High-Angle-of-Attack Technology Program (HATP), 154–155
Highly Integrated Digital Electronic Control (HIDEC) program, 142–145

Index

High-Speed Research (HSR) program, 168, 197–199
hypersonic research, 50, 203
Hyper-X research program, 201
Improved Digital Electronic Engine Controller (IDEEC) system, 156
Integrated High Performance Turbine Energy Technology (IHPTET) program, 176–179
integrated propulsion control system (IPCS) program, 142
Integrated Resilient Aircraft Control (IRAC), 154
Integrated System Research Program, 222
J85 turbojet experimental nozzle, 58
Joint Office of Noise Abatement, 81
Joint Technology Demonstrator Engine (JTDE) program, 168
Large-Scale Advanced Propfan (LAP) project, 116
Laser Doppler Velocimetry (LDV), 117
Learjet, 120, 122, **128**
multidimensional inlet, 158–159
Nacelle Acoustic Treatment program, 76–77, 81
NASP Joint Program Office, 200
National Aero-Space Plane (NASP) Project Office, 200
Navy-NASA Engine Program, 55
noise abatement, 73–87, 125, 186–197
noise abatement budget, 85–86
noise abatement retrofitting, 79, 82, 84–86
noise test, **72**
nuclear propulsion, 49–50
Office of Advanced Research and Technology (OART), 90
Office of Aerospace Technology Vehicle Systems Program, 180
"open rotor" engine, 218
overwing pod installation, 93
P.1127 Kestrel testing, 88

Performance Seeking Control (PSC) program, 144–145
Polymerization of Monomer Reactants (PMR), 168
propeller research, 127–128
propfan, 114–125, 124
Propulsion and Power Base Research and Technology Program, 172
Propulsion Systems Laboratory (PSL), 62
propulsion-controlled aircraft (PCA) program, 149, 152–154
QCSEE development, 92–93
QSRA development, 90, 93
QUESTOL development, 92
Quiet Aircraft Technology (QAT), 194–195
Quiet and Clean Turbofan Engine (QCGAT), 128–131
Quiet Engine Program (QEP), 76, 78, 81, 82, 84–86, 92
Quiet Experimental Short Takeoff and Landing (QUESTOL), 91–92
Quiet Technology Demonstrator 2 (QTD2) program, 196–197
Quiet Technology Demonstrator (QTD) program, 194–195
Quiet-Fan (Q-Fan) program, 91–92
Quiet-STOL (Q/STOL) aircraft program, 89–90
Refan program, 82–84, 129
resignation of Director of Lewis, 67
Revolutionary Turbine Accelerator (RTA), 200–201
rocket propulsion, 49–50
role, 61, 88
Separate-Flow Nozzle (SFN) Jet Noise Reduction Test Program, 194
space program, 50, 52
SR-71 Blackbird, 57, 199
STOL development, 88–92
Supersonic Cruise Aircraft Research (SCAR) Program, 55–56, 87

supersonic transport (SST), 50–55
support for, 73
synthetic fuels, 226
Technology Readiness Level (TRL), 188–189, 191
Throttles Only Control (TOC), 154
"Transforming Global Mobility" white paper, 218
transonic region testing, 88
turbofan development, 63, 73, 94, **102**
Turning Goals into Reality Awards, 175, 178
two-segment landing approach, 78–79
Ultra-Efficient Engine Technology Program (UEET), 201
Versatile Affordable Advanced Turbine Engine (VAATE), 179
V/STOL development, 87–92
Wallops Flight Facility, 119–120
wind testing, 118–119, 121–122, 128
X-15 flight research program, 50
X-51A WaveRider, 203
XB-70 Valkyrie program, **53**, 54–55
XV-6A Kestrel, 88
YF-12 Blackbird, **48**
National Aero-Space Plane (NASP), 200
National Air and Space Act, 49
National Research Council (NRC), 227
National Sonic Boom Program, 55
National Transportation Safety Board (NTSB), 147
Navy, U.S.
   Advanced Turboprop Project (ATP), 116–117
   airfoils (16-series), 10
   Bureau of Construction and Repair, 8
   F-14 aircraft, 65
   F-111B, 59–61
   F-111C, 61
   F-111/TF-30 Propulsion Program Review Committee, 60–61
   FB-111A, 61
   fuels research, 94–95

Integrated High Performance Turbine Energy Technology (IHPTET) program, 176
Naval Air Development Center, 55
Naval Appropriation Act, 1
Navy-NASA Engine Program, 55
propulsion-controlled aircraft (PCA) program, 152
Special Committee on Jet Propulsion, 27
TFX program, 58
Towing Tank, 5
turbofan preference over turbojet, 64
V/STOL development, 92
Neumann, Gerhard, 36
New York Daily News, 86
New York University, 3
Nixon, Richard E., 81–82, 103
Noebe, Ronald, 170
noise, 73–87, 92
   airport noise, 77–80, **79**, 90–93, 219
   engine noise, 90–91, **102**, 218–219, 225
   explained, 75–76
   "footprint," 93, 129
   inlet, 191–192
   "noise annoyance footprint," 187
   passenger cabin noise, 122–123, 191, 196
   speed sensor, 141
   V/STOL engine, 89
Nored, Donald, 105
North American Aviation
   A3J-1 Vigilante, 36
   B-70 aircraft series, 51
   F-86 Sabre, 32, 140
   F-86D, K, and L interceptors, 140
   F-100 Super Sabre, 33
   NA-73, 20
   P-51 Mustang, 11, 20
   SM-64 Navaho cruise missile, 37–38
   T-2 Buckeye, 63
   T-39 Sabreliner, 94
   X-15 aircraft, 37, 50, **51**

Index

XB-70 Valkyrie, **53**
XB-70/A Valkyrie, 50, 53–55
North Atlantic Treaty Organization (NATO), 95
Northrop Aircraft, 26, 37, 63, 65
Northrop Grumman, 103. *See also* Grumman Corporation
nuclear bomber aircraft, 38, 50, 54
Nuclear Energy for the Propulsion of Aircraft (NEPA), 38
nuclear propulsion, 37–38, 49–50, 54
Numerical Propulsion System Simulation (NPSS) program, 177–178, 199

O

Office National d'Etudes et Recherches Aérospatiales (ONERA), 117
Office of Aeronautics and Space Technology (OAST), 151
Office of Aerospace Technology Vehicle Systems Program, 180
O'Hare International Airport, 78
Ohio State University, 125, 128, 205n19
Operation Babylift, 146
Organization of Petroleum Exporting Countries (OPEC), 103
Orient Express, 199
Otto, Edward W., 141
Oxide Dispersion Strengthened (ODS) superalloys, 167

P

P-38 Lightning, 20
P-51 Mustang, 11, **11**, 20
P-59 Airacomet, 30
P-61C Black Widow, 37
P-80 Shooting Star, 30, 32
P.1127 Kestrel aircraft, 88, **88**
Packard Motor Car Company, 14
Pan American Airlines, 106, 108
Pappas, Drew, 151

Parker, Jack, 63–64
Pathfinder unmanned aerial vehicles, 223
Pavlecka, Vladimir H., 26
"PCA Lite"/"PCA Ultralite," 153. *See also* propulsion-controlled aircraft (PCA) program
Pereira, J. Michael, 170
Performance Seeking Control (PSC) program, 144–145, 160n11
personal air vehicle, 183, 223, **224**
Pinkel, Benjamin, 19–20
Pinkel, I. Irving, 95
Plum Brook Station, 38
Polymerization of Monomer Reactants (PMR), 168
Popular Science, 124
Porter, Lisa, 218
Power Jets, Ltd., 23
power plants. *See* engines
Pratt & Whitney Aircraft
  578-DX demonstrator engine, 123–124, **123**
  Advanced Control Technology for Integrated Vehicles (ACTIVE) program, 156
  Advanced Ducted Propulsor (ADP), 185
  Advanced Ducted Propulsor (ADP) low-speed fan, **186**, 192
  Advanced Subsonic Technology (AST) program, 188–193, 192–193
  Advanced Turboprop Project (ATP), 115
  aircraft testing, 188
  airliner market, 112
  alloys, 168, 180
  chevron studies, 193
  chevron-enhanced engines, 197
  computer software, 199
  creation of, 17, 26
  Critical Propulsion Components (CPC), 198
  Energy Efficient Engine (E^3), 110–113, 125
  Engine Alliance GP7200 engine, 180
  Engine Component Improvement (ECI) Program, 106–110

297

Engine Validation of Noise Reduction Concepts (EVNRC) program, 191
engine-component noise nomenclature, 190
F100 engine, 65–66, **67**, 178
F-100-229 engine, 168
F100-PW-229 engine, 158
F119 engine, **178**
F119-PW-100 turbofan, 145, 158
Flight Propulsion System (FPS), 110–112
fuel efficiency, 132n8
Geared TurboFan, 185–186, **221**, 222
High Stability Engine Control (HISTEC) project, 157
high-bypass-ratio engines, 93, 192
high-bypass-ratio turbofan engines, 187
High-Speed Research (HSR) program, 197–199
Hornet series, 17
Integrated High Performance Turbine Energy Technology (IHPTET) program, 176
integrated propulsion control system (IPCS) program, 142
J42 engine, 33
J48 engine, 33
J57 engine, 33–34, 141
J58 turbo-ramjet, 56–57, 104
JT3 turbojet, 34
JT3D turbofan, 63, 75, 86
JT8D refan series, 83–84
JT8D turbofan, 86, 98n34, 105–106, **108–109**, 122
JT8D turbofan (low-bypass), 107–108
JT8D turbofan (noise of), 75, 105
JT9D turbofan, 87, 105–108, **109**, 112
JT9D-7A engine, 111
Large Military Engines Division, 156
NPSS team, 177
nuclear propulsion research, 38
Polymerization of Monomer Reactants (PMR), 168

PurePower PW1000G engine, 186
PW2000 turbofan, 108
PW2037 turbofan, 123, 143
PW4000 engine series, 112–113, 125
PW4000 turbofan, **113**
PW4084 engine, 112–113
PW4098 engine, 113, 192
Quiet Engine Program (QEP), 87
Refan program, 83
relationship with NACA, 19
RL-10 liquid-fueled rocket, 49
Rolls-Royce Nene turbojet, 33
TF30 turbofan, 60, 64–66
Throttles Only Control (TOC) tests, 149–150
turbofan design, 81
turbofan market, 73
turboprop design, 26
Ultra-Efficient Engine Technology Program (UEET), 179
variable-cycle engine (VCE), 55–56
Vorbix axial series, 87
Wasp series, 14–15, 17
XRJ47 engine, 38
Pratt & Whitney Canada, 126
Pratt & Whitney Rocketdyne, **202**, 203
President's Aircraft Board, 5
Price, Nathan C., 26
"propeller research airplane," 35
Propeller Research Tunnel (PRT), 9–10, 15, **16**, 42n44
propeller-driven aircraft, 1–22, 25–32, 34–35, 40n14, 46n119
propellers, **4**, 7
  as airfoils, 3–4
  blade-element theory, 3
  constant-speed, 26
  counter-rotating (defined), 120
  cuffs, 11
  design versus airplane design, 6
  development, 3–4

Index

as dominant propulsion system, 2
efficiencies, 10–12
experimental engines, 26
first, 4
function of, 6
general aviation, 127–128
helicopter, 8
lack of data for, 5
low tip speed, 89
low-speed, 182
paddle blades, 12
propfan, 115–125, **119**
single-rotation (SR) propfan, 120, 125
transonic, 35
variable-pitch, 7–8, 91, 125
wind testing, 18, **19**
Wright-Patterson Propeller Laboratory, 117
Propulsion Systems Laboratory (PSL), 35–38, 49
   engine research, 62
   GAP FJX-2 turbofan, 183
   PSL No. 1, 35–36, **36**, **37**, 64–67
   PSL No. 2, 35, **62**, 66–67
   PSL No. 3, 66
   PSL No. 4, 66, **67**
   ramjet research, 37–38
   rocket research, 49
   turbojet tests, **36**
propulsion-controlled aircraft (PCA) program, 147–154, **148**, **151**, 160n11. *See also* "PCA Lite"/"PCA Ultralite"
Pucinski, Roman C., 76
Puffin (personal air vehicle), 223, **224**
Purifoy, Dana, 152

Q

Quesada, Elwood R., 51
Quiet, Clean, Short-Haul Experimental Engine (QCSEE), 92–93, 179
Quiet Aircraft Technology (QAT), 194, 222
Quiet and Clean Turbofan Engine (QCGAT), 128–131
Quiet Experimental Short Takeoff and Landing (QUESTOL), 87, 91–92
Quiet Short-Haul Research Aircraft (QSRA), 90, **90**, 92–93
Quiet Technology Demonstrator 2 (QTD2) program, 196–197
Quiet Technology Demonstrator (QTD) program, 194–195
Quiet-STOL (Q/STOL) aircraft program, 87, 89–90

R

R&D 100 Awards, 169
R3C-2 Racer aircraft, 25
Raburn, Vern, 184–185
RAF-6 (airfoil), 10, 12
Reaction Motors, 50
Reagan, Ronald, 199
Records, William R., 146–147
Reeder, Jack, 88
Rentschler, Frederick B., 17
Republic Aviation Corporation, **xii**, 11, 29, 35
Research & Technology, 169
rocket propulsion, 27, 31–32, 35, 38, 49–50, 85
Rocketdyne, 49, **202**, 203
rockets, 27, 37–38, 49–50, 85, 178, 201
Rockwell International, B-1 Lancer, 61
Rohr Industries, 115, 118–119, 188
Rohrbach, Carl, 115
Rolls-Royce, Ltd., 73
Rolls-Royce Corporation, 20.
   Advanced Subsonic Technology (AST) program, 188
   airliner market, 112
   business jet engines, 188
   chevron-enhanced engines, 197
   engine-component noise nomenclature, 190

299

high-bypass-ratio engines, 187, 192
Jet Engine Containment Concepts and Blade-Out Simulation, 205n19
Nene turbojet, 33
noise abatement, 193
NPSS team, 177
"open rotor" engine, 218
Pegasus engine, 87
Quiet Technology Demonstrator (QTD) program, 194
Trent 800 turbofan, 194
Trent engine series, 112
Ultra-Efficient Engine Technology Program (UEET), 179
Roosevelt, Franklin Delano, 17
Roots supercharger, 13–14, 19
Rosen, Robert, 131
Rostock airfield, 22
Rothrock, Addison M., 3, 19–20
Rowe, Brian, 124
Royal Aeronautical Society, 6
Royal Air Force (Great Britain), 22–23, 27–28
Royal Australian Air Force, 61
Royal Flight of Saudi Arabia, 152
Rutan, Burt, 184
Ryan, William F., 74

S

S1MA Continuous-Flow Atmospheric Wind Tunnel, 117–118
Sadin, Stan, 188–189
Sanders, Newell D., 34
Scaled Composites, 184, 208n60
Schey, Oscar W., 3, 19–20
Schkolnik, Gerard S., 157
Schmittman, Craig, 114
Schneider Trophy, 25
"Screw Propeller: With Special Reference to Aeroplane Propulsion, The," 5
Selfridge Air Force Base, 58

Separate-Flow Nozzle (SFN) Jet Noise Reduction Test Program, 193–194
Sewell, Charles "Chuck," 64
Shaffer, John H., 76–77, 81
shape memory alloys (SMAs), 175, 195–196
Shaw, Robert J., 179
Shivashankara, Belur, 194
Short Takeoff and Landing (STOL) program, 88–92, 89
Short Takeoff and Vertical Landing (STOVL) program, 179, 222
Sievers, G. Keith, 129
Silverstein, Abe, 30, 34, 38, 199
single-lever power control (SLPC) systems, 182
"Skunk Works," 56. *See also* Lockheed Aircraft Corporation Advanced Development Projects division
SM-64 Navaho cruise missile, 37–38
Smith, C.R., 103
Smolka, James W. "Jim," 157
sonic booms, 52–53, 55, 199, 218
Soucek, Apollo, 14
Sound Absorption Material (SAM), 83
Soviet Union, 32–33, 37–38, 49
space program, 49–52, 60, 228
Space Shuttle, 56, 149, 201, 217
Spirit of St. Louis aircraft, 25
Sputnik, 1, 49
SR-71 Blackbird, 56–58, 104, 199, 203
Stability Management Control, 158
Stack, John, 10, 12, 60, 88
stall, 55, 60–61, 65, 142–144, 158
Stanford University, 3, 5–10
Stanley, Robert M., 28
Stapleton International Airport, 146
Staten, Kenneth, 200
statorlike turbine, 76
stators, 36, 54, 76, 83, 92, 141, 190
Stepka, Francis S., 168
Stewart, Jim, 147

Index

Strategic Air Command, 54, 140
Subsonic Fixed Wing (SFW) Program, 222
SunPower Corporation, 223
Supermarine Seafire fighters, 12
Supersonic Cruise Aircraft Research (SCAR) Program, 55–56, 87
supersonic flight, 50–62, 199, 218, 228. *See also* high supersonic flight
   high supersonic (defined), 57
supersonic transport (SST), 50–51, 55, 73, 199
surface-to-air missiles, 38, 54
Swissair, 152
Systems Control, Inc., 66
Szalai, Ken, 149, 153

T

T-2 Buckeye, 63
T-33 trainer, 30
T-38 Talon, 63
T-39 Sabreliner, 94
Tactical Fighter Experimental (TFX) program, 58–61
Tan Son Nhut Air Force Base, 146
Taurus G4 electric aircraft, 224, **225**
Taylor, Burt L. III, 141
Taylor, C. Fayette, 14
Team Pipistrel-USA.com, 223–224, **225**
Technology Concept Airplane (TCA), 197–198
Technology Readiness Level (TRL), 188–189, 191–193, 197, 211n88
Teledyne Continental Motors, 127, 177, 181
Teledyne Ryan Aeronautical, 176
Textron Lycoming, 176
Textron Specialty Materials, 169
Theodorsen, Theodore, 10
Théorie Générale de l'Hélice Propulsive, 3
Thompson Products, 18
Thomson, Clarke, 2
Throttles Only Control (TOC), 148–150, 154
thrust vectoring, 154–160

Tokyo International Airport, 146
Trans World Airlines (TWA), 106
transonic flight, 32, 35, 87–88, 220
Tri-Service XV-6A Kestrel Evaluation Squadron, 88
turbulence, 55, 76

U

U-2 spyplane, 56
Ultra-Efficient Engine Technology Program (UEET), **166**, 179–180, 201, 222
Unducted Fan (UDF), 114, 121–124, 126, 219
United Aircraft Corporation, 2
United Aircraft and Transport Corporation, 17
United Airlines (UAL), 83, 106, 146–147
United Airlines (UAL) Flight 232, 153, 170
United Kingdom, 87. *See also* Great Britain
United Nations, 86, 187
United Technologies, 120–121
United Technologies Research Center (UTRC), 120–121, 193
universal asynchronous receiver/transmitter (UART), 142–143
University of Akron, 173, 205n19
University of Michigan, 3
unmanned aerial vehicles (UAVs), 223
U.S.S. *Kitty Hawk*, 90, **90**

V

Variable-Density Tunnel (VDT), 10
Versatile Affordable Advanced Turbine Engine (VAATE), 179–180, 222
Vertical and Short Takeoff and Landing (V/STOL) program, 73, 87–89, **88**, 92
Vietnam War, 65, 146, 154
Virginia Polytechnic Institute and State University (Virginia Tech), 190
VK-1 aircraft, 33
Volpe, John A., 81
von Ohain, Hans, 22–24, 203

W

Walker, Joe, 54
Ware, Marsden, 2, 13–14
Warsitz, Erich, 22
Washington News, 73–74
Washington Post, 80
Wasielewski, Eugene W., 29
Weick, Fred E., 10, 15
Westinghouse, 27, 30
Wheeler, Keith, 54
Whitcomb, Richard, 32, 127
White, Kevin H., 77–78
Whitehead, Bob, 151
Whittle, Frank, 22–23, 27–28, 203
Whittle Unit (W.U.), 23
Williams, Sam, 184
Williams International, 172, 176–179, 181–184, 205n19, 208n60
Wilson, Eugene E., 2
women, **20**, **28**
Worcester Polytechnic Institute, 4
World Health Organization, 73–74
World War II, 16–22, 25–28, 62
Wright (Wilbur) Memorial Lecture, 6
Wright, Orville, 3–5, 22, 203
Wright, Wilbur, 3–5, 22, 203
Wright Aeronautical 17, 19, 25–27, 33
Wright Field, 14
Wright Flyer, **4**
Wright-Martin Company, 17
Wright-Patterson Air Force Base, 61
    Aeronautical Systems Division, 200
    Air Force Aero Propulsion Laboratory, 65, 94, 176
    Air Force Research Laboratory, 203
    High Stability Engine Control (HISTEC) project, 157
    Integrated High Performance Turbine Energy Technology (IHPTET) program, 176
    Propeller Laboratory, 117
Wydler, John W., 78

X

X-5 research aircraft, 59
X-15 aircraft, 37, 50, **51**, 201
X-31 (Enhanced Fighter Maneuverability program), 158, **159**
X-43A research aircraft, 201, 203
X-51 WaveRider, 201
X-51A WaveRider, 203
XB-70/XB-70A Valkyrie, 50, 53, **53**, 54–55
XF3W-1 Apache, 14
XF10F-1 Jaguar, 59
XF-84H aircraft, 35
XF-88B aircraft, 35
XP-59A Airacomet, 22–23, **23**, 28–29, 29, 44n78
XP-84 Thunderjet, 29
XV-6A Kestrel aircraft, 88

Y

YF-12/YF-12A Blackbird, **48,** 56–58, **57**
YF-12C aircraft, 57–58
YF-22 (ATF program), 159
YF-102 fighter, 32

Printed in Great Britain
by Amazon